AF334485

FAR-FROM-EQUILIBRIUM DYNAMICS OF CHEMICAL SYSTEMS

PROCEEDINGS OF THE
THIRD INTERNATIONAL SYMPOSIUM

FAR-FROM-EQUILIBRIUM DYNAMICS OF CHEMICAL SYSTEMS

Editors

J. Gorecki
A. S. Cukrowski
A. L. Kawczyński
B. Nowakowski

Institute of Physical Chemistry
Polish Academy of Sciences

World Scientific
Singapore • New Jersey • London • Hong Kong

Published by

World Scientific Publishing Co. Pte. Ltd.

P O Box 128, Farrer Road, Singapore 9128

USA office: Suite 1B, 1060 Main Street, River Edge, NJ 07661

UK office: 73 Lynton Mead, Totteridge, London N20 8DH

**FAR-FROM-EQUILIBRIUM DYNAMICS OF
CHEMICAL SYSTEMS**

ISBN 981-02-1801-X

Printed in Singapore.

PREFACE

The growing interest in nonequilibrium phenomena in chemistry, encouraged us to continue the series of the International Symposia on "Far-From-Equilibrium Dynamics of Chemical Systems". The Third International Symposium was held in Borki (near Tomaszów Mazowiecki), Poland, September 6–10, 1993. This Symposium continued the idea of our previous meetings ("The Dynamics of Systems with Chemical Reactions", Swidno June 6–10, 1988 and "Far-From-Equilibrium Dynamics of Chemical Systems", Swidno September 3–7, 1990) of bringing together chemists, physicists and mathematicians who are interested in various aspects of nonlinear dynamics. This time our attention was focussed on nonlinear oscillations, autowave phenomena in reaction-diffusion systems and on the applications of a kinetic thoery to nonequilibrium effects. We hope that the participants who came to Borki (about 50 from 14 countries) enjoyed the very informal atmosphere of our meeting, interesting lectures and lively discussions. This volume contains invited lectures and selected posters presented at the Symposium.

The Symposium was organized by the Institute of Physical Chemistry of the Polish Academy of Sciences.

We would like to express our thanks to:

> Polish State Committee of Scientific Research,
> Stefan Batory's Foundation,
> Addison-Wesley International Publishing Group,
> Linegal Chemicals

for the financial support of the Third Symposium.

The Editors
Warsaw, March 1994

CONTENTS

IN MEMORIAM
PROFESSOR JAN MARIA POPIELAWSKI

A.S. Cukrowski, J. Gorecki, A.L. Kawczyński and B. Nowakowski

Institute of Physical Chemistry, Polish Academy of Sciences,
Warsaw, Poland

We commemorate Professor Jan Maria Popielawski who was the Director of the Institute of Physical Chemistry of the Polish Academy of Sciences and the head of Laboratory of Theory of Chemical Kinetics. His ideas on international scientific collaboration resulted in the organization of a series of international symposia on far-from-equilibrium dynamics of chemical systems. He died suddenly in February 1992 during the course of preparation of this Symposium which was originally scheduled for June 1992.

The participants of the Symposia on the dynamics of chemical systems held in Świdno, Poland in 1988 and 1990 remember Professor Jan Maria Popielawski, who was the main organizer of these conferences. The organization of this Symposium was begun by Professor Popielawski. Unfortunately, he died suddenly on the 7 th of February, 1992. We feel obliged to commemorate him and his scientific achievements in this short note.

Professor Jan M. Popielawski was born on the 6th of August, 1939 in Warsaw He graduated from the Department of Chemistry of Warsaw University in 1961. He then joined the group of Prof. B. Baranowski at the Institute of Physical Chemistry of the Polish Academy of Sciences in Warsaw.

Jan M. Popielawski's Ph.D. thesis was concerned with theoretical and experimental research on irreversible processes in an adsorbed phase. He received his Ph.D. degree in 1965. In the next year Jan M. Popielawski went to the group of Prof. S. Rice at the James Franck Institute at the University of Chicago for his postdoctoral stay. During this stay he started research on the statistical theory of transport processes and of electron excitations in simple liquids. After coming back to the Institute of Physical Chemistry Dr. Popielawski started theoretical studies on the electronic properties in disordered systems. These resarches were the basis for his habilitation degree completed in 1972. His further career was closely related to the Institute of Physical Chemistry. Dr. Popielawski was appointed as an associate professor in 1972. Between 1974 and 1982 he was a head of the Graduate School of the Institute and lectured on statistical mechanics. Prof. Jan M.Popielawski introduced many young scientists to the research work. Five of them completed their Ph.D. theses under his supervision. He continued the researches with Prof. Baranowski up to 1980, when he organized the Laboratory of the Theory of Chemical Kinetics. Since then his research interest was mainly concerned with nonequilibrium effects and with the applications of stochastic methods in chemistry. Together with his collaborators from Warsaw, scientists from Leipzig (Dr. W. Stiller, Dr. R. Schmidt, and Dr. S. Fritzsche) and Minneapolis (Prof. J.S. Dahler and Dr. Lihong Qin) he performed extensive studies on nonequilibrium effects in a thermally acivated chemical reaction. He obtained important results concerned with nonequilibrium contributions to the rate constant and the interaction between a chemical reaction and transport processes.

Prof. Popielawski was the Deputy Director of the Institute from 1983 to 1986. In 1986 he was nominated as a full professor of the Institute. Jan M. Popielawski was always active in the democratic movement in Poland. After the political turn over , he was designated as the Director of the Institute in the first free election in 1990. Beside his work at the Institute he was a very active member of the Polish Chemical Sociaty, in which he held many prominent positions.

Prof. Popielawski always understood the importance of

international scientific collaboration. He was a fellow of the Humboldt Foundation (Germany) and an international member of Bunsengesellschaft für Physikalische Chemie. In 1987 he spent his sabbatical year as a visiting professor at the James Franck Institute of the University of Chicago. He had common research projects with many centers abroad (Prof. H. Kehlen from Merseburg, Prof. F. Hensel from Marburg, Prof. F. de Pasquale from Rome). To encourage collaboration between the Polish and foreign scientists working in the field of chemical kinetics he organized a series of Świdno Conferences. These meetings created the unique, unformal atmosphere, which led to many joint projects.

His untimely death at the age of 52 ended such a fruitful life. Nevertheless, his ideas are still among the people who had the honour to collaborate with him and these ideas are being developed. We hope that the success of the series of Symposia on chemical dynamics manifests the vitality of his ideas.

Obituary notices about Professor Jan M. Popielawski have appeared in:

1). G. Nicolis and S. Rice, *Physics Today* (Dec. 1992) p. 108,

2). A. Fuliński, *Acta Physica Polonica* B23 (1992) p. 175,

3). P. Modrak and M. Janik-Czachor, *Wiadomości Chemiczne* 46 (1992) p. 601.

4). *Ber. Bunsenges. Phys. Chem.* 96 (1992) p. 634.

THE PUBLICATIONS OF PROFESSOR JAN MARIA POPIELAWSKI

1. B. Baranowski, J. Popielawski, "Thermodynamics of irreversible processes in surface systems, p. I.", Bull. Polon. Acad. Sci. ser. sci. chim., 11 (1962) 445.

2. B. Baranowski, J. Popielawski, "Thermodynamic of irreversible processes in two-phase systems, p. III", Rocznikl Chem, 36 (1962) 36.

3. B. Baranowski, J. Popielawski, "Thermodynamics of irreversible processes in surface systems, p. I.", Bull. Polon. Acad. Sci. ser.

sci. chim., $\underline{11}$ (1963) 33.

4. B. Baranowski, J. Popielawski, "Thermodynamics of irreversible processes in surface systems, p. III"., Bull. Polon. Acad. Sci. ser. sci. chim., $\underline{11}$ (1963) 39.

5. B. Baranowski, J. Popielawski, "Thermodynamics of irreversible processes in surface systems, p. IV"., Bull. Polon. Acad. Sci. ser. sci. chim., $\underline{11}$ (1963) 253.

6. B. Baranowski, J. Popielawski, "Active transport in biological systems as the consequence of diffusional cross-effects", Roczniki Chem., $\underline{38}$ (1964) 483.

7. B. Baranowski, J. Popielawski, "Statistical mechanics of transport processes in adsorbed gases", Molec. Phys., $\underline{9}$ (1965) 59.

8. J. Popielawski, "Mikroskop emisji polowej i jego zastosowania fizykochemiczne", Wiad. Chem. $\underline{19}$ (1965) 49, (in Polish).

9. J. Popielawski, "The kinetic theory of transport processes in adsorbed gases", Molec. Phys., $\underline{10}$ (1966) 483.

10. B. Baranowski, J. Popielawski, "Interaction between diffusion and viscous flow in the mixtures of rarefied gases with Maxwellian intermolecular forces", Acta Phys. Polon., $\underline{12}$ (1966) 821.

11. J. Popielawski, "On the solution of kinetic equation for adsorbed gases", Molec. Phys., $\underline{12}$ (1967) 97.

12. J. Popielawski, "The surface flow of adsorbed carbon dioxide in active carbon", J. Catalysis, $\underline{7}$ (1967) 263.

13. J. Popielawski, S.A. Rice, N. Hurd, "Functional integral representation of non-equilibrium statistical mechanics", J. Chem. Phys., $\underline{46}$ (1967) 3707.

14. J. Popielawski, S.A. Rice "Theory of excitones in liquids. III Nonresonant broadening of impurity spectra in simple liquids", J. Chem. Phys, $\underline{47}$ (1967) 2292.

15. J. Popielawski, "On the kinetic theory of chemical reactions in surface gases", Molec. Phys., $\underline{14}$ (1967) 341.

16. J. Popielawski, "Procesy transportowe w zaadsobowanych gazach", Wiad. Chem. $\underline{21}$ (1967) 437, (in Polish).

17. J. Popielawski, "Electrical conductivity in dense metallic vapours", Chem. Phys. Letters, $\underline{2}$ (1968) 71.

18. J. Popielawski, "The kinetic theory of desorption and chemical reactions in surface gases", Molec. Phys., $\underline{17}$ (1969) 341.

19. J. Popielawski, "The thermodynamic theory of irreversible processes in surface systems", Surface Sci. $\underline{19}$ (1970) 355.

20. J. Popielawski, "The statistical mechanics of free electrons in some disordered systems", J. Chem. Phys, $\underline{52}$ (1970) 1.

21. J. Popielawski, "On the application of tight binding method to disordered systems", J. Chem. Phys, $\underline{53}$ (1970) 957.

22. M. Sitarski, J. Popielawski, "On the kinetic theory of diffusion, self-diffusion and thermal diffusion in adsorbed gases", Molec. Phys., $\underline{13}$ (1970) 741.

23. J. Popielawski, "Semiconductor model of electric conductivity in dense metallic vapours", J. Chem. Phys, $\underline{57}$ (1972) 929.

24. M. Sitarski, J. Popielawski, "The kinetic theory of diffusion and thermal diffusion in adsorbed gases", Molec. Phys., $\underline{23}$ (1972) 365.

25. J. Popielawski, "On the application of Van Hove theory to disordered systems", Physica $\underline{65}$ (1973) 203.

26. H. Kehlen, B. Baranowski, J. Popielawski, "Zur Thermodynamik irreversibler Process in einem Zwei-phasen System. I. Bilanzgleichnungen und Entropieproduktion", Z. physik. Chem. (Leipzig) $\underline{254}$ (1973) 337, (in German).

27. H. Kehlen, B. Baranowski, J. Popielawski, "Zur Thermodynamik irreversibler Process in Grenz-flachen System", Z. physik. Chem. (N.F.) $\underline{86}$ (1973) 282, (in German).

28. M. Sitarski, J. Popielawski, "On the derivation of kinetic equations for absorbed gases", Molec. Phys. $\underline{28}$ (1974) 353.

29. J. Popielawski, "On the single site electron theory of correlation disordered systems with short-range correlations", Physica $\underline{72}$ (1974) 101.

30. J. Popielawski, "On the single site theory of electric conductivity in some disordered systems", Physica $\underline{78}$ (1974) 97.

31. J. Popielawski, "Electron theory of disordered systems with short range correlations", II. Arbeitstagung Statistische Physik, Leipzig 1974, s.101.

32. J. Popielawski, "On the Schwartz-Ehrenreich closure relations for

correlation functions in simple liquids", Physica $\underline{71A}$ (1975) 145.

33. H. Kehlen, B. Baranowski, J. Popielawski, "Zur Thermodynamik irreversibler Process in einem Zwei-phasen System. II. Phänomenologische Gleichnungen", Z. physik. Chem. (Leipzig) $\underline{256}$ (1975) 713, (in German).

34. J. Popielawski, J. Gryko, "The semiconductor model of electric conductivity in supercritical mercury vapours", J. Chem. Phys., $\underline{66}$ (1977) 2257.

35. J. Gryko, J. Popielawski, "A comment on the application of the Cohen-Lekner theory of excess electron mobility in liquid krypton" Phys. Rev., $\underline{16A}$ (1977) 1333.

36. J. Popielawski, "On the optical criterion for the Mott transition", Phys. Stat. Sol. (b), $\underline{88}$ (1978) 241.

37. J. Popielawski, "On the equivalence of the theories of liquid metals due to Ashcroft and Schaich and due to Gyoffy", Philos. Mag. $\underline{39}$ (1979) 61.

38. J. Popielawski, H. Uhtmann, F. Hensel, "The shape of absorption edge in compressed fluid memory vapour", Ber. Bunsenges. Phys. Chem. $\underline{83}$ (1979)123.

39. J. Gryko, J. Popielawski, "On theory of the electro drift mobility in compressed argon", Phys. Rev. $\underline{A21}$ (1980) 1717.

40. J. Gryko, J. Popielawski, "The theory of drift velocity of excess electron in argon", Greifswalde Physikalilische Hefte (DDR) $\underline{5}$ (1980) 22.

41. J. Gryko, J. Popielawski,"Comment on the Braglia-Dellasca theory of density dependence of electron drift velocity in gases", Phys. Rev. $\underline{A24}$ (1981) 1129.

42. H. Uhtmann, J. Popielawski, F. Hensel, "Radiation emitted by a slightly ionized non-ideal high pressure plasma", Ber. Bunsenges. Phys. Chem. $\underline{85}$ (1981) 555.

43. J. Popielawski, "The theory of non-equilibrium electron processes in disordered systems", Acta Phys. Polon. $\underline{A59}$ (1981) 623.

44. J. Górecki, J. Popielawski, "On the application of the long mean free path approximation to the theory of electron transport properties in liquid noble metals", J. Phys. F. (Met. Phys.) $\underline{13}$

(1983) 1197.

45. J. Górecki, J. Popielawski, "On the applicability of the nearly free electron model to resistivity calculations for liquid metals", J. Phys. F. (Met. Phys.) 13 (1983) 2107.

46. A.S. Cukrowski, J. Popielawski, "The dependence of the rate constant of chemical reaction on the density derived from the the generalized Enskog equation for dense gases", J. Chem. Phys. 78 (1983) 1197.

47. J. Górecki, J. Popielawski, "On the comparison of the Rubio-Ashcroft-Shaich and the Rousseau-Stoddart-March formulae", J. Phys. F. (Met. Phys.) 14 (1984) L49.

48. J. Popielawski, "On the theory of interaction between chemical reaction and viscous flow in dilute gases", J. Chem. Phys. 83 (1985) 790.

49. J. Górecki, J. Popielawski, "On the stochastic theory of adiabatic thermal explosion in small systems - numerical results", J. Stat. Phys. 44 (1986) 961.

50. A.S. Cukrowski, J. Popielawski, "The effect of viscous flow and thermal flux on the rate of chemical reaction in dilute gases", Chem. Phys. 109 (1986) 215.

51. A.S. Cukrowski, J. Popielawski, " The effect of chemical reaction on viscosity in dense gases", Acta Phys. Polon. A70 (1986) 321.

52. A.S. Cukrowski, J. Popielawski, "The theory of the effect of chemical reaction on viscosity coefficient in dilute gases - comparison of different approaches", Acta Phys. Polon. A71 (1987) 853.

53. F. De Pasquale, J. Górecki, J. Popielawski, "On the stability of chain reaction with respect to global and local fluctuations", J. Phys. A20 (1987) 5231.

54. A. S. Cukrowski, J. Popielawski, R. Schmidt, W. Stiller, "Non-equilibrium contribution to the rate of chemical reaction in the Lorentz gas. A comparison of perturbation and numerical solutions of the Boltzmann equation", J. Chem. Phys. 89 (1988) 197.

55. J. Górecki, F. De Pasquale, J. Popielawski, "The range of spatial correlations in stochastic systems in small noise", Proc. Intern.

Conf. Spatial Inhomogeneities and Transient Behaviour in Chemical Kinetics, Brussels, Belgium 1987. Eds. P. Gray, G. Nicolis. Manchester Univ. Press, Manchester, G.B. 1988.

56. B. Nowakowski, J. Popielawski, "Nonisothermal condensation on spherical aerosol particles from the Grad solution of the Boltzmann equation", J. Colloid Interface Sci., 122 (1988) 299.

57. J. Popielawski, A.S. Rice, "A generalized regular solution model of liquid supported monolayer of long chain amphifile molecules", J. Chem. Phys. 88 (1988) 1279.

58. J. Popielawski, A.S. Rice, "Hyperthermal scattering of atoms from disordered surfaces", Langmuir 4 (1988) 681.

59. A.S. Cukrowski, S. Fritzsche, J. Popielawski, "The effect of chemical reaction on the relaxation of energy in binary mixtures of dilute gases composed of hard spheres", Proc. Intern. Symp., Dynamics of Systems with Chemical Reactions, Świdno, Poland, 1988. Ed. J. Popielawski, World Scientific, Singapore 1989.

60. J. Popielawski, "The theory of nonequilibrium effects in chemical kinetics", Proc. Intern. Symp., Dynamics of Systems with Chemical Reactions, Świdno, Poland, 1988. Ed. J. Popielawski, World Scientific, Singapore 1989.

61. J. Popielawski Ed., Proceedings of the international Symposium "The dynamics of systems with chemical reactions", Swidno, Poland, June 6-10 1988, World Scientific, Singapore 1989.

62. J. Popielawski Ed., "Teoria kinetyki chemicznej, podstawy molekularno statystyczne i wybrane zastosowania", selected materials from the international Symposium "The dynamics of systems with chemical reactions", Swidno, Poland, June 6-10 1988, Selected materials , Institute of Phys. Chem., Polish Acad Sci., Warsaw, 1989.

63. J. Popielawski, F. Hensel, "A model of charged clusters for dielectric anomaly near the critical region of fluid mercury", Proc. 12 AIRAPT-27 EHPRG Intern. Conf. on High Pressure Technology, Padeborn, FRG, High Pressure Rev. 4 (1990) 586.

64. J. Popielawski, A.S. Cukrowski, "Extension of the Lorentz theory of electric conductivity in gases to systems with chemical

reactions", Acta Phys. Polon. $\underline{A78}$ (1990) 815.

65. W. Stiller, R. Schmidt, J. Popielawski, A. S. Cukrowski, "Nonequilibrium kinetics of reaction H + Br_2 in xenon within Lorentz gas model", J. Chem. Phys. $\underline{93}$ (1990) 2445.

66. A.S. Cukrowski, J. Górecki, J. Popielawski, S. Fritzsche, "Investigation of biomolecular chemical reaction with negative Arrhenius activation energy", Proc. Intern. Symp., Far-from-Equilibrium Dynamics of Chemical Reactions, Świdno, Poland, 1991. Eds. J. Popielawski, J. Górecki, World Scientific, Singapore 1991.

67. J. Popielawski, J. Górecki, F. De Pasquale, "The small noise expansion in the theory of spacial correlations in chemical reacting systems", Proc. Intern. Symp., Far-from-Equilibrium Dynamics of Chemical Reactions, Świdno, Poland, 1991. Eds. J. Popielawski, J. Górecki, World Scientific, Singapore 1991.

68. J. Popielawski, J. Górecki,F. De Pasquale, "The travelling front between stable and unstable states in a system with quadratic chemical dynamics" Proc. Intern. Symp., Far-from-Equilibrium Dynamics of Chemical Reactions, Swidno, Poland, 1991. Eds. J. Popielawski, J. Górecki, World Scientific, Singapore 1991.

69. A.S. Cukrowski, S. Fritzsche, J. Popielawski, "Non-equilibrium chemical and thermal effects in a bimolecular chemical reaction in dilute gas", Proc. Intern. Symp., Far-from-Equilibrium Dynamics of Chemical Reactions, Świdno, Poland, 1991. Eds. J. Popielawski, J. Górecki, World Scientific, Singapore 1991.

70. J. Górecki, J. Popielawski, A.S. Cukrowski, "Molecular dynamics study in the influence of nonequilibrium effects on the rate of chemical reactions", Phys. Rev. $\underline{A}$ $\underline{44}$ (1991) 3791.

71. A.S. Cukrowski , J. Popielawski, W. Stiller , R. Schmidt "Remarks on nonequilibrium contributions to the rate of chemical reaction in the Lorentz gas", J. Chem.Phys. $\underline{95}$ (1991) 6192.

72. J. Popielawski, J. Gorecki, Eds., Proceedings of the second international Symposium "Far-from-equilibrium dynamics of chemical systems", Swidno, Poland, September 3-7 1990, World Scientific, Singapore 1991.

73. A.S. Cukrowski., A.L. Kawczyński, J. Popielawski, W. Stiller , R. Schmidt, "Interaction between two chemical reactions in gases – the nonequilibrium effects from the Boltzmann and Fokker-Planck equations within a Lorentz gas model", Chem. Phys. $\underline{159}$ (1992) 37.

74. J. Popielawski , A.S Cukrowski, S. Fritzsche, "Perturbation of the thermal equilibrium by a simple chemical reaction", Physica $\underline{A188}$ (1992) 344.

75. A.S. Cukrowski, S. Fritzsche, J. Popielawski, "The theory of translational energy relaxation in binary mixtures of dilute gases with chemical reaction", Acta Phys. Polon. $\underline{A82}$ (1992) 1005.

76. A.S. Cukrowski, J. Popielawski, Lihong Qin, J.S. Dahler, " A simplified theoretical analysis of nonequilibrium effects in bimolecular gas phase reactions", University of Minnesota Supercomputer Institute Research Report UMSI 92/193 September 1992, J. Chem. Phys. $\underline{97}$ (1992) 9086.

77. F. De Pasquale, J. Gorecki, J. Popielawski, "On the stochastic correlations in a randomly perturbed chemical front", J. Phys. A. Math. Gen. $\underline{25}$ (1992) 433.

78. A.S. Cukrowski, S. Fritzsche, J. Popielawski, "Nonequilibrium effects in a bomolecular chemical reaction in a dilute gas", Acta Phys. Polon. $\underline{A84}$ (1993) 369.

79. B. Nowakowski, J. Popielawski, "The kinetic theory of the effect of chemical reaction on diffusion of a trace gas", J. Chem. Phys., scheduled for May 15, 1994 issue.

80. B. Nowakowski, J. Popielawski, "The kinetic theory of the effect of chemical reaction on diffusion", this volume.

Periodic Perturbations of an Oscillatory Chemical System

Gregory Markman[1] and Kedma Bar–Eli[2]

[1] *Department of Mathematics, University of Rostov Engelsa 105 Rostov on Don, Russia*

[2] *Sackler Faculty of Exact Sciences, School of Chemistry Tel–Aviv University, Ramat Aviv 69978, Israel*

Abstract

The Oregonator model in a CSTR is subjected to periodic modulation of the input and output flows of reagents. Three cases are investigated. In all cases the system has 3 steady states (one of which is always a saddle). In case 1, two steady states are stable. In case 2, SS2 is stable while SS1 is stable in a certain range and the system oscillates via a subcritical Hopf bifurcation in another. In case 3 similar oscillations occur but they end via saddle–loop bifurcation. Small harmonic oscillations occur if the modulation span only the stable steady state. When modulation period and amplitude are such that the modulated flow stays partly in the stable region and partly in the unstable region, the oscillations become synchronized with the modulation period. There is a range of modulation amplitude (or period) in which there is a rational ratio between the oscillations period and that of the modulation. Between any two such steps of synchronization there is always another step in which the period is the sum of the two periods, and the pattern is a combination of the patterns on its two sides. As the period increases its step size decreases. A simple power relationship exists between the step size and its period. These synchronized oscillations are always periodic and no chaotic oscillations have been observed.

1 Introduction

The Belousov- Zhabotinsky (BZ) [1, 2, 3] reaction has been a subject to many investigations. Many of the experimental results have been explained via the Field-Körös-Noyes (FKN) mechanism [4]. A simplified version of the above mechanism which includes flows to and from the system, and thus is particularily useful for investigating CSTR's, was used by De Kepper and Bar-Eli [5] and by Showalter et al. [6]. De Kepper and Bar-Eli [5] compared the simplified and complete versions to the behavior regarding the various control parameters.

Zhabotinsky [7] investigated the behavior of the BZ reaction under the influence of a periodic light modulation. Dulos [8] did similar experiments on the Briggs–Rauscher oscillator [9]. Synchronization of the oscillations with the modulation period was found in both cases. Ito [10] and Kai and Tomita [11] studied the effect of small periodic perturbations on the limit cycle of the Brusselator model near its Hopf bifurcation point. The effect of small sinusoidal perturbations on a general limit cycle has been taken by Rehmus and Ross [12] and specific models were examined by Rehmus et al. [13]. Bar-Eli [14, 15] invesigated the peristaltic effect (which is equivalent to periodically perturbing the flow) on the BZ and other systems. Weiner et al. [16, 17] modulated the minimal bromate oscillator [18] - [23] by a delayed signal from another, similar, oscillator.

All these results have prompted us to investigate more extensively the influence of modulation of a parameter of the oscillating system on the system characteristics. The simplest parameter to use is of course the flow to and from the system. This flow can be very conveniently controlled and thus the results reported here can be easily compared to experiments.

2 The Model

The model used here is the one used earlier by De Kepper and Bar-Eli [5]. This model is the same as the Oregonator model described by Field and Noyes [24] except that terms describing the flow of reactants to and from the vessel are included. The relevant equations are:

$$\dot{x} = k_1 Ay - k_2 xy + k_3 Ax - 2k_4 x^2 - k_0(t)x \tag{1}$$
$$\dot{y} = -k_1 Ay - k_2 xy + k_5 fz + k_0(t)(y_0 - y) \tag{2}$$
$$\dot{z} = k_3 Ax - k_5 z - k_0(t)z \tag{3}$$

The rate of flow term, $k_0(t)$, is modulated according to, (similarily to the earlier work [14, 15])

$$k_0(t) = k_0(1 + \epsilon \sin \omega t) \tag{4}$$

In our case the inflow consists only of the species y_0 i.e. bromide ions, while all the species namely, bromous acid (x), bromide (y) and ceric ions (z) are included in the out flow.

The system can be non-dimensionalized using the following transformation of the variables:
$x = k_1 A x'/k_2 \quad y = k_3 A y'/k_2 \quad z = k_1 k_3 A A z'/(k_2 k_5) \quad t = t'(k_1 A) \quad y_0 = k_2 y_0'/(k_3 A) \quad \omega = \omega'/(k_1 A) \quad k_0 = k_0'/(k_1 A)$, the constants $a = k_3/k_1 \quad b = 2k_4/k_2 \quad c = k_5/(k_1 A)$ and omitting the primes, the following equations are obtained :

$$\dot{x} = a(x + y - xy) - bx^2 - k_0(t)x \tag{5}$$
$$\dot{y} = -y + fz - xy + k_0(t)(y_0 - y) \tag{6}$$
$$\dot{z} = c(x - z) - k_0(t)z \tag{7}$$

By adding the equations $\dot{u} = v, \quad \dot{v} = -\omega^2 u$ with the initial conditions $u(0) = 0, \quad v(0) = 2\pi/T_{in}$ (with T_{in} being the period of the modulating force) eq. 4 becomes

$$k_0(t) = k_0(1 + \epsilon u) \tag{8}$$

The larger set is more convenient to use, since one does not have to compute the values of $\sin \omega t$ for large values of t.

3 The Autonomous Equations

The modulation of the system is investigated in three cases in which $a = 550.$ $b = .05$ $c = 25.$ are kept constant while the other parametrs vary. When the value of $\epsilon = 0$, the system becomes an autonomous one as dealt earlier by De Kepper and Bar-Eli [5].

In all three cases described, there is a region where three steady states exist. These are denoted as: SS1 where the value of x is the greatest, SS2 where the value of x is the smallest, and SS3 in the middle which behaves always as a saddle with one positive eigenvalue. Typical values are $x(SS1) \approx z(SS1) \approx 5000$ $y(SS1) \approx .5$ $x, y, z \approx 1 - 10$ for SS2 and SS3, with $x(SS2) \approx z(SS2) < x(SS3) \approx z(SS3)$ $y(SS2 > y(SS3)$.

For each case the values of y_0 and f are fixed while the value of k_0 - the flow rate - is used as a bifurcation parameter. The description of the three cases, together with the flow values for the critical points $A - D$ is given in table I and schematically in figure 1. From the above we can tabulate the ranges of existence of the stable steady states or stable oscillations or combinations of them, as seen in table II.

Typical oscillations, which occur in the indicated ranges, are shown in figure 2. The sytem spends most of the time at low values of x near SS2 and only a comparatively short times at the peak values near $x \approx 10^4$. In case 2 the period

14

is near 2 (see below), while in case 3, the time spent near SS2 increases as the flow k_0 approaches point A, since the oscillations approach saddle–loop bifurcation [32, 33].

The behavior of the system is investigated as a function of the three free parameters namely, k_0 , ϵ and T_{in}.

4 Case 1

In this case the system will be stable at either SS1 or SS2. When the modulation is activated, one obtains small harmonic oscillations around either SS1 or SS2, depending on the initial conditions. The period of these harmonic oscillations is the same as that of the modulation.

If the initial point is at SS2 and the value of the flow rate k_0 is near point A, the saddle-node point, and ϵ is such that during the oscillation the system spends some time below the point A where only SS1 exists, then the system may jump to SS1 and continue its harmonic oscillations around it. Just before the "jump" to the other steady state, the simple harmonic oscillations are slightly modified and become more pointed at the top (high x's) and more flat at the bottom (low x's), as shown in figure 3.

The occurence of the "jump" to SS1 occurs only if T_{in} is large enough, i.e. the modulation is fairly slow. Table III shows the critical flow needed to transfer from the vicinity of SS2 to the vicinity of SS1, as a function of the modulation period.

The transfer of a system from one steady state to the other has been studied extensively. Thus Bar-Eli and Geiseler [25] perturbed a bistable bromate - cerium ions in sulfuric acid and measured the kinetics of the transition from one steady state to the other. The duration of the "successful" perturbation i.e. the one that can cause the transition, must be with appropriate amplitude and duration. When the perturbing amplitude becomes smaller, its duration must become longer.

Showalter et al. [26, 27, 28] studied extensively, both theoretically and experimentally the transitions of the bistable iodate - arseneous system located near the saddle - node critical point. They perturbed the system beyond the critical point and found that as the perturbation amplitude becomes smaller the transition to the other steady state takes longer. They also found the relationship, initially pointed out by Dewel et al. [29, 30] namely, $\Delta k_0 \tau^2 = const$ where Δk_0 is the deviation from the critical point (the hysteresis limit–i.e.point A) and τ is the duration of the transfer to the available stable steady state. The system located near the hysteresis limit is *critically slowed down* [29, 30, 31] since it must change its kinetics near the saddle -node point due to one of the eigenvalues approaching zero.

The explanation to our results depicted in table III seems obvious. As the system leaves the A-B region it starts going towards SS1. If the time spent below the point A is too short, the system will revert to SS2, since it does not have enough time to reach the separatrix and traverse it; as the flow increases the system spends smaller portion of the cycle below the point A, and thus we need longer modulation times in order to have enough time to traverse the separatrix.

From eq. (4), the time to reach the critical point A from the initial point $k_0(0)$ is

$$\frac{t_{c1}}{T_{in}} = 2\pi + \frac{\arcsin \frac{k_0(A) - k_0}{\epsilon k_0}}{2\pi} \tag{9}$$

and the total time spent below point A is:

$$\Delta t_c = 2(t_{c1} - t_{min}) \tag{10}$$

where $t_{min} = T_{in}\frac{3}{4}$ is the time needed for the flow to reach its minimum.

A plot of Δt_c^2 vs. $1/\Delta k_0$ gives a straight line with slope of 0.12926 and intercept 0.087 (shown in figure 4) Thus, either from this plot or from the last column of table III, (average .135) we obtain that the minimum flow deviation (from point A) needed to transfer the system from SS2 to SS1 is inversly proportional to the square of the time spent below point A. This result is in complete agreement with the results cited above ([26]– [29]) : in order for the transition to the upper, stable SS1 to occur, the system, during its modulated perturbation, must spend enough time below critical point A.

Similar phenomenon will occur, of course, when the system is near point B ; in this case the system, initially at SS1, will, under similar conditions, go over to SS2.

5   Case 2

Autonomous oscillations (existing between 0 and D) have periods between 2.232 to 1.948 as k_0 is changed from 0 to $k_0(D)$ (tables I and II) . When the modulation is activated various results can occur depending on the values of the parameters and the initial point.

a) Small harmonic oscillations (with the modulation period) around SS2 (SS1) occur when $k_0(A) < k_{0min}$ $(k_0(C) < k_{0min} < k_{0max} < k_0(B))$ and the initial point is near SS2(SS1).

b) When $k_{0max} < k_0(D)$, the system oscillates in a quasiperiodic manner with $2.13 > T_{out} > 2.02$ (similarily to the autonomous ones- figure 2) independent of the initial point.

c) $k_0 > k_0(C)$. For certain values of ϵ, as it is shown in table IV, the oscillations are synchronized with the modulation with period $T_{out} = 2$ provided the initial point is near the autonomous limit cycle. It is seen, that the limits of synchronization are maximized when k_0 is near point D. For points beyond these limits of ϵ the oscillations are quasiperiodic as described in (b) above.

d) When $k_{0min} < k_0(C) < k_0(D) < k_{0max}$ the system crosses the Hopf bifurcation point at C; the small harmonic oscillations around SS1 may be intercepted by small fast oscillations (with "Hopf" period - in our case $f \approx .08$). These oscillations grow, and if enough time is available (i.e. T_{in} is long), the system goes over to large oscillations.

Figures 5,6 and 7 show these cases. In figure 5, T_{in} is long enough for the fast "Hopf" oscillations to be seen in the midst of the simple harmonic ones. In figure 6, T_{in} is longer and the system goes over the separatrix to the large oscillations. Computation errors resulting from the very short period of the "Hopf" oscillations may be responsible for the seemingly erratic behavior seen in the figure. Figure 7 shows the system with very large T_{in}. Series of large oscillations with a span of slow harmonic oscillation around SS1, each about half the time are seen. In between of thees two types of oscillations, the growing and decaying Hopf oscillations are observed.

6   Case 3

6.1   No Modulation

Oscillations exist for $k_0 < k_0(A)$ (even for negative, nonphysical, flows) with period increasing to infinity towards $k_0(A)$ as $1/\sqrt{\Delta}$, where $\Delta = k_0 - k_0(A)$ [32, 33, 34], where they cease at a saddle-loop.

6.2   Change with flow

When $k_{0max} = k_0(1 + \epsilon) < k_0(A)$ large, quasiperiodic oscillations, similar to the autonomous ones, are obtained. Thus when $k_0 = 0.23$ and $k_{0max} = .46 < k_0(A)$, large oscillations with periods changing from 2.30 to 2.64 are obtained. They are very similar to those obtained in the Case 2 above.

When $k_{0max} > k_0(A)$, the resulting oscillations are a combination of large ($L \approx 10^4$ similar to the autonomous ones) and small ($S \approx 1 - 10$) peaks always synchronized with the modulation. Thus when $k_0 = 0.24 < k_{0max} = .48 > k_0(A)$ oscillations with $T_{out} = 5$ and pattern $L^2 S$ are obtained.

6.3 Change with ϵ

The behavior of the system with change of ϵ has been done mainly under the constant flow and period conditions namely, $k_0(A) < k_0 = .475 < k_0(C)$ and $T_{in} = 1$. Under these conditions the flow is partly above and partly below the critical point A.

Typical oscillations are shown in figure 8. The oscillations have the pattern $LSLS^2$ with reduced period of $P = T_{out}/T_{in} = 7$. The period is thus seen to be synchronized [35] with the modulation period i.e. the latter is an exact multiple of the former. Note the similarity to the autonomous oscillations, figure 2, and those cited above for $k_0 = .24$ (although with different pattern): the large oscillations look the same while the small ones (in the vicinity of SS2) differ in size.

As ϵ increases the pattern changes from S type oscillations to mixture of large and small and finally to LS type oscillations. Calculations with the nonphysical values of $\epsilon > 1$, show that the patterns continue changing and finally end at L pattern with period 2. Between LS and L patterns, there are combination patterns and periods as explained below. The results for $\epsilon < 1$ are shown in table V, while those for $\epsilon > 1$ (nonphysical) are given in table VI.

The oscillations are arranged in steps. In other words there is a range of ϵ in which the same pattern and period persists, thus making the plot of $P = T_{out}/T_{in}$ vs ϵ a step like plot. Between any two steps of certain pattern, and period say P_1 and P_2, there exist another step with a pattern which is a combination of the two namely, P_{12} with period which is the sum of the two. Thus between period 4 step $.392 < \epsilon < .550$ with pattern LS^2, and pattern 3 step between $.672 < \epsilon < 1$ with pattern LS there exists a period 7 step with pattern $LSLS^2$ (shown in figure 8), period 10 $((LS)^2 LS^2)$ etc. In this way very long periods with complicated patterns may result, but careful examination shows that no chaotic, or even quasiperiodic oscillations occur. Note also that the patterns of type $L^n S$, beyond $\epsilon > 1$, are just combinations of LS on one side and L on the other. Thus the two tables V and VI, can be looked upon as one table starting with S pattern with period 1 and ending with L pattern with period 2, with all the combination patterns and periods in between. These results are depicted in figure 9 where the data in table V are shown.

It should be born in mind that slight changes in the pattern may occur between the beginning of the step to its end due to small changes in the maxima of the small oscillations as ϵ changes. As can be seen in figure 10 which shows the change of the small maxima with ϵ. As ϵ increases the maxima of the small i.e. S type oscillations increase, and eventually a new small maximum appears, thus changing the pattern but not the period. In the period 5 step, for example, the pattern at the beginning of the step is LS^2 while at the end it is LS^3 with no change in the period.

When the period of modulation becomes very large the resulting oscillations always follow the modulated ones thus $T_{in} = T_{out}$ but the pattern changes as shown in the table VII. In this case the system stays about half the time below point A and performs large oscillations there (with slowly changing period i.e. slowly changing times between the large peaks), and about half the time near SS2. When ϵ is very small, the time spent below point A may be too short for even a single large oscillation to be executed.

Comparing this case to that previously described ($T_{in} = 1$, figure 9), one sees that with long modulation periods only patterns of the type $L^n S$ are obtained. These patternss are not seen in figure 9, but they do appear in table VI for the region $\epsilon > 1$. Thus the two sets of results for $T_{in} = 1$ and $T_{in} = 100$ differ only in the periods but not in the patterns.

As we have seen in section 6.2, the flow $k_0 = .475$ with which these results were obtained, need not be above $k_0(A)$; the only demand is that during the modulation the system will be partly in the stable and partly in the oscillating region. Using $k_0 = .4735 < k_0(A)($ but very near it) very similar results to those obtained in table V and figure 9 are obtained. In particular, the steps, patterns and periods and combinations of them are nearly the same. The only difference is that the position of the steps move slightly to lower values of ϵ's as k_0 decreases. Thus, for example, for $k_0 = .475$ pattern LS^7 with period 10 is obtained at $\epsilon = .145$ (see table V), while the same pattern and period are obtained for $k_0 = .4735$ at $\epsilon = .115$. In a similar way, using table VI, pattern $L^2 S$ with period 5 is obtained at $1.7 < \epsilon < 1.85$, while from section 6.2, this pattern is obtained at the lower flow of $k_0 = .24$ with $\epsilon = 1$.

6.4   Change of modulation period

When the amplitude of the modulation is kept constant and its period is being changed i.e. when, as in our case $\epsilon = 0.6$, very similar results to the previous ones are obtained. The results are summarized in tables (VIII–XI) and in figures 11, 12. The main finding is, again, that the period, or rather the reduced period $P = \frac{T_{out}}{T_{in}}$, and the pattern, are step functions of the parameter - in this case - the modulation period; moreover, between two periods (and patterns) there exist another period which is the sum of the two and pattern which is a combination of the two.

In table VIII the behavior at high values of T_{in} is seen. The pattern is always of the type $L^n S$, with n decreasing with T_{in}. Note the similarity to the results of table VII.

Further reduction of T_{in} results in patterns of the type $L^n S$, but this time the period is $n+1$ and finally, when the modulating period is very short, S pattern is observed as shown in table IX. As in the previous case ϵ changes, the range of existence of each step with the same pattern decreases as its period increases. We shall discuss this point below.

Patterns of the type LS^n are given in table X with periods $2n+3$ and combination patterns between periods 5 and 4, 4 and 3, and 3 and 2 are given in table XI.

In all of these tables, similar behavior is observed: smaller range of pattern and period as the latter increases and combination patterns and periods (synchronized [35] with the modulation) made up of those on its two sides. No quasiperidicity or chaos were obtained, although very long periods and complicated patterns (spanning very small ranges) are clearly seen.

In figure 11 a plot of the reduced period vs. the modulation period is given, together with the obsereved patterns ohowing the data in tables VIII, IX ,X. The pattern may change along the step, in a similar way to that of figure 10 above. Thus period 2 step is L on the left and LS on the right.

Since each step is made up of the two lower steps on its sides, we have period 3 LS coming from $L+S$ on the left, and L^2S coming from $LS+L$ on the right, and similarily for higher period steps.

Figure 12 shows a detail of the above figure near $T_{in} = 2.00$ indicating the various combinations of the patterns LS (period 2) and L^2S (period 3). Thus on the left combinations of the type $(LS)^n L^2S$, $P = 2n+3$ are obtained, while on the right type $LS(L^2S)^n$ $P = 3n+2$ are shown.

The change in maximum of the small amplitude oscillation is seen in figure 13 where the maxima are plotted vs. the modulating period. The small oscillation of the pattern L^2S with period of 3, increases slowly at first, and then very sharply, until it reaches some critical value, at which point it disappears, a large amplitude oscillation comes instead, thus changing the period (to 4) and the pattern (to L^3S with a small oscillation that starts to grow again It is obvious that as the small oscillation grows, it passes, towards the end of the step, a separatrix after which it become a large oscillation which consumes more time.

6.5   Self Similarity

In figure 9 two series of steps are seen. On the left the LS^n series, which is made up of an S step added each time to LS^{n-1} and thus increasing the period by 1. On the right, the steps $(LS)^n LS^2$ are made by adding each time a set LS and increasing the period by 2. In a similar way we see the "build-up" of LS^n and L^nS (by adding

20

S's or L's respectively), in figure 11. Again, in figure 12 the same phenomena are seen on a smaller scale: $(LS)^n L^2 S$ are made by adding (LS)'s (period increase by 2), while $LS(L^2 S)^n$ are made by adding $(L^2 S)$'s (period increase by 3). Each new step occupies a smaller range than the previous one. It is suggested that similar contraction rules hold. Plotting the period (P) vs. range (R) of the three sets LS^n, $L^n S$ and $(LS)^n L^2 S$, figure 14, the power law

$$PR^a = b \qquad (11)$$

is obtained.

The step size (range) decreases very sharply as the period (or n) increases in all three cases. The values of a and b for the various cases are given in table XII

The data for figure 14 are given in table XIII.

7   Discussion

The three cases investigated in this work have each its own characteristics.

Case 1 is the simplest one; having only stable steady states, the system can only oscillate with the same period as the modulation and with amplitude which is dictated by the change in the steady state values in the vicinity of the modulated flow. Similar results were obtained by Buchholz and Schneider [36], who modulated a stable BZ reaction located near a focus point. The system can also transfer from one steady state to the other one provided the modulated flow stays long enough below (above) the lower (the upper) critical point. The kinetics of the "jump" can be easily explained and is in an agreement with earlier results ([26]– [29]) namely, that the minimum flow deviation from the critical point, needed to transfer the system from one steady state to the other one, is inversely proportional to the square of the time spent below the critical point.

Case 2 seems to be very complicated: harmonic oscillations near stable steady states, "jumps" from one steady state to other, quasiperiodicity, transfer from small to large oscillations and interference with the movement near the focus of the Hopf bifurcation are all possible. It seems that further investigation is needed here.

The main finding in case 3 is the fact that the oscillations are synchronized [35] with the modulation i.e. the ratio between the modulation period and the resulting oscillation period is always an integer. The plot of the oscillation period vs. the modulation period (or amplitude) is a step like plot, with step size decreasing with the increase of the period. Each step has its own pattern made up of a certain combination of small and large oscillations and is a composition of the two steps surrounding it.

The pattern and period of the BZ oscillations can thus be controlled by the experimentalist by properly perturbing the flow to the CSTR. By changing either the period or the amplitude of the perturbation, one can achieve a variety of periods and patterns. It is desirable to test these predictions in a carefully designed experiment.

Table I: Bifurcation points (values of k_0) for the various cases

case	y_0	f	A[a]	B[b]	C[c]	D[d]
1	4.	.5045	1.774430068	222.3895935	–	–
2	4.	.533	1.736235858	221.4339273	.4883548	1.070279435
3	10.8	.533	.4735965056	124.4243564	.5224043335	–

[a] A:lower saddle-node SS2 and SS3
[b] B:upper saddle-node SS1 and SS3
[c] Hopf bifurcation on SS1
[d] "end" of oscillations in case 2

Table II: Range of stable steady states and/or oscillations

case	SS1	SS1+SS2	SS2	osc.	osc. +SS1
1	0-A	A-B	B-∞	-	-
2	D-A	A-B	B-∞	0-C	C-D
3	-	C-B	A - C ; B-∞	0-A	-

Table III: Values of k_0 and T_{in} where transition to SS1 occurs

$\epsilon = 0.5$ initial point near SS2

k_0	k_{0min}	$k_0(A) - k_{0min}$	T_{in}	$\frac{t_{c1}}{T_{in}}$	$\frac{\Delta t_c}{T_{in}}$	Δt_c	$\Delta k_0(\Delta t_c)^2$
2.	1.	.77443007	.895	.96379	.42758	.381	.1124
3.2	1.6	.17443007	6.325	.8250	.1500	.95	.157
3.4	1.7	.07443007	14.565	.79727	.09454	1.3769	.141
3.5	1.75	.02443007	43.465	.77662	.053249	2.314	.131

Table IV: Limits of synchronization as function of k_0 and ϵ ; $T_{in} = 1$

k_0	.5	.7	.9	1.05	1.2	1.4	1.8	2.0
ϵ_{min}	.78	.40	.25	.17	.35	.55	.80	.92
ϵ_{max}	1.	1.	1.	1.	1.	1.	1.	1.

Table V: Pattern and reduced period as a function of ϵ

$$k_0 = .475 \quad T_{in} = 1$$

ϵ	pattern	T_{out}/T_{in}
0.05 - .089	S	1
0.097	LS^{24}	27
0.1	LS^{20}	23
0.105	LS^{16}	19
0.107	LS^{15}	18
0.115	LS^{12}	15
0.118	LS^{11}	14
0.124	LS^{10}	13
0.130	LS^9	12
0.136	LS^8	11
0.145	LS^7	10
0.160	LS^6	9
0.175 - 0.180	LS^5	8
0.200	LS^4	7
0.220	$LS^3 LS^4$	13
0.223	$(LS^3)^2 LS^4$	19
0.225	$(LS^3)^3 LS^4$	25
0.226	$((LS^3)^4)LS^4$	31
0.230 - 0.255	LS^3	6
0.282 - 0.300	$LS^2...LS^3$	5
0.392 - 0.550	$LS...LS^2$	4
0.600	$LSLS^2$	7
0.630	$(LS)^2 LS^2$	10
0.650	$(LS)^3 LS^2$	13
0.655	$(LS)^4 LS^2$	16
0.660	$(LS)^5 LS^2$	19
0.663	$(LS)^6 LS^2$	22
0.664	$(LS)^7 LS^2$	25
0.665	$((LS)^7 LS^2)^2 (LS)^8 LS^2$	78
0.666	$(LS)^8 LS^2$	28
0.667	$(LS)^9 LS^2$	31
0.669	$(LS)^{14} S$	43
0.671	$(LS)^{27} S$	82
0.672 - 1	LS	3

Table VI: Periods and pattern for the nonphysical regime $\epsilon > 1$

$$k_0 = .475 \quad T_{in} = 1$$

ϵ	period	pattern
1 - 1.55	3	LS
1.6	8	LSL^2S
1.7 - 1.85	5	L^2S
1.9	12	L^2SL^3S
1.95 - 2	7	L^3S
3	2	L

Table VII: Pattern change with ϵ

in all items $T_{out} = T_{in} = 100$

ϵ	period	pattern
0.003	1	S
0.0035 - 0.0045	1	LS
0.0048 - 0.006	1	L^2S
0.007 - 0.009	1	L^3S
0.01	1	L^4S
0.1	1	$L^{11}S$
0.2	1	$L^{14}S$
0.9 - 1.0	1	$L^{20}S$

Table VIII: T_{out} vs. T_{in}

$$\epsilon = 0.6, \; k_0 = .475$$

T_{in}	$\frac{T_{out}}{T_{in}}$	pattern
100	1	$L^{18}S$
50	1	$L^{10}S$
25	1	L^5S
20	1	L^4S
10 - 8	1	L^2S
7 - 5	1	LS
4.5 - 2.4108	1	L

Table IX: T_{out} vs. T_{in} Patterns : $L^n S$

$\epsilon = 0.6,\ k_0 = .475$

T_{in}	$\frac{T_{out}}{T_{in}}$	pattern
2.3998 - 2.3987	17	$L^{16}S$
2.3986 - 2.3972	16	$L^{15}S$
2.3971 - 2.3955	15	$L^{14}S$
2.3953 - 2.3933	14	$L^{13}S$
2.3931 - 2.3907	13	$L^{12}S$
2.3904 - 2.3873	12	$L^{11}S$
2.387 - 2.3831	11	$L^{10}S$
2.3827 - 2.3776	10	$L^9 S$
2.3771 - 2.3702	9	$L^8 S$
2.3696 - 2.36	8	$L^7 S$
2.3592 - 2.3452	7	$L^6 S$
2.344 - 2.3224	6	$L^5 S$
2.3206 - 2.2846	5	$L^4 S$
2.2813 - 2.2131	4	$L^3 S$
2.2063 - 2.0427	3	$L^2 S$
2.0202 - 1.381	2	$LS...L$

Table X: Period vs. T_{in}

T_{in}	$\frac{T_{out}}{T_{in}}$	pattern
1.2689 - 1.033	3	LS
.9753 - .8568	4	$LS^2...LS$
.8184 - .7477	5	LS^2
.7187 - .6716	6	LS^3
.6484 - .6146	7	$LS^4...LS^3$
.5953 - .5699	8	LS^4
.5534 - .5336	9	LS^5
.5193 - .5035	10	LS^6
.4908 - .4780	11	$LS^7...LS^6$
0.162 - 0.05	1	S

Table XI: Details of period vs. T_{in}

T_{in}	$\frac{T_{out}}{T_{in}}$	pattern	remarks
2.2845	34	$L^3S(L^4S)^6$	$5 > P > 4$
2.2843	19	$L^3S(L^4S)^3$	$5 > P > 4$
2.2841 - 2.284	14	$L^3S(L^4S)^2$	$5 > P > 4$
2.283	9	L^3SL^4S	$5 > P > 4$
2.282	13	$(L^3S)^2L^4S$	$5 > P > 4$
2.2819	17	$(L^3S)^3L^4S$	$5 > P > 4$
2.2817	21	$(L^3S)^4L^4S$	$5 > P > 4$
2.212	11	$L^2S(L^3S)^2$	$4 > P > 3$
2.211 - 2.209	7	L^2SL^3S	$4 > P > 3$
2.208	10	$(L^2S)^2L^3S$	$4 > P > 3$
2.207	13	$(L^2S)^3L^3S$	$4 > P > 3$
2.2065	28	$(L^2S)^8L^3S$	$4 > P > 3$
2.042	20	$LS(L^2S)^6$	$3 > P > 2$
2.041	11	$LS(L^2S)^3$	$3 > P > 2$
2.039 - 2.038	8	$LS(L^2S)^2$	$3 > P > 2$
2.036 - 2.027	5	LSL^2S	$3 > P > 2$
2.025	7	$(LS)^2LS^2$	$3 > P > 2$
2.022	9	$(LS)^3LS^2$	$3 > P > 2$
2.021	15	$(LS)^6LS^2$	$3 > P > 2$
2.0205	23	$(LS)^{10}LS^2$	$3 > P > 2$

Table XII: a and b of eq. 11

pattern	a	b
LS^n	.439	1.60
L^nS	.348	1.61
$(LS)^nL^2S$	.258	1.45

Table XIII: Reduced period vs. range of step

$$\epsilon = 0.6 \quad k_0 = .475 \quad y_0 = 10.8$$

pattern $L^n S$

begin	end	period	range
1.381	2.0202	2	.6393
2.0427	2.2063	3	.1637
2.2131	2.2813	4	.0683
2.2846	2.3206	5	.0361
2.3224	2.344	6	.0217
2.3452	2.3592	7	.0141
2.36	2.3696	8	.0097
2.3702	2.3771	9	.0070
2.3776	2.3827	10	.0052
2.3831	2.387	11	.0040
2.3873	2.3904	12	.0032
2.3907	2.3931	13	.0025
2.3933	2.3953	14	.0021
2.3955	2.3971	15	.0017
2.3972	2.3986	16	.0015
2.3987	2.3998	17	.0012

pattern LS^n

begin	end	period	range
1.381	2.0202	2	.6393
1.033	1.2689	3	.236
.8568	.975	4	.1186
.7477	.8184	5	.0708
.6716	.7187	6	.0472
.6146	.6484	7	.0339
.5699	.5953	8	.0254
.5336	.5534	9	.0199
.5035	.5193	10	.0159
0.4780	0.4908	11	.0129

pattern $(LS)^n L^2 S$

begin	end	period	range
2.02634	2.03687	5	.01054
2.02321	2.02545	7	.00225
2.0220	2.02288	9	.00089
2.0214	2.02184	11	.00045
2.02106	2.02131	13	.00026
2.02085	2.02100	15	.00016

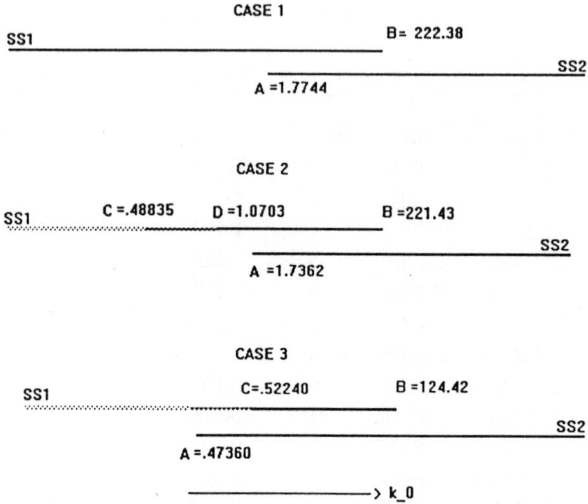

Figure 1: A schematic plot of the stable and oscillatory steady states. SS3 (not shown) connects points A and B in all cases. This figure is a pictorial representation of tables I and II.

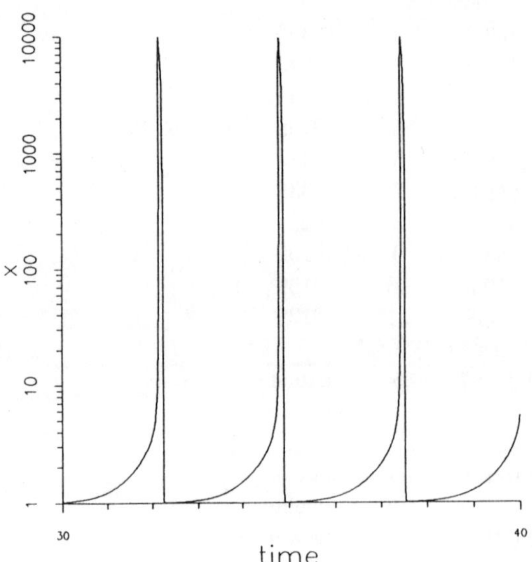

Figure 2: Typical autonomous ($\epsilon = 0$) oscillations (plotted for case 3) $y_0 = 10.8$ $f = .533$ $k_0 = .3$ Plot of x vs time is shown.

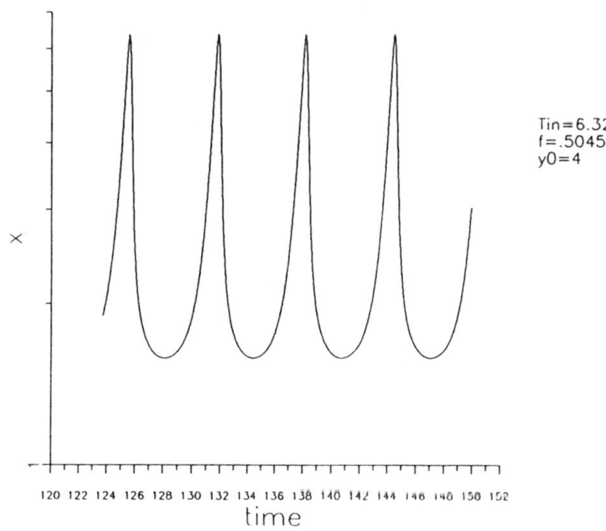

Figure 3: Case 1: nearly harmonic oscillations (near SS2), just before the "jump" to SS1. Note that the maxima are sharp while the minima are flat compared to real harmonic oscillations. $T_{in} = 6.32$ $k_0 = 3.2$ $\epsilon = .5$

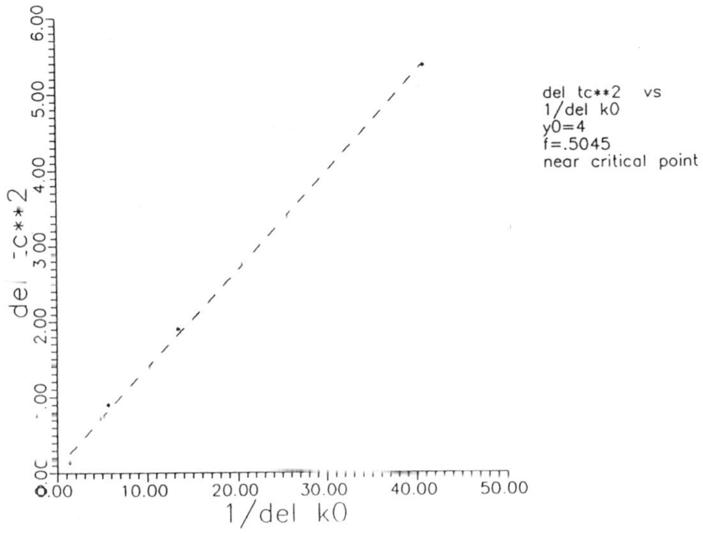

Figure 4: Case 1: plot of Δt_c^2 (the time spent during an oscillation below point A) vs. $1/\Delta k_0 = 1/(k_0 - k_0(A))$. $\epsilon = 0.5$. Initial point near SS2 .

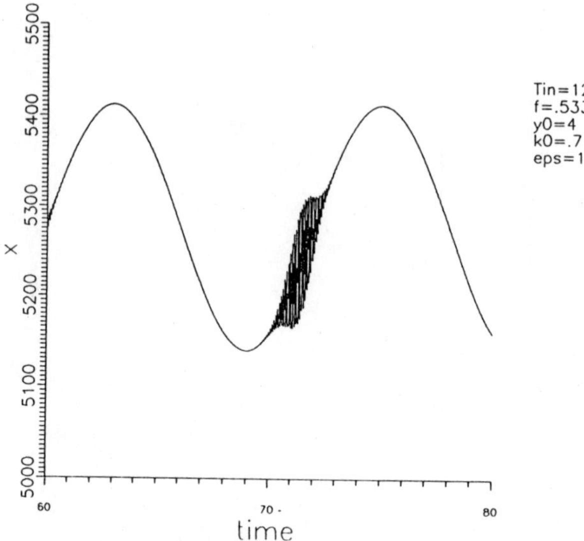

Figure 5: Case 2: small harmonic oscillations around SS1 intercepted by fast increasing and decreasing Hopf oscillations. $k_0 = .7$ $\epsilon = 1$ $T_{in} = 12$

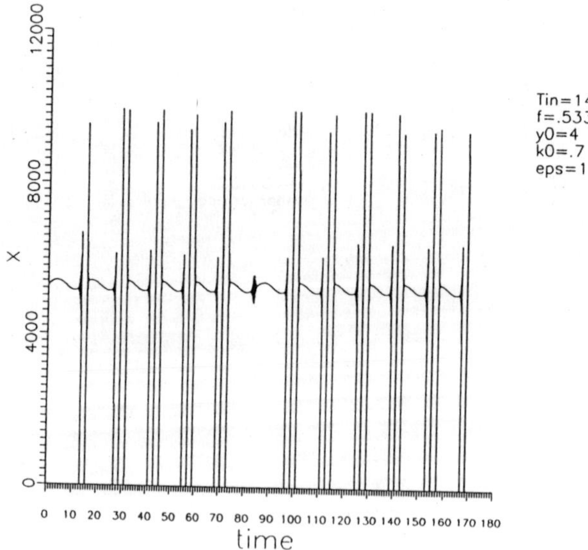

Figure 6: Case 2: small harmonic oscillations transfered to erratic large ones. $k_0 = .7$ $\epsilon = 1$ $T_{in} = 14$

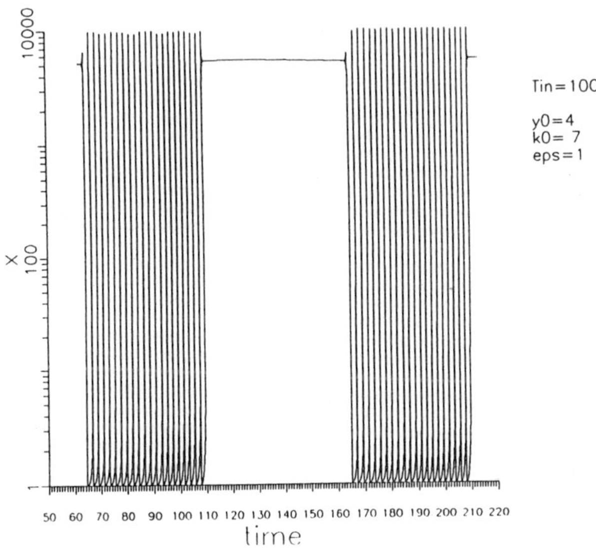

Figure 7: Case 2: series of large oscillations is intercepted by a steady state. Hopf type oscillations are seen in between. $k_0 = .7$ $\epsilon = 1$ $T_{in} = 100$

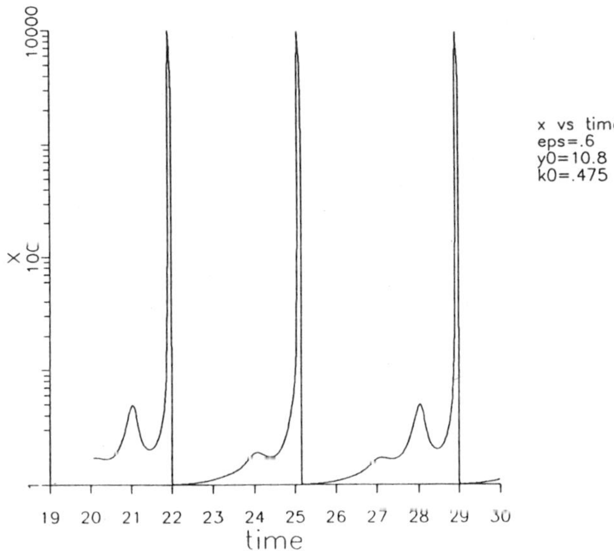

Figure 8: Case 3: plot of x vs. time for $k_0 = .475$ $T_{in} = 1$ $\epsilon = .6$ with pattern $LSLS^2$ and reduced period of 7.

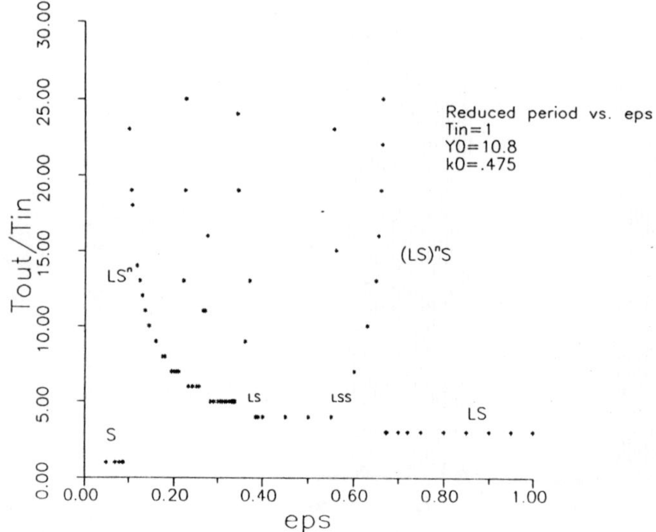

Figure 9: Case 3: plot of period vs. ϵ for $k_0 = .475$ $T_{in} = 1$. Patterns are indicated on the plot.

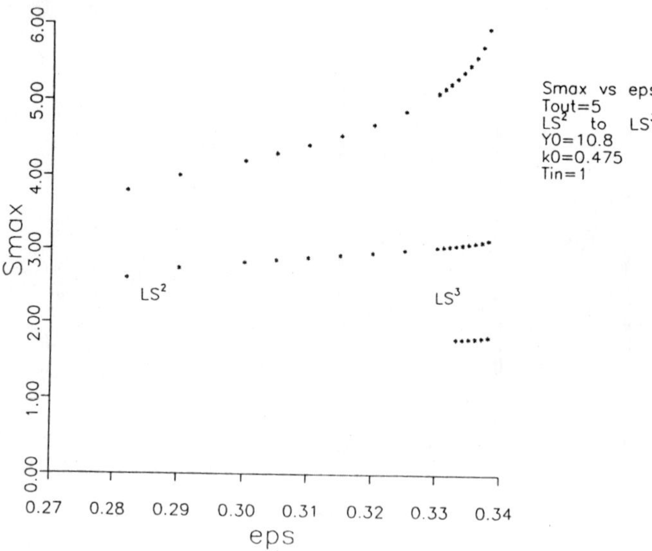

Figure 10: Case 3: plot of the small oscillations maxima vs. ϵ for period 5 step. The pattern is LS^2 for small ϵ, and LS^3 for large ones. Similar behavior is noted in figure 9 for the period 4 step.

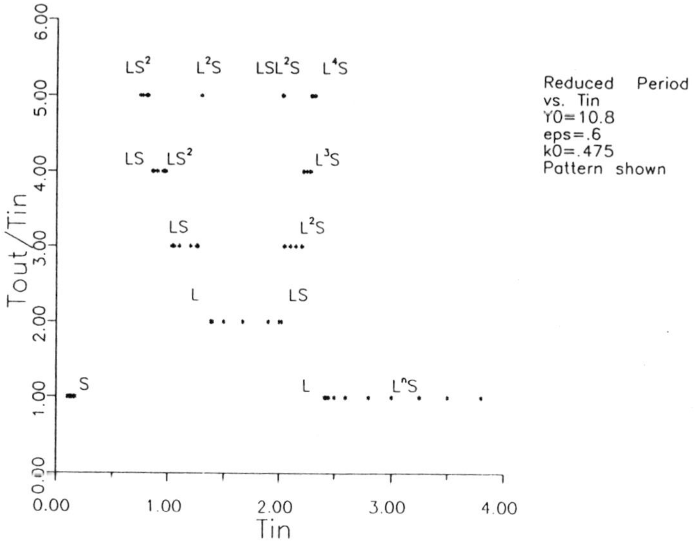

Figure 11: Case 3: plot of the reduced period vs. modulating period $\epsilon = .6$

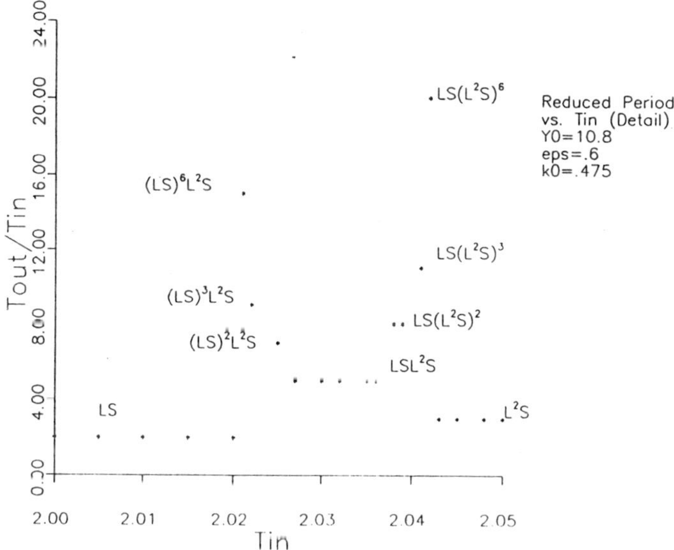

Figure 12: Case 3: plot of the reduced period vs. modulating period for the patterns located between L^2S, period $P = 3$ and LS, period $P = 2$. Combination periods and patterns $(LS)^n L^2 S$ and $LS(L^2 S)^n$ are clearly seen.

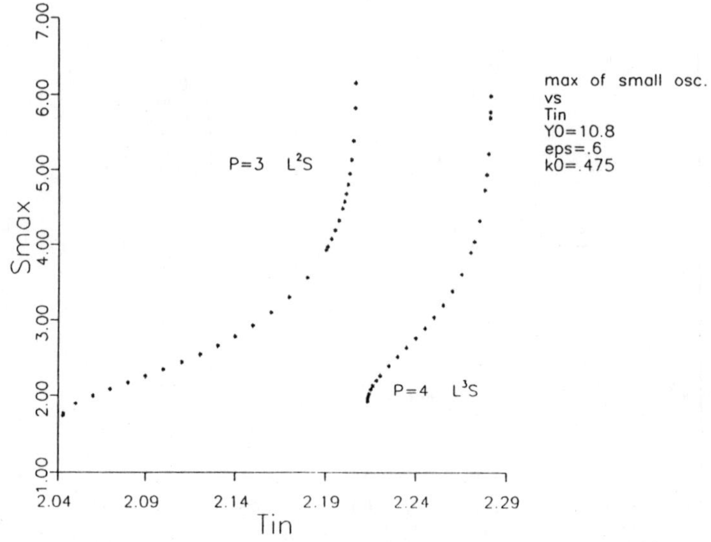

Figure 13: Case 3: plot of the maxima of the small oscillation vs. the modulating period for the case of L^2S, period $P = 3$ and L^3S , period $P = 4$. Note the sharp increase of S_{max} just before the point where the step ends.

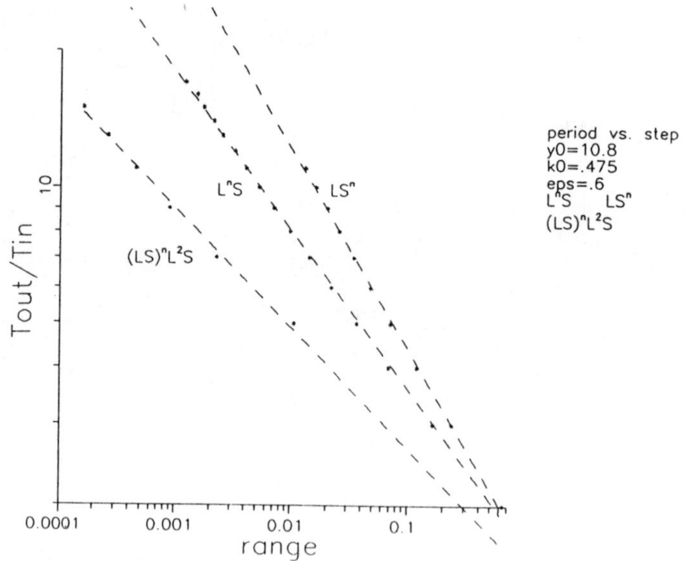

Figure 14: Case 3: $\log P$ vs. $\log R$ (the range of the period step) for the case of LS^n , L^nS and $(LS)^nL^2S$ patterns.

References

[1] B.P. Belousov; Sb.Ref. Radiats. Med. 145 (1958)

[2] A.M. Zhabotinsky; Dokl. Akad. Nauk. USSR $\underline{157}$ 392 (1969)

[3] A.M. Zhabotinsky; Biofizica $\underline{9}$ 306 (1964)

[4] R.J.Field, E. Körös , R.M. Noyes; J.Am.Chem.Soc. $\underline{94}$ 8649 (1972)

[5] P.De Kepper, K.Bar-Eli; J.Phys.Chem. $\underline{87}$ 480 (1983)

[6] K. Showalter, R.M. Noyes , K. Bar-Eli; J. Chem. Phys. $\underline{69}$ 2514 (1978)

[7] A.M. Zhabotinsky; Kontsentratsionnie avtokolebaniya, Nauka, Moskva (1974)

[8] E. Dulos in "Nonlinear Phenomena in Chemical Dynamics", C. Vidal and A. Pacault, Eds. Springer - Verlag, Berlin, 1981 pp. 140 -146

[9] T.S. Briggs, W.C. Rauscher; J. Chem. Educ. $\underline{50}$ 7 (1973)

[10] A. Ito; Prog. Theor. Phys. $\underline{61}$ 45 (1979)

[11] T. Kai, K. Tomita; Prog. Theor. Phys. $\underline{61}$ 54 (1979)

[12] P. Rehmus, J. Ross; J. Chem. Phys. $\underline{78}$ 3747 (1983)

[13] P. Rehmus, E.C. Zimmermann, J. Ross, H.L. Frisch; J. Chem. Phys. $\underline{78}$ 7241 (1983)

[14] K. Bar-Eli; J. Phys. Chem. $\underline{89}$ 2852 (1985)

[15] K. Bar-Eli in " Temporal Order" L. Rensing and N.I. Jaeger Eds., Springer Verlag, Berlin 1985 p. 126

[16] J. Weiner, F.W. Schneider, K. Bar-Eli; J Phys. Chem. $\underline{93}$ 2704 (1989)

[17] J. Weiner, R.Holz, F.W. Schneider, K. Bar-Eli; J Phys. Chem. $\underline{96}$ 8915 (1992)

[18] K. Bar-Eli; J. Phys. Chem. $\underline{89}$ 2855 (1985)

[19] K. Bar-Eli, W. Geiseler; J. Phys. Chem. $\underline{87}$ 3769 (1983)

[20] K. Bar-Eli, R.J. Field, J. Phys. Chem. $\underline{94}$ 3660 (1990)

[21] C.E. Dateo, M. Orban , P. De Kepper, I.R. Epstein; J. Am. Chem. Soc. $\underline{104}$ 504 (1982)

[22] W. Geiseler; Ber. Bunden-Ges. Phys. Chem. $\underline{86}$ 721 (1982)

[23] W.Geiseler; J. Phys. Chem. <u>86</u> 4394 (1982)

[24] R.J. Field, R.M. Noyes; J. Chem. Phys. <u>60</u> 1877 (1974)

[25] K. Bar-Eli, W.Geiseler; J. Phys. Chem. <u>87</u> 1352 (1983)

[26] N. Ganapathisubramanian, J.S. Reckley, K. Showalter; J. Chem. Phys. <u>91</u> 938 (1989)

[27] G.A. Papsin, A. Hanna, K. Showalter; J. Phys. Chem. <u>85</u> 2575 (1981)

[28] T. Pifer, N. Ganapathisubramanian, K. Showalter; J. Chem. Phys. <u>83</u> 1101 (1983)

[29] G. Dewel, P. Borckmans, D. Walgreaf; J. Phys. Chem. <u>88</u> 5442 (1984)

[30] G. Dewel, P. Borckmans, D. Walgreaf; J. Phys. Chem. <u>89</u> 4670 (1985)

[31] M. Heinrichs, F.W. Schneider; J. Phys. Chem. <u>85</u> 2112 (1981)

[32] K. Bar-Eli , M. Brøns; J. Phys. Chem. <u>94</u> 7170 (1990)

[33] C. Kaas-Petersen, S.K. Scott; Physica <u>D32</u> 461 (1988)

[34] V. Gaspar, K. Showalter; J. Chem. Phys. <u>88</u> 778 (1988)

[35] N. Minorsky "Nonlinear Oscillators" D. Van Nostrand Company Inc., Princeton N.J. (1962) p.438 (and references there).

[36] F. Buchholz, F.W. Schneider; J. Am. Chem. Soc. <u>105</u> 7450 (1983)

SYMMETRY BREAKING AND FRACTAL DEPENDENCE ON INITIAL CONDITIONS IN DYNAMICAL SYSTEMS

A. Okniński[*], R. Rynio[*], J. Peinke[+]

[*] University of Technology (Politechnika Świętokrzyska),
Phys. Dept., Al. 1000 lecia PP 7, 25-314 Kielce, Poland

[+] C.N.R.S. - C.R.T.B.T., B.P. 166, 25 Avenue de Martyrs,
F - 38042 Grenoble Cedex 9, France

ABSTRACT

It is shown that symmetry caused degeneracy, leading to the presence of two co-existing attractors, may lead to fractal dependence on initial conditions. Breaking of a discrete symmetry leads to fractal dependence on initial conditions in 1D noninvertible mappings as well as in 2D invertible mappings. It is demonstrated that in the case of dynamical systems with symmetry it is the symmetry breaking which determines the basin boundary structure. We suggest that a generalization for systems of ordinary differential equations is possible.

1. INTRODUCTION

A typical nonlinear dynamical systems can have several co-existing attractors. Presence of co-existing attractors may lead to complicated structure of attractor basin boundaries.

Recently, basin boundaries of a broad class of nonlinear systems having co-existent attractors have been studied [1-8]. It was demonstrated that the basin boundaries of 1D maps can be classified as being smooth, quasifractal, or fractal [7,8].

A mechanism generating fractal boundaries in the case of coexistence of two attractors (i.e. in bistable systems) in 2D and 3D maps was studied [9,10]. It was shown that chaotic forcing applied to a bistable system can generically lead to a nowhere differentiable boundary [9,10] (c.f. also [11] for a nongeneric example). Fractal boundaries may also be created by 1D maps using qualitatively equivalent approach of generating functions [12,13]. It may be shown that bistability plus transient chaos are already sufficient to generate fractal boundaries [12,13] due to the presence of chaotic repellers [14].

Coexisting attractors may generically appear in dynamical systems due to the presence of symmetry. Let us note that in all 1D dynamical systems reported above to have qusifractal or fractal boundaries multiple attractors can be associated with presence of symmetry.

We thus propose a general treatment of 1D nonlinear dynamical systems with symmetry. In the next Section properties of 1D maps with the simplest symmetry possible are discussed and several examples of such maps are given. In Section 3 it is demonstrated that co-existing attractors appear in 1D dynamical systems in the process of symmetry breaking. The structure of the attractor basin boundaries is related with broken symmetry extending the work of Kocarev [7,8]. It is shown in Section 4 that apparently unrelated systems can be shown to possess analogous symmetries and can be thus predicted to display analogous behavior. For example, symmetry can be shown to determine charecter of intermittent chaos and hence dynamical systems with the same symmetry exhibit intermittent chaos of the same form. In the next Section possible applications to 2D noninvertible maps and systems of ordinary diffrential equations are discussed. It is stressed that symmetry induced bistability may lead to fractal boundaries without chaotic forcing. Other consequences of symmetry for 1D maps are also discussed.

2. ONE DIMENSIONAL NONINVERTIBLE MAPS WITH SYMMETRY

Let us consider a map f:

$$x_{N+1} = f(x_N, a), \qquad f : [A, B] \longrightarrow [A, B], \qquad (1)$$

a **being** the control parameter, which has a symmetry s:

$$s \circ f = f \circ s , \qquad s : [A, B] \longrightarrow [A, B], \qquad (2a)$$

and we shall investigate the simplest case

$$s^2 \equiv s \circ s = 1. \tag{2b}$$

Due to the property (2) the dynamical system (1) has two kinds of M - cycles. There can be a M-cycle A symmetric under s:

$$\{x^A_{*_1}, \ldots, x^A_{*_M}\} = s \; \{x^A_{*_1}, \ldots, x^A_{*_M}\} . \tag{3a}$$

In what follows we shall use a shorthand notation to denote M-cycle, $\{x^A_{*_M}\} \equiv \{x^A_{*_1}, \ldots, x^A_{*_M}\}$.

In the case when a M - cycle B is not symmetric under s then there exists another M - cycle C such that:

$$s \; \{x^B_{*_M}\} = \{x^C_{*_M}\}, \quad s \; \{x^C_{*_M}\} = \{x^B_{*_M}\}. \tag{3b}$$

The case (3b) corresponds to degeneration of dynamics - there are two solutions for a given value of the control parameter a.

Let us consider a maps of form (1)

$$f(x, a) = (1-a) \, x + a \, x^3,$$

$$s(x) = -x, \quad [A, B] = [-1, 1], \tag{4}$$

Let us note that obviously $s \circ s = 1$.

For $a \in [0, 2]$ there is a stable 1-cycle $\{x_{*_1}\} = \{0\}$ which is symmetric under s. For $a \in [2, 3]$ there is a stable 2-cycle $\{x^A_{*_2}\} = \{- (a-2)/a^{1/2}, + (a-2)/a^{1/2}\}$ symmetric under s. Then for $a > 3$ the symmetry is broken and for $a \in [3, 3.2355]$ there are two stable unsymmetrical 2-cycles: $\{x^B_{*_2}\} = \{\alpha, \beta\}$, $\{x^C_{*_2}\} = \{-\alpha, -\beta\}$ such that $s[\{x^B_{*_2}\}] = \{x^C_{*_2}\}$, $s[\{x^C_{*_2}\}] = \{x^B_{*_2}\}$ [15, 16].

Beginning of bifurcation diagram for the map (4) for $a - 3.02$ and two initial values: x_0 and $s(x_0) = x_0$ is shown in Fig. 1.

3. TWO DIMENSIONAL INVERTIBLE MAPS WITH SYMMETRY

Let us consider a 2D map of form:

$$(x_{N+1}, y_{N+1}) = F (x_N, y_N) = [f(x_N, a) + \beta y_N, x_N]. \tag{5}$$

Let f have properties defined in Eqs. (1,2) and be given by Eq. (4), β

$\neq 0$ being a control parameter. Then the map (5) can be written in explicit form as

$$x_{N+1} = (1-a)\, x \;+\; a\, x^3 \;+\; \beta\, y_N, \tag{6a}$$

$$y_{N+1} = x_N. \tag{6b}$$

We note that for $\beta \neq 0$ the map (6) is invertible and has symmetry S, $S(x,\, y) = (-x,\, -y)$, $S \circ S = 1$, $S \circ F = F \circ S$. The map (6) can be also written in form of 1D map:

$$x_{N+1} = (1-a)x_N \;+\; ax_N^3 \;+\; \beta x_{N-1} \equiv g(x_N,\, x_{N-1};\; a,\, \beta). \tag{7}$$

The map g has symmetry s, $s(x) = -x$.

For $a \in [0,\, 2-\beta]$ there is a stable fixed point $x_{*_1} = 0$ symmetric under s. For $a \in [2-\beta,\, 3-2\beta]$ there is a 2-cycle, $\{x_{*_2}^A\} = \{-\,(a+\beta-2)/a^{\,1/2},\; +\,(a+\beta-2)/a^{\,1/2}\}$, stable and symmetric. Then for $a > 3-2\beta$ the symmetry is broken and there are two stable unsymmetrical 2-cycles $\{x_{*_2}^B\} = \{\alpha,\, \beta\}$ and $\{x_{*_2}^C\} = \{-\alpha,\, -\beta\}$, such that $s[\{x_{*_2}^B\}] = \{x_{*_2}^C\}$ and $s[\{x_{*_2}^C\}] = \{x_{*_2}^B\}$.

Bifurcation diagram for the map (7) is shown in Fig. 2 for $\beta = 0.3$ and two different initial values: x_0, x_1, and $s(x_0) = -\,x_0$, $s(x_1) = -\,x_1$.

4. FRACTAL DEPENDENCE ON INITIAL CONDITIONS

We shall investigate consequences of symmetry for the dynamics. Let us note first that after the symmetry is broken there are two attractors and there must be two attractor basins with respect to the initial conditions. It has been shown by Kocarev that in many cases the basin structure is the closure of the stable manifold of an unstable periodic point p, $W_s(p)$ [7,8]. The stable manifold is defined as the set of points x such that $f^k(x) = p$ for finite k and $f^k = f(f^{k-1})$. It should be, however, investigated which unstable periodic points $p_1,\, \ldots,\, p_k$ contribute to the basin boundary while the boundary is given by the closure of the sum of sets $W_s(p_k)$.

We shall show that in the case of dynamical systems with symmetry it is the process of symmetry breaking which determines the basin boundary structure. Namely, the basin boundary equals the closure of the sum of sets $W_s(p_k)$, where p_1, ..., p_k are all unstable periodic points, symmetric under symmetry s.

In the case of the map f there can be two co-existing basins of attraction. The basin boundary is the sum of stable manifolds $W_s(\{x_{*_1}\})$, $W_s(\{x_{*_2}^A\})$, of 1-cycle x_*^1 and 2-cycle $\{x_{*_2}^A\}$, s − invariant, $x_{*_1} = 0$, $\{x_{*_2}^A\} = \{-A, A\}$, $A = (a-2)/a^{1/2}$. $W_s(\{x_{*_1}\})$ consists of points converging to ± 1, while $W_s(\{x_{*_2}^A\})$ consists of points converging to 0.

It follows from the symmetry s that for a > 3 $x_{*_1} = 0$ must divide two basins. Indeed, if sequence x_0, x_1, ... converges to, say, 2-cycle $(\alpha, -\beta)$, then the sequence $-x_0$, $-x_1$, ... must converge to $(-\alpha, \beta)$. The preimages of $x_{*_1} = 0$, $f^{-1}(0)$, $f^{-2}(0)$, ... must also divide two basins of the two attractors. Therefore, the basins boundary is $W_s(\{x_{*_1}\})$.

Also for a > 3 the 2-cycle $\{x_{*_2}^A\} = \{-A, A\}$ becomes unstable. It can be analogously shown that the points −A, A and their preimages divide the interval [−1, 1] into basins of initial conditions from which 2-cycle $(\alpha, -\beta)$ or $(-\alpha, \beta)$ is obtained. It follows from the bifurcation diagram (Fig. 1), that other unstable fixed points do not belong to the basin boundary. In Fig. 3 values of $x_{250, 251}$ for a = 3.02 were plotted.

It can be seen that there is a fractal structure in the neighbourhood of the point $x_{*_1} = 0$. This fractal structure is formed by preimages of the points $\{x_{*_2}^A\}$ converging to 0: $\{x_{*_2}^A\} = +0.581161$, $f^{-1}(\{x_{*_2}^A\}) = \pm 0.354061$, $f^{-1}[f^{-1}(\{x_{*_2}^A\})] = \pm 0.184697$ (all values computed for a = 3.02). Then this fractal structure is transported by the map f^{-1} from the nighbourhood of 0 towards the ends of the interval [−1, 1] where $f^{-1}(0) = \pm 0.817847$, $f^{-1}[f^{-1}(0)] = \pm 0.973211$ (computed for a = 3.02).

The same mechanism leads to fractal dependence on initial conditions in dynamical system (7). In Fig. 4 dependence on initial condition is shown for a = 2.6 and $\beta = 0.3$. Dark regions correspond to initial conditions x_0, x_1 escaping to infinity, white and shaded regions denote initial conditions yielding two unsymmetrical 2-cycles.

5. DISCUSSION

We have demonstrated that bistability, manifested by presence of two co-existing merging attractors in 1D maps, in presence of three repellers is sufficient to generate a fractal boundary. Bistability in the investigated system (6) was due to the discrete symmetry s, $s(x) = -x$.

It is interesting that generalization for systems with continuous time seems to be possible. Map of form (6) which was shown by Holmes [24] to be a model of the Poincaré return map for the Duffing equation

$$\dot{x} = -\delta x + \beta y - \alpha y^3 + f \cos(\omega t), \qquad (8a)$$

$$\dot{y} = x, \qquad (8b)$$

has the discrete symmetry Σ

$$x \longrightarrow x, \quad y \longrightarrow y, \quad t \longrightarrow t + T/2, \qquad (9)$$

$T = 2\pi/\omega$, analogous to symmetry $S(x, y)$ of the map (6), $S(x, y) = (-x, -y)$. Since Eqs. (8) and (6) have similar symmetries, fractal dependence on initial conditions can be predicted in the model (8) after the symmetry is broken.

ACKNOWLEDGEMENT

One of the authors (AO) wishes to express his gratitude to T. Tél for discussion.

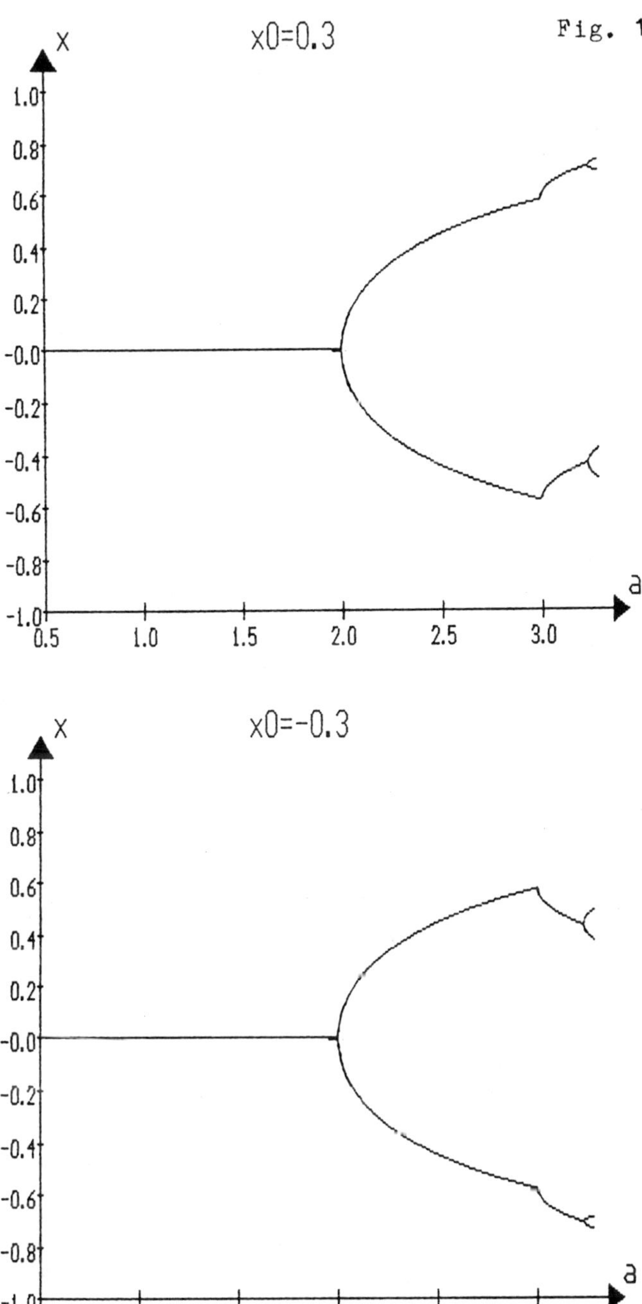

x0=0.3
Fig. 1
X
1.0
0.8
0.6
0.4
0.2
-0.0
-0.2
-0.4
-0.6
-0.8
-1.0
0.5
1.0
1.5
2.0
2.5
3.0
a
x0=-0.3
X
1.0
0.8
0.6
0.4
0.2
-0.0
-0.2
-0.4
-0.6
-0.8
-1.0
0.5
1.0
1.5
2.0
2.5
3.0
a

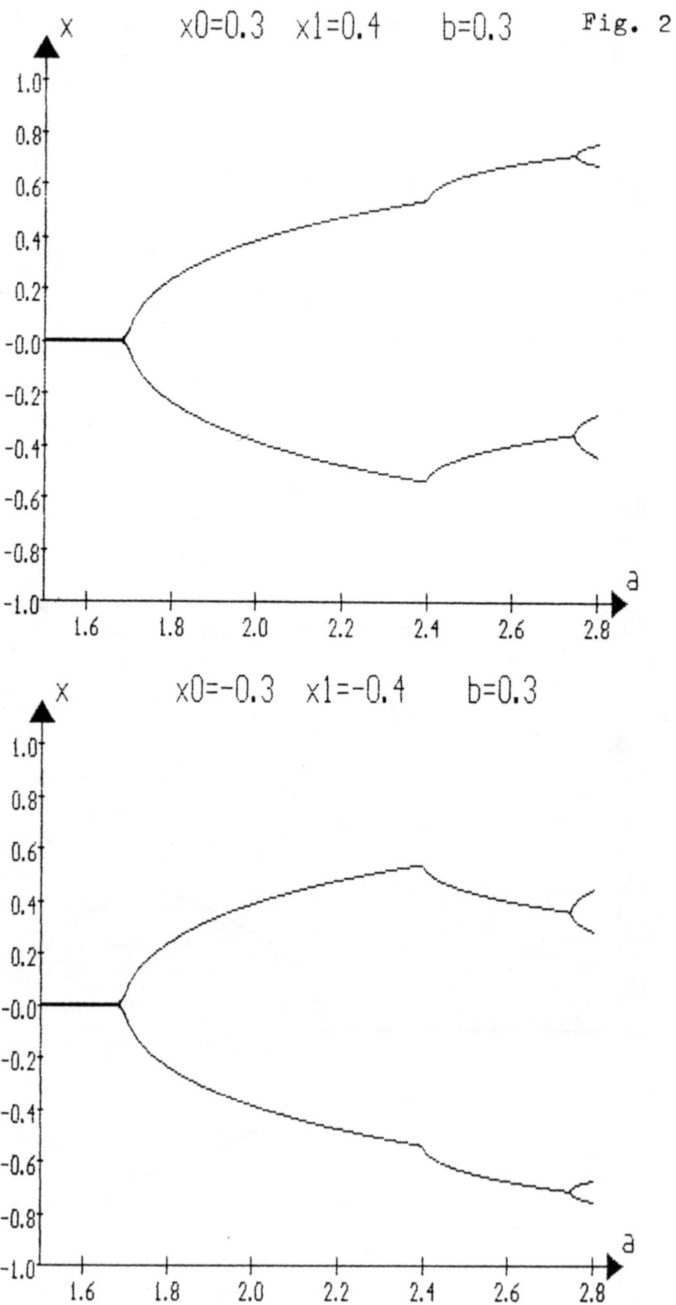
x
x0=0.3   x1=0.4   b=0.3   Fig. 2
1.0
0.8
0.6
0.4
0.2
-0.0
-0.2
-0.4
-0.6
-0.8
-1.0
1.6 1.8 2.0 2.2 2.4 2.6 2.8
a
x
x0=-0.3   x1=-0.4   b=0.3
1.0
0.8
0.6
0.4
0.2
-0.0
-0.2
-0.4
-0.6
-0.8
-1.0
1.6 1.8 2.0 2.2 2.4 2.6 2.8
a

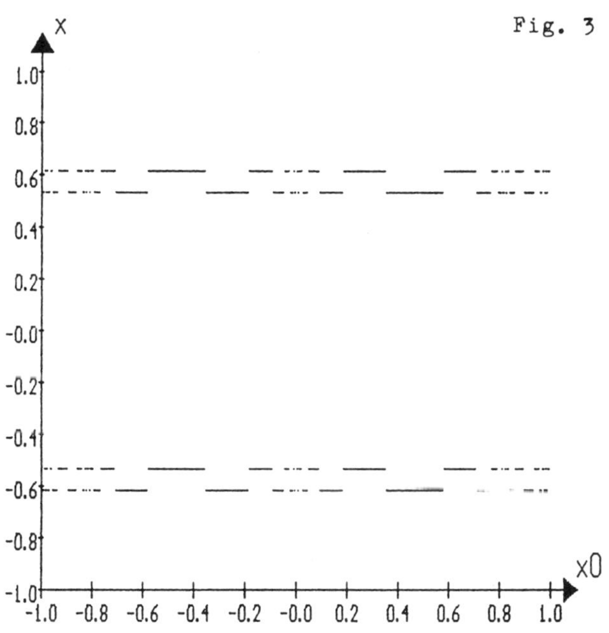
Fig. 3
X
xO

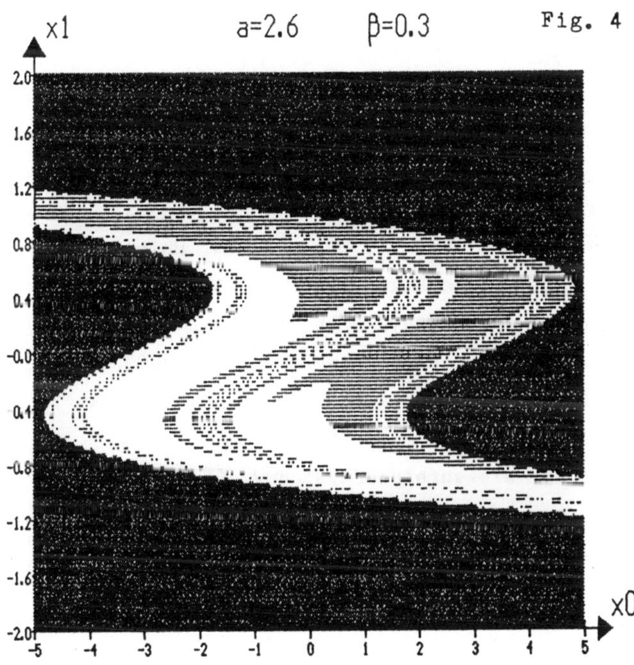
x1
a=2.6   β=0.3   Fig. 4
xO

REFERENCES

1. C. Gerbogi, E. Ott, J.A. Yorke, Physica **7D** (1983) 181.
2. C. Grebogi, E. Ott, J.A. Yorke, Phys. Rev. Lett. **50** (1983) 935; **51** (1983) 942 (E).
3. C. Grebogi, S.W. McDonald, E. Ott, and J.A. Yorke, Phys. Lett. **99A** (1983) 415.
4. S.W. McDonald, C. Grebogi, E. Ott, and J.A. Yorke, Physica **17D** (1985) 125.
5. S.W. McDonald, C. Grebogi, E. Ott, and J.A. Yorke, Phys. Lett. **107A** (1985) 51.
6. C. Grebogi, E. Ott, and J.A. Yorke, Phys. Rev. Lett. **56** (1986) 1011.
7. Lj. Kocarev, Phys. Lett. **121A** (1987) 274.
8. Lj. Kocarev, Phys. Lett. **125A** (1987) 389.
9. O.E. Rössler, J.L. Hudson, and M. Klein, J. Phys. Chem. **93** (1989) 2858.
10. J. Peinke, M. Klein, A. Kittel, G. Baier, J. Parisi, R. Stoop, J.L. Hudson, O.E. Rössler, Europhys. Lett. **14** (1991) 615.
11. C. Grebogi, E. Ott, J.A. Yorke, Physica **7D** (1983) 181.
12. A. Okninski, J. Stat. Phys. **52** (1988) 577.
13. J. Peinke, M. Klein, J. Parisi, A. Kittel, O.E. Rössler Acta Phys. Polon. **B23** (1992) 409.
14. T. Tél, *Directions in Chaos*, vol. 3, ed. Bai-lin Hao, Singapore 1990, World Scientific, pp. 149-221.
15. T.D. Rogers, D.C. Whitley, Math. Modell. **4** (1983) 9.
16. J. Testa, G.A. Held, Phys. Rev **28A** (1983) 3085.
17. P. Chossat and M. Golubitsky, SIAM J. Math. Anal. **19** (1988) 1259.
18. J.W. Swift, K. Wiesenfeld, Phys. Rev. Lett. **52** (1984) 705.
19. U. Parlitz, W. Lauterborn, Phys. Rev. **36A** (1987) 1428.

20. R. Mettin, Chaos in Dissipative Systems, WE-Heraeus-Seminar, Trassenheide 1992.

21. J. Grasman, H. Niemeijer, E.J.M. Veling, Physica **13D** (1984) 195.

22. J.D. Crawford, E. Knobloch, and H. Riecke, Physica **44D** (1990) 340.

23. F.M. Izrailev, M.I. Rabinovich, A.G. Ugodnikov, Phys. Lett., **86A** (1981) 321.

24. P.J. Holmes, Phil. Trans. Royal Soc. **A292** (1979) 419.

CHAOS IN CHEMICAL REACTIONS - A REVIEW

ANDRZEJ L. KAWCZYŃSKI

Institute of Physical Chemistry,
Polish Academy of Sciences,
01224 Warszawa

Deterministic chaos is a widespread phenomenon which has been recently observed in various nonlinear dynamical (including chemical) systems. Selected information from the theory of nonlinear dynamical systems, necessary to understand chaotic phenomena, are shortly described. Chaotic behavior in far from equilibrium open chemical systems is reviewed. Consequences of deterministic chaos for chemical kinetics are discussed.

Introduction

How can concentrations change if chemical reactions occur in homogeneous (well stirred) systems? In recent years the answer to this fundamental question in chemical kinetics has been changed dramatically and this paper is an attempt to present these changes. As a matter of fact results concerning nonlinear phenomena in many branches of science have been accumulated in the past. The turning-point arrived in late seventies when it became clear that the chaotic behavior in relatively simple, completely deterministic, nonlinear systems was possible to occur. As a consequence an explosion of interest in the nonlinear dynamics has been observed. It became clear that different systems as for example the human heart and the forced pendulum behave in a similar way and their behavior can be described by similar laws of dynamics. The theory of nonlinear dynamics has become a very successful tool in studies of a very broad class of phenomena. Monographs, conference proceedings, collections of papers and reviews concerning the results on the nonlinear dynamic phenomena are available in almost all branches of science, beginning from mathematics and physics and ending on economy. Many of them are cited in the monographs [1-3] and those especially important in chemical kinetics are cited in this paper or may be found in recent monographs [5),83)]. There are two monographs in Polish [4),82)] concerning nonlinear phenomena in chemical kinetics.

Chemical kinetics used to deal normally with systems in which monotonic changes of concentrations are observed. Oscillatory changes of concentrations, like the Bray reaction [6], in which concentrations of intermediates change periodically in time were exceptions. These oscillations were explained by physical effects rather than by chemical reactions. The Bray reaction as well as the Lotka – Volterra models [7-9], describing in a formal way some model chemical systems and exhibiting oscillatory behavior, did not initiate a general interest. In the sixties, a new oscillating reaction was discovered by Belousov [10], and then Zhabotinsky [11] and co workers proved that oscillations of intermediates were caused by chemical reactions. Simultaneously the theory of dissipative structures has been developed by Prigogine and his co workers [12,13], forming a thermodynamical basis for oscillatory behavior in the far-from-equilibrium systems.

The changes of concentrations in homogeneous (well stirred) chemical systems are described by kinetic equations which are ordinary differential equations. General conditions for the existence of periodic solutions (limit cycles) were well known at that time from the qualitative theory of ordinary differential equations. This theory was well developed for 2D systems, that is for systems described by 2 ordinary differential equations [14]. It was a common believe that also higher-dimensional systems, namely those described by 3 and more equations, have asymptotically attracting trajectories (attractors) of the same type as those for 2D systems. These attractors for the 2D systems can be fixed points and limit cycles where fixed points correspond to stationary solutions and limit cycles to periodical ones.

The fact that nonlinear dynamical systems can have very

complicated solutions was known to Poincaré [15] but this had no practical influence on the theory of dynamical systems up to 1963 when Lorenz [16] published the results of numerical solutions of a simplified model of the Bénard convection. The solutions of the completely deterministic system of three ordinary differential equations are oscillating but not periodic. These solutions are unstable and undergo irregular changes, like fluctuations, with no source of randomness from the outside. A term **chaotic** was coined for such solutions and this term means **nonperiodical oscillatory behavior that arises from the nonlinear nature of deterministic kinetic equations**. The deterministic chaos should be distinguished from noisy behavior arising from random driving forces. Chaotic solutions for systems with so small dimension i. e. for systems described by 3 or more ordinary differential equations were a great surprise. In 1971 Ruelle and Takens [17] used the concept of asymptotic chaotic trajectory to a model of hydrodynamic turbulence. They introduced the term "strange attractor" for an asymptotic chaotic orbit. In 1975 Li and Yorke [18] published a paper entitled "Period three implies chaos" with apparently random results of some mappings. They seemed to be the first in using the term "chaos" for characterization of the behavior of deterministic systems. In 1978 Feigenbaum [19] discovered the universal properties of scaling in one-dimensional mappings. After some induction period the situation changed drastically and the interest in chaotic behavior gained considerable popularity. Dynamical systems with chaotic behavior form a long list, beginning from simple physical systems as driven pendulum, through more complicated systems in classical and quantum mechanics, hydrodynamics, optics, solid state physics, electronics, chemistry, electrochemistry, biochemistry up to biological and ecological systems [1-5]. At that time chemical kinetics, mainly through its achievements in dissipative structures, was well prepared to play an active role in the progress of dynamical systems. Indeed, the first paper on chaotic behavior in a model chemical system was published in 1976 [20] and one year later an experimental evidence of chaotic changes in concentrations was reported [21,22]. Since that time there is a growing interest in chaotic behavior in chemical systems.

This paper is organized as follows: The second chapter contains a short mathematical background necessary to understand the experimental results. In the further chapter some experimental results are presented. Experiments with the Belousov - Zhabotinsky (BZ) reaction are described in detail and other systems are only mentioned. The fourth chapter contains models for the BZ reaction and an example of a model based on enzymatic reactions is given. Some consequences of the deterministic chaos in chemical kinetics are shortly discussed in the last chapter.

Mathematical background

Kinetic equations in homogeneous phenomenological chemical kinetics are ordinary differential equations. For open chemical systems with steady inflows and outflows - continuously stirred tank reactors (CSTR) and for some class of closed chemical systems, kinetic equations are autonomous (with constant coefficients) ordinary differential equations:

$$\frac{dc}{dt} = f(c,a) \tag{1}$$

where $c = [c_1, c_2, \ldots, c_n]^T$ refers to concentrations of reacting species and $a = [a_1, a_2, \ldots, a_k]^T$ denotes such parameters as rate constants, flow rates of feed streams and temperature. If inflows and/or outflows change in time, the corresponding coefficients in kinetic equations also depend on time and the equations become non autonomous.

Chemical systems usually belong to dissipative systems. This means that they globally contract a phase space and evolve to some asymptotic trajectories. They should be distinguished from conservative (hamiltonian) systems in which the phase space is conserved. In the sequel only dissipative systems are considered. Results concerning the hamiltonian systems can be found in monographs 23,1,3). It is well known [4] that for 1D systems (one kinetic

52

equation) asymptotic trajectories are always only fixed points (
stationary states) but a 1D nonlinear system can have more than one
stationary state. Every attracting stationary state is surrounded by
an interval (a basin of attraction) from which all the trajectories
tend to this state. The basins of attraction are separated by
repelling asymptotic trajectories, which in 1D systems are unstable
stationary states. The concept of the basin of attraction can be
easily extended to more dimensional systems. In 1D nonlinear systems
a bistability (multistability) as well as hysteresis phenomena can
appear but the asymptotic behavior of such systems is very simple.
This is not always the case for ordinary differential equations with
two variables [14] that is for the 2D systems. In addition to fixed
points they can have also other types of attractors, which have the
form of closed curves in a phase plane, such that in some their
neighborhood there are no other closed curves. These are called
"limit cycles" and their solutions are periodic in time. Also changes
of attractors caused by changes of parameters (at least for one of
them which is called the bifurcation parameter) are well known [14,4].
For example in the Hopf bifurcation, the attracting (stable)
stationary state becomes repelling (unstable) and an attracting limit
cycle appears with an amplitude growing from zero (see Fig.1).

Fig.   1   Schematic   illustration   of   the   Hopf   bifurcation.   An
attracting (stable) stationary state (focus) becomes repelling (
unstable) with changes of a bifurcation parameter and an attracting (
stable) limit cycle appears.   At a critical value of the bifurcation
parameter the stationary state (center) is densely surrounded by
closed orbits.

Surprisingly   in   3D   systems,   (systems   described   by   three
differential equations) fixed points and limit cycles do not exhaust

all possible attractors. Besides there are so-called strange attractors [17] (see Fig.2). By a strange attractor we mean an asymptotic trajectory with the following properties:

1) all points belonging to the basin of attraction are attracted to the attractor

2) any point belonging to the strange attractor will be, after some time, close to each other point belonging to the attractor.

3) any two points close to each other at some moment will be separated, after some time, to a macroscopic distance comparable with the size of the attractor.

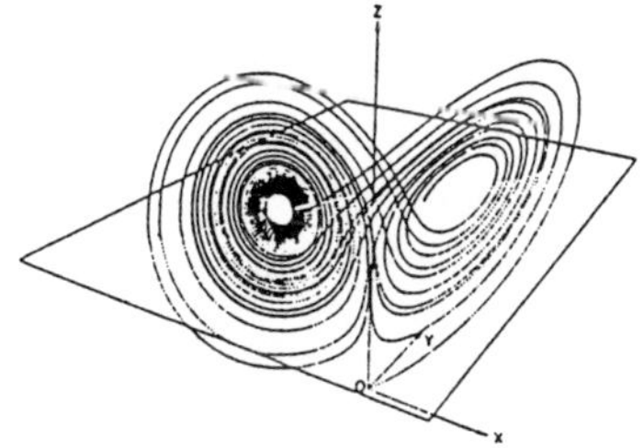

Fig. 2 A segment of the trajectory for the Lorenz model: $\overset{\circ}{x} = \sigma\,(y - x);\ \overset{\circ}{y} = -xz + rx - y;\ \overset{\circ}{z} = xy - bz;$. (From [3], page 379 and reference herein)

The last property, known as a strong dependence on the initial conditions, makes the attractor strange, and the behavior of systems with strange attractors is practically unpredictable.

A measure of the average exponential divergence or contraction of trajectories is the Lyapunov exponent [24,25]: Let us consider two trajectories which at time t = 0 are at c_o and at $c_o + \Delta c_o$. The vector Δc_o will change in time to Δc and the length of Δc is denoted by:

$$d\,(c_o,t\,) = \|\,\Delta c\,(c_o,t)\,\| \qquad\qquad (2)$$

where $\|\,.\,\|$ denotes a norm.

The largest Lyapunov exponent λ can be calculated from:

54

$$\lambda(c_o, t) = \lim_{\substack{t \Rightarrow \infty \\ d(0) \Rightarrow 0}} \frac{1}{t} \, \ln \frac{d(c_o, t)}{d(c_o, 0)} \qquad (3)$$

where $d(c_o, 0)$ means the length of the initial vector Δc_o.

Systems with strange attractors are characterized by positive Lyapunov exponents. Lyapunov exponents characterize the dynamic properties of strange attractors. They allow us to estimate an interval of time after which trajectories close to a chosen one with a precision of Δc diverge to distances comparable with the size of the strange attractor.

The "geometrical" aspects of strange attractors are characterized by their dimensions. Strange attractors usually have fractal dimensions. In the 3D phase space they are something in between 1D curves and 2D surfaces (see Fig.2). As the topological dimension for trajectories is always equal to 1, the "metric" dimensions can be higher than 1 and be fractal. The simplest of the "metric" dimensions is the capacity dimension [25,26] which can be defined in the following way: Let $N(l)$ denote the minimum number of d-dimensional cubes of a side l needed to cover some set in d-dimensional Euclidean space. If, for $l \Rightarrow 0$ the number of N is given by:

$$N(l) \propto l^{-d_c} \qquad (4)$$

then the capacity dimension is defined by:

$$d_c = \lim_{l \Rightarrow 0} \frac{\log N(l)}{\log (1/l)} \qquad (5)$$

For typical objects such as lines, squares, cubes etc., the metric dimension is equal to the topological dimension but for such objects as the Cantor set, the Koch curve, the Sierpinsky carpet etc. the metric dimension is fractal and is larger than the topological one. Sets with metric dimensions larger than the topological dimension have been called fractals [27]. Fractals have an inherent scale invariance. Upon repeated magnification parts of a fractal look identical with the whole object. They are self - similar [27].

The capacity dimension does not take account of "inhomogeneities" of the objects. In order to characterize inhomogeneities of strange attractors an infinite set of information dimensions has been introduced [28]. Let us chop a strange attractor to points corresponding to $t = 0, \tau, 2\tau, \ldots, N\tau$. Region in a d-dimensional phase space occupied by the strange attractor is divided into d-dimensional cubes with an edge length l. The probability of finding a point that belongs to the strange attractor in i-th cube ($i = 1, 2, \ldots, M(l)$) may be determined by:

$$p_i = \lim_{N \Rightarrow \omega} \frac{N_i}{N} \tag{6}$$

where N_i denotes a number of attractor points belonging to the i-th cube. The infinite set of information dimensions is defined by:

$$d_f = \lim_{l \to 0} \frac{1}{f-1} \frac{\lg \sum_{i=1}^{M(l)} p_i^f}{\lg(l)} \tag{7}$$

For $f = 0$, d_0 is the capacity dimension. For $f \Rightarrow 1$:

$$d_1 = \lim_{l \to 0} \frac{\sum_{i=1}^{M(l)} p_i \lg p_i}{\lg(l)} \tag{8}$$

and, if all p_i are identical, d_1 also becomes equal to the capacity dimension. If p_i are different, what means that the attractor is inhomogeneous, the difference $d_0 - d_1$ is a measure of a nonuniformity of the attractor. It can be shown that

$$d_{f'} \leq d_f \quad \text{for} \quad f' \geq f. \tag{9}$$

For $f = 2$, d_2 is determined by the probability that two points of the attractor lie within a l^d cell and is simply related to a correlation integral. d_2 is therefore referred to as the correlation dimension. The dependence of the correlation integral on the scale unit l can be used for separation of the deterministic chaos from an external white noise [25]. The log - log dependence for stochastic processes with the white noise gives a value of the slope equal to 3 whereas the deterministic process with the strange attractor gives the slope equal to the correlation dimension d_2 which in 3D systems is between 1 and 2. Higher order dimensions give more information about the nonuniformity of the attractor.

The theory of ordinary differential equations for two variable systems is well developed and one can easily predict or exclude the existence of limit cycles using the Bendixson - Poincaré theorem in which simple properties of the right-hand sides of the equations are of importance. For three- and more-variable systems it is much more difficult to predict the existence of strange attractors. Usually the corresponding equations must be solved numerically and the solutions obtained are analyzed. It is not simple to prove whether the numerical solutions are chaotic and some other methods must be involved. One of the most important means is the Poincaré transformation. In n-dimensional phase space it is possible to find a n-1 hyperplane intersected transversally by the trajectory. A point transformation is introduced, which transforms each point of intersection of the trajectory with the hyperplane into the next point of intersection (see Fig. 3).

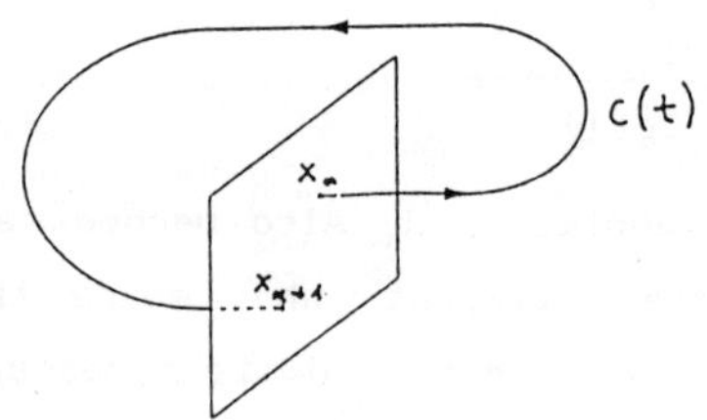

Fig. 3   Schematic illustration of the Poincaré transformation. The point x_n of intersection of a trajectory c(t) with the plane is transformed to a next one x_{n+1}.

If a 2D point transformation is expanding in one direction and squeezing in another one, with a subsequent folding, this results in the transformation of some 2D region into itself, as for instance in the Henon transformation [29], then it exhibits a strong dependence on initial conditions and the corresponding trajectory is chaotic.

The situation is much simpler if the Poincaré transformation contains sets of points which can be approximated by one-dimensional curves. Then it becomes possible to introduce a transformation of these sets to themselves and after appropriate normalization to obtain the 1D return (Poincaré) maps. The theories of such maps are much better developed than the theory of ordinary differential equations for 3D and more dimensional systems and an important information concerning the chaotic behavior can be extracted from obtained maps. It should be noticed that at present there is no direct way which would allow us to predict the form of the map from the properties of the equations. Numerical solutions cannot be avoided. There are some types of maps with known properties which describe some generic routes leading to a chaotic behavior. Three simplest types of such maps are shortly described below.

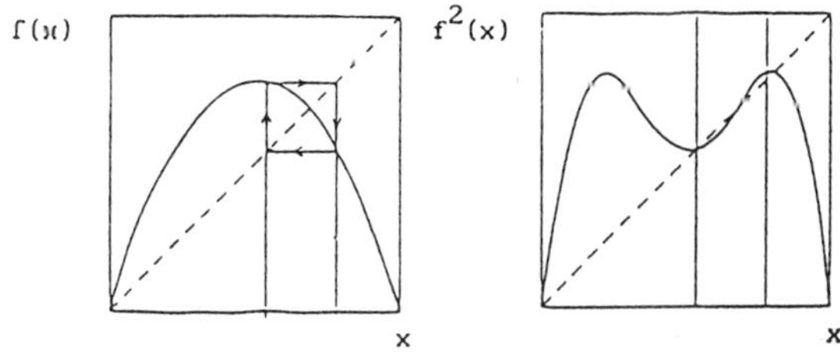

Fig. 4 (a) The logistic map for μ at which the 2-periodic orbit is attracting. (b) The f^2 iterate has two attracting fixed points.

A typical map (see Fig. 4) iterates a point from the interval [0,1] to a next point from the same interval by means of the relation:

$$x_{n+1} = f(x_n) \tag{10}$$

There is usually a point for which $f(x^*) = x^*$. This is the fixed point of the map, in which the graph of the map intersects the diagonal of the square built up on the side of the unit interval. There can be a sequence of points $x_1^*, x_2^*, \ldots, x_k^*$ transformed successively by the map:

$$x_2^* = f(x_1^*), \ x_3^* = f(x_2^*), \ldots, \ x_1^* = f(x_k^*) \tag{11}$$

to produce a k-periodic orbit. Each of these points is the fixed point of the $f^k = f(f(\ldots f(.)$ iterate of the map: $x_i^* = f^k(x_i^*)$ (i = 1,2,...,k) and at each x_i^* the k-th iterate of the map intersects the diagonal. Fixed points of the map represents simple limit cycles of the corresponding differential equations, whereas the k-periodic orbits are equivalent to limit cycles with k loops in the period. Fixed points as well as k-periodic orbits can be attracting (stable) or repelling (unstable) what means that points from their neighborhood, after having passed many iterations, approach them or run away. The stability of a fixed point or a k-periodic orbit is defined by the value of the derivative of the corresponding transformation. If at the fixed point

$$\left|\frac{df}{dx}\right|_{x^*} < 1 \quad \text{or} \quad \left|\frac{df^k}{dx}\right|_{x_i^*} < 1 \tag{12}$$

then the point is stable, otherwise it is unstable. The derivative of the k-iterate of the map is given by:

$$\frac{df^k}{dx}\bigg|_{x_i^*} = \frac{df}{dx}\bigg|_{x_1^*} * \frac{df}{dx}\bigg|_{x_2^*} \cdots \frac{df}{dx}\bigg|_{x_k^*} \tag{13}$$

Stable fixed points correspond to stable limit cycles and the unstable ones determine unstable limit cycles.

Systems of two differential equations always give monotonic Poincaré maps containing only fixed points. Maps for 3D and more dimensional systems can have maps with extreme and chaotic orbits can appear.

1) The best known example is the logistic map (see Fig. 4) belonging to smooth unimodal (with one extreme) types of maps [30]:

$$x_{n+1} = f(x_n) = 4\mu \, x_n \, (1 - x_n) \qquad 0 \le \mu \le 1 \qquad (14)$$

Maps of this type have been obtained for a broad class of nonlinear systems [31]. When the control parameter μ is increased the period-doubling route to chaos, known also as the Feigenbaum cascade, is observed [19,30,32]. The point $x = 0$ is of attracting type for $\mu \le 0.25$ and for higher values of μ it becomes repelling. Another fixed point appears (the parabola graph crosses the diagonal) and this point is attracting up to $\mu = 0.75$. At this μ the fixed point becomes repelling and a 2-periodic attracting orbit appears. This orbit is stable in some interval of μ and next it becomes repelling and at the higher values of μ a 4-periodic orbit becomes attracting.

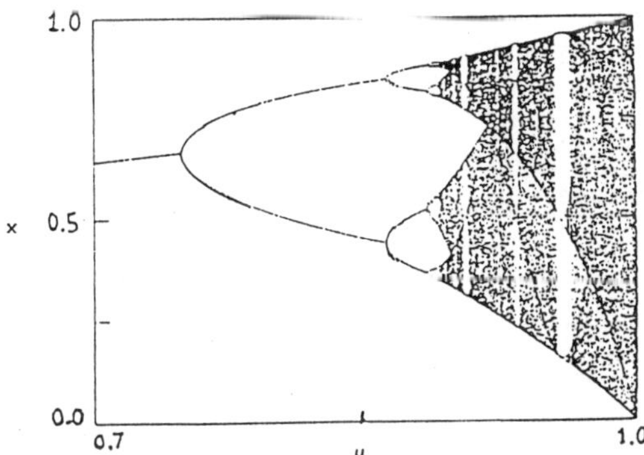

Fig. 5 Iterates of the logistic map as a function of μ for μ belonging to [0.7,1.0]. A few hundred of initial iterations has been omitted to see the asymptotic behavior. Three initial period-doublings of the fixed point are clearly visible. Windows with attracting 6-periodic, followed by 5-periodic and next by 3-periodic orbits being elements of the universal sequence can also be seen.

With increasing μ subsequent an attracting 2^n-periodic orbit arises whereas a 2^{n-1}-periodic orbit becomes unstable. The values of μ_n, at which a 2^{n-1}-periodic orbit becomes unstable and an attracting 2^n-periodic orbit appears, are scaled according to:

60

$$\delta_n = \frac{\mu_{n+1} - \mu_n}{\mu_{n+2} - \mu_{n+1}} \qquad (15)$$

with δ_n quickly approaching to δ = 4.669201... .   After infinite sequence of period-doublings a chaotic orbit arises.   Consecutive iterations of almost all points do not tend to any periodic orbit but they wander in some subintervals of [0,1].   By increasing μ one can see other periodic orbits which appear in the universal sequence [33,30], which means that they appear in such a way that a given periodic orbit is preceded and followed by other well determined orbits.   A p-periodic orbit successively bifurcates to $p*2$, $p*2^2$,..., $p*2^n$-periodic orbits with the same scaling as described above for the fixed point.   At μ = 4 the chaotic orbit is distributed over the whole interval [0,1].   This is the period-doubling route to chaos which is universal in the sense that it appears for all smoothed unimodal maps with the parabolic extreme.

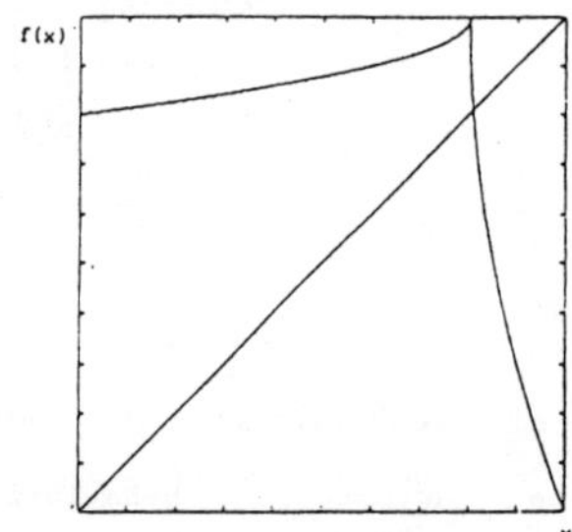

Fig. 6   An example of a cusp-shaped map.   Shift of the left branch of the map towards the tangency with the diagonal is accompanied by the period-adding phenomenon.

2) Another simple example is the so called cusp-shaped map (see Fig. 6) defined by:

$$x_{n+1} = \begin{cases} f_1(x_n) & \text{for} \quad 0 \le x_n \le a \\ f_2(x_n) & \text{for} \quad a \le x_n \le 1 \end{cases} \qquad (16)$$

where f_1 and f_2 are continuous functions of x and $f_1(0)$ = b (with 0 < b < 1), $f_1(a)$ = 1, $f_2(a)$ = 1 and $f_2(1)$ = 0. The cusp-shaped maps also are unimodal but they have a discontinuous derivative at the maximum. The first example of a dynamical system with this type of map seems to be the Lorenz system [16]. Such maps have been also found in chemical systems [34].

For cusp-shaped maps a typical phenomenon, which appears when one of the branches of the map tends to the tangency with the diagonal, is the period-adding [35-37]. As the control parameter changes successive attracting periodic orbits appear with period values of 2,3,..., up to infinity (see Figs. 7 and 8).

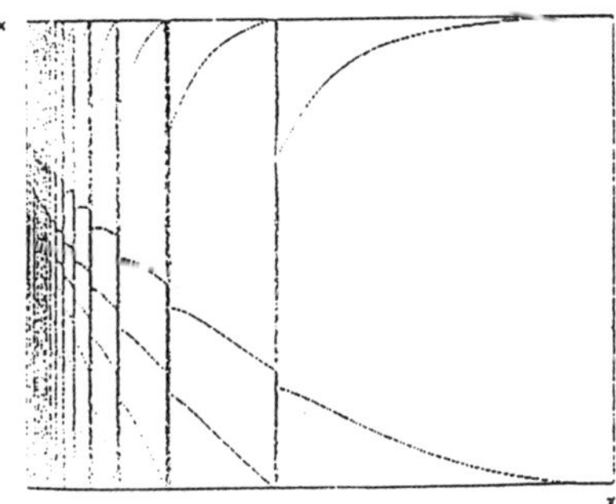

Fig. 7   Period-adding with chaotic windows for the cusp-shaped map consisting of two hyperbolic curves (17a - b). The bifurcation parameter varies in the range from 0.5 (left) to 0.0 (right). (From 35)).

These orbits are attracting on some intervals of the control parameter. For the map in which f_1 and f_2 are hyperbolas:

$$f_1(x) = \frac{\phi}{\beta}\left(1 - \frac{\phi - \beta}{\phi - x} \right) \qquad \text{if } x \leq \beta \tag{17a}$$

and

$$f_2(x) = \gamma^2\left(\frac{1}{\beta - \eta} - \frac{1}{x - \eta} \right) \qquad \text{if } x \geq \beta \tag{17b}$$

where $0 < \beta < 1$, $\phi > \beta > \eta \geq 0$ and $\gamma^2 < (\beta - \eta)(1 - \eta)/(1 - \beta)$ and

62

β, ϕ, η are fixed, intervals of the control parameter γ, on which periodic orbits are attracting, are separated by intervals on which the orbits are chaotic (see Fig. 7). In this case chaotic orbits appear without any warning, such as bifurcations leading to more complex periodic orbits. Maps with nonpositive Schwarzian derivative exclude the coexistence of attractors at the same value of the control parameter. For maps with positive Schwarzian derivative

$$Sf(x) = \frac{f'''(x)}{f'(x)} - \frac{3}{2}\left(\frac{f''(x)}{f'(x)}\right) \tag{18}$$

different attractors can coexist for the same value of the control parameter. It is a necessary but not the sufficient condition for a coexistence of attractors.

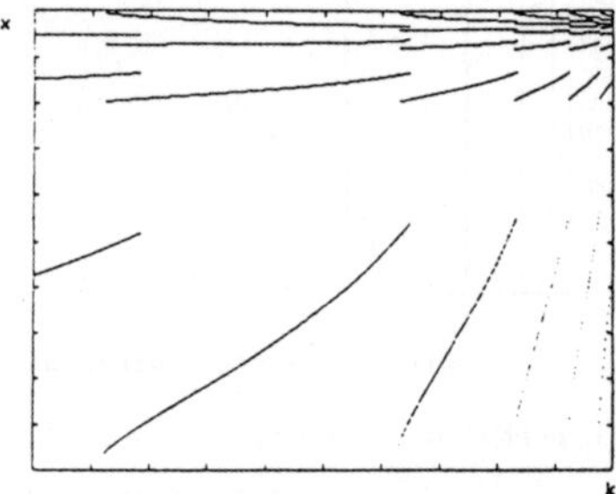

Fig. 8   Period-adding for the cusp-shaped map consisting of two square root curves (19a - b) exhibiting the coexistence of successive attracting periodic orbits. In the subintervals of k where the successive orbits coexist an orbit is attracted to one of them depending on initial conditions. (From [37]).

The coexistence of two successive different periodic orbits (see Fig. 8) as well as the coexistence of a chaotic orbit with a 2-periodic attracting orbit has been found in the map, in which f_1 and f_2 are square root curves for which the Schwarzian derivative is positive [37]

$$f_1(x) = 1 - l\sqrt{1 - x/k} \qquad \text{if } x \leq k \qquad (19a)$$

and

$$f_2(x) = 1 - l\sqrt{(x - k)/(1 - k)} \qquad \text{if } x \geq k \qquad (19b)$$

where $l = 0.2$ and the control parameter k varies from 0 to 1. Successive attracting periodic orbits, $2, 3, 4, \ldots$ up to infinity, coexist on subsequent subintervals of k and are not separated by chaotic orbits. The appearance of the successive periodic orbits for the cusp-shaped maps can be scaled according to:

$$\varepsilon_{n+1} - \varepsilon_n = \text{constant } [n^{-2} - (n + 1)^{-2}] \qquad \text{for large n} \qquad (20)$$

where ε_{n+1} and ε_n denote the values of the control parameter at which (n+1) and n periodic orbits appear. When the branch of a map crosses the diagonal a saddle - node (tangent) bifurcation occurs in which both the attracting and the repelling fixed points appear. By changing the value of the parameter in the opposite direction one can see the intermittency route to chaos [38]. Stable limit cycle, corresponding to the attracting fixed point (node), vanishes when the map moves locally above the diagonal and a chaotic orbit, corresponding to the shifted position of the map, appears. It should be noted, however that in some situations like those with the square root map (19a-b) an orbit, which appears when the branch of the map is just above the tangency, is not chaotic but still periodic with a very long period.

3) Circle maps are obtained by transformation of the circumference of a circle onto itself. The initial angle θ_o which determines the position of a point on the circumference of a circle is transformed by $f(\theta)$ to a next angle θ_1 and so on. The angles are usually normalized to the interval [0,1] (instead of [0, 2π]) and $f(0) = f(1)$. The circle maps appear in a natural way in the periodically driven pendulum and in many problems with coupled oscillators [39,40], for which trajectories evolve on the surface of a 2D torus . Circle maps are usually discontinuous (see Fig. 9), though continuous ones are

64

not excluded. They can be invertible or noninvertible. The sine map:

$$\theta_{n+1} = \theta_n + \Omega - \frac{k}{2\pi} \sin (2\pi \, \theta_n) \qquad (\text{mod } 1) \qquad (21)$$

plays a generic role for this class of maps. Maps in which the
function $\sin (2\pi\theta)$ is replaced by other functions with single cubic
inflection point yield the same qualitative results. The sine map is
a two-parameter (Ω and k) map.

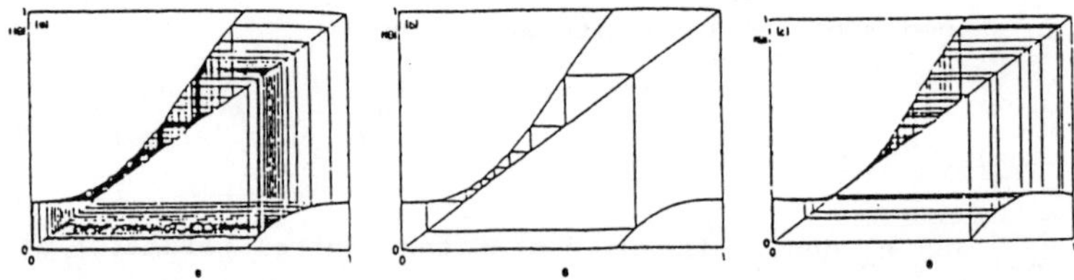

Fig.   9   An   example   of   circle   maps   —   the   sine   map.
Quasi-periodic   (a)   k < 1,   periodic   (b)   k = 1 and   chaotic   orbits
(c) k > 1 are seen.   (From [3], page 320)

The value of Ω represents an average number of rotations of the
transformed angle determined by one full rotation and for the sine map
with k = 0 it is equal to the winding (rotation) number which in
general is defined by:

$$W = \lim_{n \Rightarrow \infty} \frac{\theta_n - \theta_o}{n} \qquad (22)$$

For k belonging to [0,1] the sine map is invertible and for k > 1 it
becomes noninvertible (see Fig. 9). A maximum and a minimum appear
on each of its two branches. A characteristic phenomenon for the sine
map is the frequency- (phase-) locking. For k = 0, the iterates of
the sine map are periodic if Ω is rational or quasi-periodic if Ω is
irrational. For k > 0 but lower than 1, the rotation number W is no
longer equal to Ω. Finite subintervals of Ω with lengths growing with
increasing k appear over which the iterates of the map achieve each
rational W = p/q and a set of θ_i is the periodic orbit with period q.
The iterates are locked to a given rational W over the interval of Ω.
When k reaches 1 the whole interval of Ω is covered by subintervals

corresponding to rational W and the set corresponding to irrational W forms a fractal with measure 0. The locked intervals on the parameter plane k versus Ω form slightly distorted triangles known as the Arnold tongues with their apices lying on the k = 0 axis (see Fig. 10).

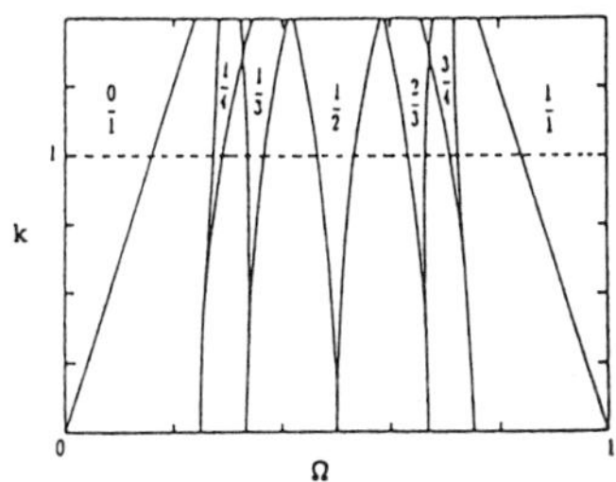

Fig. 10   Schematic Arnold tongues showing the phase-locking phenomenon for the sine map.

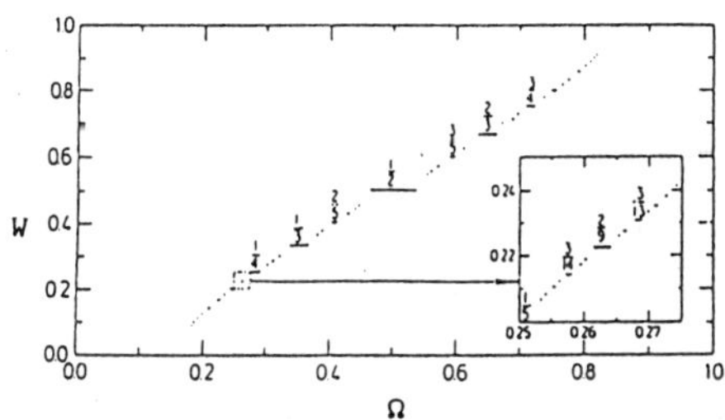

Fig. 11   A devil's staircase for the sine map (k = 1).   The steps show the regions of Ω in which W is constant.   Inset presents an enlarged view of the indicated fragment of the whole picture, showing the self-similarity.   Fractions above the steps indicate W for a few of wider steps. (From [3], page 302).

For every two Arnold tongues with the rotation numbers p/q and r/s there is another Arnold tongue with the winding number given by the Farcy sum:

$$\frac{p}{q} \oplus \frac{r}{s} = \frac{p + r}{q + s}$$

(23)

For fixed k, the dependence of W on Ω forms the devil's staircase, which is a monotonic, increasing, continuous function with plateaus of finite length at every rational W (see Fig. 11). The segments of the devil's staircase are self-similar. The enlarged segments looks identical to the whole staircase. For k > 1, when the sine map becomes noninvertible (see Fig. 9), the Arnold tongues begin to overlap. Some of them persist but others disappear. Chaotic orbits and hysteresis effects appear but the dynamics is not completely clear. As the map has a smooth maximum and a smooth minimum the period-doubling route to chaos can appear. Also the period-adding effects accompanied by approaching the tangency bifurcation should be visible but the dynamics is not completely clear.

It is worth to emphasize that the scaling properties for the 1D maps show that it is possible to have an infinite number of bifurcations in a quite small range of the control parameter. This fact has important experimental consequences. One of the most important is that only some number of bifurcations and corresponding periodic orbits can be observed experimentally.

There are also other problems concerning the experimental evidence and identification of strange attractors in a given dynamical system. Fluctuations and external noise always occur in experimental measurements. Fluctuations or noise can switch the system from a basin of one attractor to another one thus leading to a stochastic behavior. In chemical systems (CSTRs) there are also problems with assuring the homogeneity by a sufficiently effective stirring. Therefore any experimental proof of chaos should be based on much deeper analysis then a direct registration of concentration changes in time. It is necessary to calculate the Lyapunov exponents, information dimensions and to construct 1D return maps. This becomes possible if an attractor trajectory in a phase space is known. One should remember, however, that in chemical systems only one or two concentrations of chemical components can usually be measured directly so such measurements do not give trajectories in 3 and more dimensions. Fortunately, it is possible to reconstruct a trajectory and, in particular, also a strange attractor from experimental changes of one variable in time using the Ruelle - Takens procedure [41]. Let us assume that a strange attractor exists in a N-dimensional phase space which means that reactions between N independent chemical components govern the dynamics of the system. Let us assume, further, that we know the time evolution of concentration of one component $c_i(t)$. For $t_k = k \Delta t$ ($k = 0,1,2,...$) one can construct a point in a M-dimensional phase space with components $c_i(t_k)$, $c_i(t_k + T)$, $c_i(t_k + 2T),..., c_i(t_k + (M-1)T)$ where T is a time delay. In this way for successive values of k it is possible to obtain the trajectory in a M-dimensional phase space. It was proven that the reconstructed attractor has the same properties as the "original" one, and in particular it has the same values of the Lyapunov's exponents, if $M \geq 2N + 1$. Of course, the reconstructed attractors do not contain information on the mechanism of chemical reactions because the evolution of $c_i(t + mT)$ does not usually correspond to the changes of concentration of some chemical component in time but the profits from the reconstruction procedure are obvious. After having reconstructed the strange attractor one can construct 1D return maps, calculate the Lyapunov exponents and the dimensions. It is possible to construct 1D return maps for different values of the control parameters in experiments and if the type of the map belongs to those described above one should observe the bifurcation sequences predicted for the particular class of maps. If this is the case, then it is a very strong experimental proof that the behavior of the system is governed by the strange attractor and the measured chaotic evolutions constitute the deterministic chaos.

It is evident that 1D maps obtained for real systems are not always so simple as those described above. They are sometimes "mixtures" of maps in the sense that, for example, they are not parabolic on the whole interval but e. g., cubic with two extremes. Then one should observe the period-doubling route to chaos connected with a shift of parabolic maximum, "mixed" with the period-adding phenomenon connected with attaining the tangency with the diagonal.

Experimental examples

The first known example of a chemical system with chaotic behavior was the peroxidase reaction [21]. Molecular oxygen is reduced in this reaction to water by NADH (reduced nicotinamide adenine dinucleotide) in the presence of the enzyme horse radish peroxidase:

$$O_2 + 2NADH + 2H^+ \Rightarrow 2H_2O + 2NAD^+ .$$

where NAD^+ is oxidized nicotinamide adenine dinucleotide. Both substrates are fed continuously to a stirred, buffered reaction mixture containing a fixed amount of the enzyme. The dynamics of the reaction, which is described by a more complex model than the above scheme, is monitored by the oxygen concentration. In this reaction a bistability, periodic oscillations and a chaotic behavior have been observed. By changing the amounts of the enzyme in the system one can achieve periodic and chaotic oscillations with different numbers of small and large peaks. The dependence of a next amplitude on the size of a previous one in the chaotic regime is in a qualitative agreement with the model of this reaction (see the paper by Olsen and Degn in [3] p. 512). In this case the points do not lie on a curve but they fit to a set of curves obtained from numerical solution of the model.

The most extensively studied system is the Belousov [10]- Zhabotinsky [11] reaction conducted in CSTR (a continuous flow stirred tank reactor). The reactants are pumped into a well stirred reactor at a given flow rate while the reaction mixture flows out with the same rate. An intense (ideal) stirring should assure the uniformity of concentrations of all the species in the reactor volume. The system is maintained far from equilibrium and true asymptotic states (stationary states, limit cycles or strange attractors) can be reached. By varying of flow rates of the reactants or their concentrations in feed streams it is possible to perform a detailed study of the bifurcations . CSTRs are therefore convenient and versatile tools for studying the dynamics of chemical systems. The BZ reaction consists in oxidation and bromination of an organic compound

(malonic acid and many others) by bromate ions, catalyzed by metal ions (Ce(III) / Ce(IV) and many others). On running this reaction in CSTR it is possible to observe bistability, excitability, simple and complex periodic oscillations, quasiperiodic oscillations, coexistence of different periodic oscillations as well as chaos.

The first experimental evidence of chaos in the BZ reaction[22] was published independently and almost simultaneously with the observations of chaos in the peroxidase reaction. Since then this reaction has been intensively investigated and it is now the best known example of a nonlinear chemical system. Its dynamical behavior has been studied in dependence on the control parameters: flow rates, reagent concentrations in the feed streams, acidity, stirring intensity, temperature, etc. The flow rate is the most convenient bifurcation parameter at fixed values of the other parameters. For ideal stirring conditions the flow rate is equivalent to the residence time which is the ratio of the CSTR volume to the total flux of reagents. Under typical experimental conditions the BZ reaction exhibits an oxidized stationary state at short residence times (high flow rates) and a reduced stationary state at long residence times (low flow rates). At intermediate residence times from minutes to hours, periodic, quasi-periodic and chaotic oscillations (see Fig. 12) are often observed.

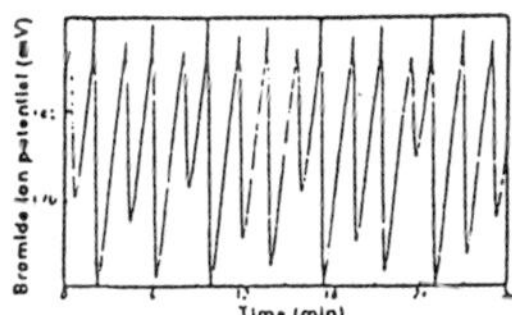

Fig. 12   Chaotic changes of the potential of the bromide selective electrode in time observed in the BZ reaction. (From 48)).

The most important experiments with the chaotic dynamics in the BZ reaction in CSTR were performed at the Virginia University (Virginia experiments [22,42,43]), at the University of Texas at Austin (Texas experiments [44-53]) and at the Bordeaux University (Bordeaux experiments [54-59]). In these experiments detailed studies of the dynamics as well as bifurcations in the BZ reaction have been performed with the residence time as the bifurcation parameter but the concentrations of reagents in feed streams, temperature and acidity were different. The main results of these investigations on deterministic chaos are as follows:

1) asymptotic trajectories of the system reconstructed from changes in concentration of the catalyst or of bromide ion in time according to the Ruelle - Takens procedure appear to be strange attractors (see Fig. 13) with positive value of the largest Lyapunov exponent.

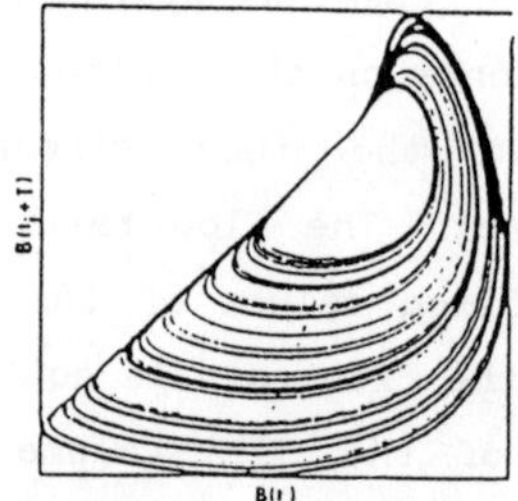

Fig. 13   A two-dimensional projection of the strange attractor reconstructed from time changes of the potential of the bromide selective electrode. (From [48]).

2) 1D return maps constructed from the experimental data have a form of one dimensional curves (see Fig. 14). They are not broad, randomly distributed clouds of points as one could expect if the chaotic behavior was the consequence of stochastic effects like a random noise or internal fluctuations.

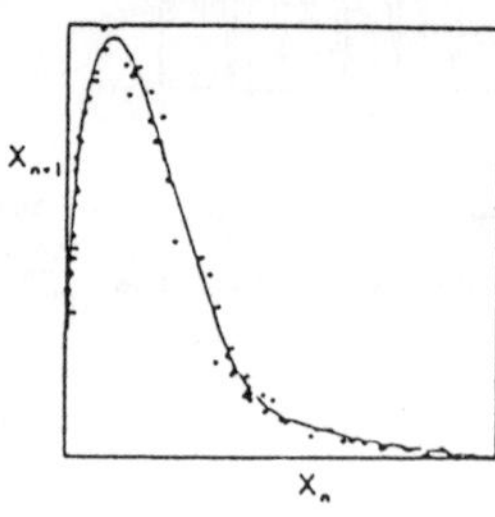

Fig. 14   A 1D map obtained from the reconstructed strange attractor. (From [48]).

3) the bifurcation sequences observed experimentally in some cases

are identical to those found for the generic maps described in the previous chapter.

The period-doubling route to chaos (see Fig. 15) followed by the periodic orbit windows, with periods predicted by the universal sequence, has been observed in experiments [45] with the flow rate as the bifurcation parameter (see Fig. 16). Due to finite accuracy in changes of the flow rate only 2 - 3 period-doublings are seen usually in experiments. Here also only the attracting orbits from the universal sequence with the widest windows have a chance to be detected in experiments. However the behavior of the system under given conditions is in agreement with the theory of smooth unimodal 1D maps.

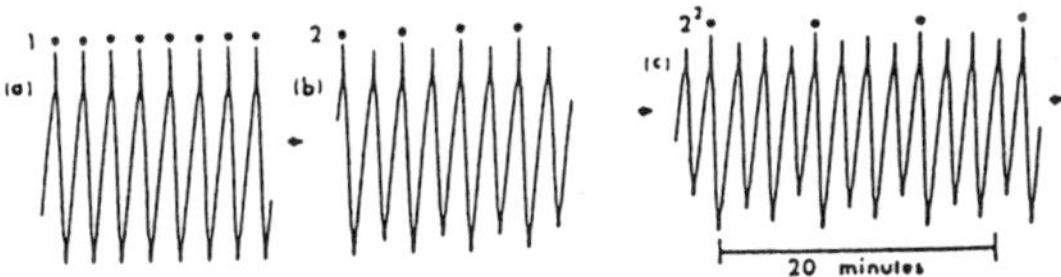

Fig. 15 First two elements of the period-doubling sequence observed in the BZ reaction. Simple periodical oscillations (a) change to 2-periodic oscillations (b) and next to 4-periodic oscillations (c) when the flow rate is varied. The dots above time series are separated by one period. (From [31]).

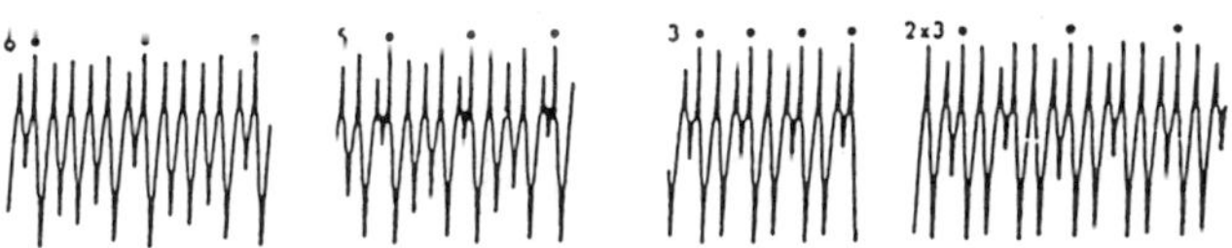

Fig. 16 Time changes with periods 6, 5 and 3 showing elements of the universal sequence observed in the BZ reaction. Amplitudes of peaks in the periods are those as predicted by the universal sequence for smooth unimodal maps. The period-doubling for period 3 is also clearly seen. The dots above the time changes correspond to one period. (From [31]).

The period-adding with chaotic windows called also the "mixed mode" oscillations caused by changes of the flow rate has been found in the Virginia experiments and in the Texas experiments [48] (see Fig. 17). At that time the period-adding phenomenon for 1D cusp - shaped maps was not known but the experimental results are in agreement with the predictions for maps of such a type.

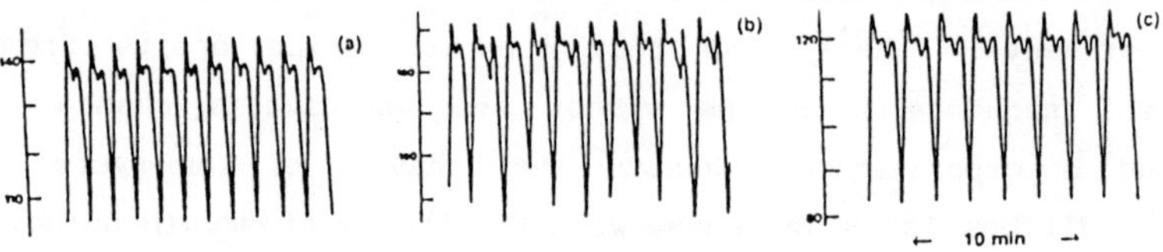

Fig. 17 Changes of the potential of the bromide selective electrode in time observed in the BZ reaction illustrating the period-adding phenomenon. (a) Periodic oscillations with one large amplitude followed by one small amplitude. (b) Chaotic oscillations with one large amplitude followed by either one or two small amplitudes in an unpredictable way. (c) Periodic oscillations with one large amplitude followed by two small amplitudes (From [42]). Similar chaotic changes were observed between others windows with periodic amplitudes [31].

The intermittency route to chaos was observed in the Bordeaux experiments [57] and the Texas experiments [47]. Small amplitude periodical (limit cycle type) oscillations become chaotic above a critical value of the flow rate (see Fig. 18).

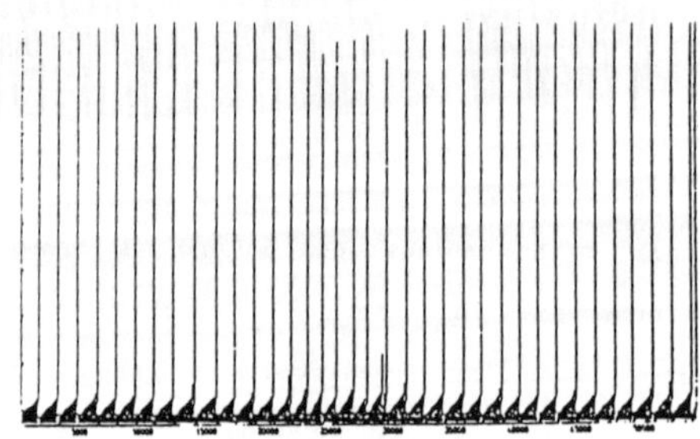

Fig. 18 Time series illustrating the intermittency route to chaos observed in the BZ reaction. (From [47]).

These changes can be described by the tangent bifurcation for 1D maps in which a saddle and a node disappear due to a local shift of the map above the diagonal as it was predicted by the Pomeau and Manneville theory.

A quasi-periodical type time series (see Fig. 19) was observed in the Texas experiments [47]. The behavior of the system is well described by trajectories moving on 2D torus but the shape of the circle map constructed is different from that of the sine map. This may be the reason why the frequency-locking have not been mentioned in those experiments.

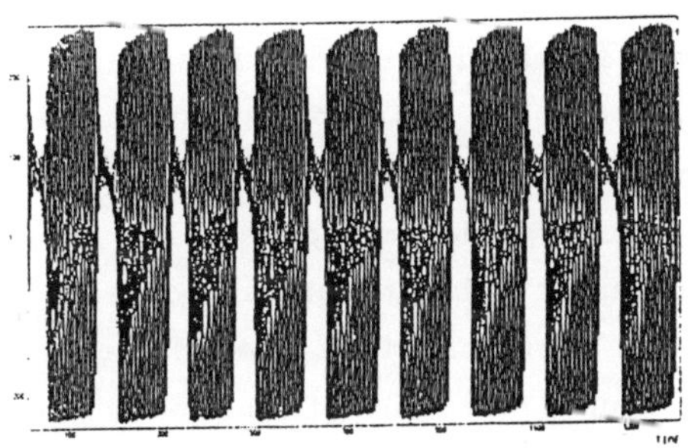

Fig. 19   Quasi-periodical   time   changes   in   the   BZ   reaction   (
From [47]).

The frequency-locking phenomenon with the Farey tree ordering, along with other orderings based also on the Farey arithmetic, has been observed, however, in the Texas experiments [49,50] with the manganese catalyzed BZ reaction and the flow rate used as the bifurcation parameter. In these experiments no quasi-periodical windows in the flow rate have been found as one might expect from the sine map. The devil's staircase constructed from the experimental results does not contain quasi-periodical windows. The shape of the circle map obtained from the experimental data is far from that of the sine map.

Bifurcation sequences observed experimentally in the BZ reaction are not always "pure" in the sense that the sequences, characteristic for one type of the map are mixed with others as should be, for example, for maps with moving smooth maximum and simultaneously approaching the tangent bifurcation. It should also be emphasized that the dynamics of the system not always can be described by 1D maps.

The chaotic behavior has been observed also in other chemical systems but it is not so well documented as in the BZ reaction. For example in the chlorite - thiosulfate reaction [60] the period-adding with chaotic windows has been found as one might expect for systems described by cusp - shaped maps.

Models

First of all one should distinguish between "theoretical" models which show new properties of solutions (like the Lotka - Volterra model, the Lorenz model etc.) or new phenomena like unknown sequences of bifurcations (forced pendulum type models, the Lorenz model and various types of 1D maps or 2D dimensional point transformations like the Henon model), and therefore they are interesting by themselves, and "experimental" models the aim of which is the description and understanding of experimental observations. Until recently physics was the main source of theoretical models but now the situation seems to change. Models from chemical kinetics and other branches of science become also important. It should be stressed, however, that experimental models should be as simple as possible to processed in terms of the nonlinear theory of dynamical systems or numerical calculations. It is not difficult to solve numerically a given, many-variable model which contains many parameters for given initial conditions but the solution is adequate only for this case. The dependence on initial conditions (many attractors can coexist) and the dependence of solutions on parameters should also be explained.

Modeling of nonlinear phenomena in chemical kinetics has its own peculiarities. In many physical systems the mechanistic details of the models are known as these systems are constructed from simple subsystems with known properties and known dynamics. In chemical kinetics a chemical knowledge seldom allows us to predict a mechanism and elementary reactions occurring in a real system. Usually there are many elementary reactions possible and moreover all their rate constants should be known. Thus, as a rule, chemical models contain many variables and many parameters such as rate constants. Beside the questions concerning the many-dimensional systems, problems with the many parameters are very important. Each rate constant is known with some experimental accuracy which is usually a few percent. If one wants to know, how does the dynamics of a given model depend on the values of rate constants, one should study it for some set of chosen values of the rate constants around the experimental, averaged values. Let us assume that we choose n values for each rate constant. Then, if there are p rate constants in the model considered, one has to study n^p different systems. Even in the case if we take only n = 3, that is the "average" and plus and minus experimental error values for the rate constants, it is easy to reach astronomical numbers. Elementary reactions important for a mechanism are often accompanied by other elementary reactions. The rate constants found from independent measurements, when an elementary reaction occurs separately, can be perturbed by other reactions in far-from-equilibrium conditions, especially when they compete for the same intermediates, so the values found are not certain. In an ideal situation one should know, from experiments, the time dependences of all the variables important in the mechanism. This occurs rarely, however. Only the concentrations of one or two chemical components are measured independently.

All these remarks try to explain that the modeling of a complex chemical system is difficult and that the models described below for the BZ reaction, although they do not explain all details of the experiments, are important achievements.

There were some difficulties in modeling the chaotic behavior of the BZ reaction. The models based on the Oregonator [61], which is the simplified version of the Field - Körös - Noyes mechanism (FKN) [62], did not give chaotic solutions [63-66]. Other efforts to model the chaotic behavior gave chaotic solutions which, in many cases, did not agree well with experimental data [44,67-69].

A very interesting attempt, based on the coupling of two

oscillators , was able to describe many details in chaotic behavior [70]. Small sinusoidal oscillations appear through the Hopf bifurcation from the stationary state with a reduced catalyst. Then the secondary Hopf bifurcation yields a quasi-periodical behavior, described by the motion of trajectories on the surface of an attracting 2D torus. The attracting torus breaks then to a fractal object. The seven-variable Oregonator-like model describes successfully the period-doubling and the period-adding routes to chaos, the quasi-periodicity with the devil's staircase for changes of the control parameter playing the role of the flow rate. This model, however, takes no account of new results in mechanistic understanding of the BZ reaction [71,72].

Recently a new class of models, realistic in the chemical sense, has been proposed by Field and co workers, that, as their authors declare, describe satisfactory all the known experimental results . These models have been based on the FKN mechanism of the BZ reaction and in their first, most detailed version [73], contain 19 reactions and 11 components:

$$HOBr + Br^- + H^+ \Rightarrow Br_2 + H_2O$$

$$Br_2 + H_2O \Rightarrow HOBr + Br^- + H^+$$

$$HBrO_2 + Br^- + H^+ \Rightarrow 2HOBr$$

$$Br^- + BrO_3^- + 2H^+ \Rightarrow HOBr + HBrO_2$$

$$HOBr + HBrO_2 \Rightarrow Br^- + BrO_3^- + 2H^+$$

$$2HBrO_2 \Rightarrow BrO_3^- + HOBr + H^+$$

$$BrO_3^- + HBrO_2 + H^+ \Rightarrow 2BrO_2^\bullet + H_2O$$

$$2BrO_2^\bullet + H_2O \Rightarrow BrO_3^- + HBrO_2 + H^+$$

$$Ce(III) + BrO_2^\bullet + H^+ \Rightarrow HBrO_2 + Ce(IV)$$

$$HBrO_2 + Ce(IV) \Rightarrow Ce(III) + BrO_2^\bullet + H^+$$

$$MA + Br_2 \Rightarrow BrMA + Br^- + H^+$$

$$MA + HOBr \Rightarrow BrMA + H_2O$$

$$MA + Ce(IV) \Rightarrow MA^{\bullet} + Ce(III) + H^{+}$$

$$MA^{\bullet} + BrMA \Rightarrow MA + Br^{-} + products$$

$$MA^{\bullet} + Br_{2} \Rightarrow BrMA + Br^{\bullet}$$

$$MA^{\bullet} + HOBr \Rightarrow Br^{\bullet} + products$$

$$2MA^{\bullet} \Rightarrow MA + products$$

$$Br^{\bullet} + MA \Rightarrow Br^{-} + MA^{\bullet} + products$$

The concentrations of H_2O , BrO_3^{-} and H^{+} are assumed constant and are included in the appropriate rate constants. MA , $MA^{\bullet}$, BrMA and "products" denote malonic acid, malonic radical, bromomalonic acid and unspecified products of the appropriate reactions respectively. Kinetic equations for CSTR, describing the above reactions according to the mass action law, have been solved numerically for a range of flow rates taken from experiments and the assumed values of rate constants for appropriate reactions. The numerical results gave both the shape of oscillations and the bifurcation sequences in which chaos appears in a very good agreement with experiments. This version of the model was further simplified to less variable models [74] for which the numerical solutions were still in a good agreement with experimental data. The simplest version obtained is the 3D model [74,75], which also gives chaotic behavior near bifurcations between oscillatory and reduced and oxidized state of the system, as it is observed in experiments, though the agreement is not so good as in the more variable models.

As an illustration of a chaotic behavior possible in coupled enzymatic systems one can present the results concerning a model with two coupled enzymatic reactions [34]. Each of these reactions is inhibited both by an excess of its own substrate (V or P) and the common product (U). The inhibition of such type is quite common in enzymatic systems important in biochemical pathways. Third enzymatic reaction which consumes the product (U) is of the Michaelis - Menten type and is not very important for the model. The considered system is open for both the reactants and the product but closed for the enzymes thus imitating systems with semipermeable membranes. The model seems to be the first one in which the period-adding phenomena were observed [76,77]. The reaction scheme consisting only of elementary reactions reads as follows:

$$V_o \Leftrightarrow V$$

$$V + E \Leftrightarrow VE \Rightarrow U + E$$

$$V + VE \Leftrightarrow V_2E$$

$$U + E \Leftrightarrow E$$

$$U + VE \Leftrightarrow VEU$$

$$U + V_2E \Leftrightarrow V_2EU$$

$$P_o \Leftrightarrow P$$

$$P + E' \Leftrightarrow PE' \Rightarrow U + E'$$

$$P + PE' \Leftrightarrow P_2E'$$

$$U + E' \Leftrightarrow E'U$$

$$U + PE' \Leftrightarrow PE'U$$

$$U + P_2E' \Leftrightarrow P_2E'U$$

$$U + E_1 \Leftrightarrow UE_1 \Rightarrow R + E_1$$

$$P \Leftrightarrow V$$

The first and the seventh reactions describe the inflow and the outflow of reagents into and from the system if concentrations of V_o and P_o are assumed constant. Concentrations of the enzymes are usually much smaller than concentrations of the substrates and the products, thus using the Tikhonov theorem, one can reduce the number of kinetic equations by elimination of concentrations of all the enzymes and all their complexes as fast variables. In a slow time scale, in which concentrations of all the enzymes and all their complexes are equal to their quasi-stationary values, the behavior of the system is described by 3 kinetic equations which in dimensionless concentrations of the substrates, the product and time have the form:

$$\frac{dv}{dt} = A_1 - A_2 v - \frac{v}{(1 + v + A_3 v^2)(1 + u)}$$

$$\frac{du}{dt} = \varepsilon_2 \left(\frac{v}{(1 + v + A_3 v^2)(1 + u)} + B \left(\frac{p}{(1 + p + B_3 p^2)(1 + Ku)} \right) + Dp - \frac{Cu}{L + u} \right)$$

$$\frac{dp}{dt} = \varepsilon_3 \left(B_1 - B_2 p - B \left(\frac{p}{(1 + p + B_3 p^2)(1 + Ku)} \right) \right)$$

where v, u and p are dimensionless concentrations of V, U and P respectively, t is dimensionless time and other symbols are parameters involving rate constants, concentrations of substrates in feed stream and total concentrations of the enzymes.

For ε_3 as the bifurcation parameter and for especially chosen values for all the remaining parameters very detailed numerical calculations show very rich sequences of bifurcations. A limit cycle with a large "radius", which exists for large values of the control parameter, looses its stability and then, at some interval of ε_3, the period-adding phenomenon appears [76]. In this interval subsequent subintervals of the bifurcation parameter, in which successive complex limit cycles are attracting, are separated by windows where trajectories are chaotic. In some subintervals the period-adding phenomenon is accompanied by the period-doubling route to chaos. In another interval of the bifurcation parameter, for lower vales of ε_3, another period-adding appears [77]. In this case the successive complex attracting limit cycles coexist in some subintervals of the bifurcation parameter from that range. The 1D maps constructed for the system in both these ranges of ε_3 exhibit the cusp-shaped form and belong to the two types discussed in the previous chapter.

Discussion

Only selected, well documented examples of homogeneous chemical(CSTR) systems exhibiting deterministic chaos are described. The list of other chemical systems in which chaotic behavior has been observed is quite long. Deterministic chaos has been observed in various electrochemical systems [5]. It is not so surprising because periodic (simple and complex) oscillations have been known in electrochemistry for a long time. One of the simplest and first electrochemical system in which chaotic oscillations have been observed seems to be the $Cu(s)\,|\,CuSO_4\,+\,H_2SO_4(aq)\,|\,Cu(s)$ system at constant voltage conditions [78,79]. In this system, at a proper potential difference, chaotic current oscillations appear. When the applied potential difference is reduced, the chaotic current oscillations become 2-periodic what means that each lager amplitude is followed by a smaller one. Further decrease of the applied potential difference causes that current oscillations become monoperiodic. For the monoperiodic oscillations a dependence of consecutive current amplitudes on previous ones (the 1D Poincaré map):

$$A_{n+1} = f(A_n) \qquad (\text{ for } n = 1, 2, \ldots)$$

gives a fixed point $A_{n+1} = A_n$. For the 2-periodic current oscillations one obtains a pair of fixed points $A_{n+2} = A_n$.

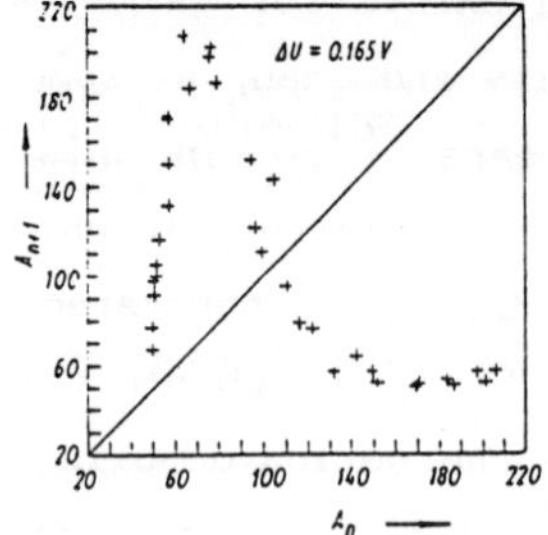

Fig. 20 A dependence of next amplitude of current oscillations on previous one observed in the $Cu(s)\,|\,CuSO_4\,+\,H_2SO_4(aq)\,|\,Cu(s)$ system at constant voltage conditions (From [79]).

However, for larger values of the applied potential difference 1D Poincare maps consist of sets of points dispersed along smooth curves with one maximum what is characteristic for chaotic oscillations (see Fig. 20). The shape of maps changes with the applied potential difference. The descending branch becomes steeper with increase of the applied potential difference whereas the ascending branch seems to stay at the same position. In this system only three regimes (monoperiodic, 2-periodic and chaotic) of current oscillations have been found. If other complex periodic oscillations are really absent, the route to chaos different from those described above was found but the investigations have not been completed.

There are known so called modulated structures exhibiting commensurate and incommensurate phases with the devil's staircase phenomena [80] which are important in the solid state physical chemistry.

Beside the chemical systems the chaotic behavior was found in many physical, biological, economical systems etc. Probably simple dynamical models could explain many historical and contemporary problems in social behaviors. All these facts are not so surprising if one takes into account that at least at the level of models the same type of the description namely nonlinear ordinary differential equations are used. This is the reason why the turning-point in the theory of dynamical system caused by the discovery of strange attractors in as low as 3 dimensional systems had and still has so important and inspiring consequences in many branches of science. In many systems chaos appears according to well determined sequences of bifurcations (routes to chaos) such as period doubling, period-adding, intermittency and quasi-periodicity. In these case there are some kinds of order in which chaos appears. Such an ordering makes that the existence of chaotic behavior can be easier to prove experimentally. Of course these routes do not exhaust possible ways in which deterministic chaos may appear.

The discovery of low dimensional deterministic chaos has also important consequences in experimental methodology. Earlier experimental observations of the same dynamical system could be compared by an agreement of time changes or trajectories of appropriate variables. For systems with strange attractors this way is useless. It is impossible to reproduce initial conditions and values of parameters with infinite accuracy. Small changes in initial values cause that the behavior of the same system is close to each other only in short time intervals. For long time intervals time changes or trajectories observed diverge, thus the results are not reproducible. In systems with chaotic behavior results must be compared on the level of 1D maps, Lyapunov exponents and metric dimensions.

In this paper non autonomous systems have been completely neglected. It is worth to mention that in chemical kinetics there is a growing interest in such systems. Many experiments in flow reactors with flow rates changing in time according to a given formula are performed. In such cases called driven systems or periodically driven systems, if flow rates are periodic functions of time, one can also observe nonlinear simple and complex oscillations and chaos. Moreover, experiments with turning back of a part of feed streams with some time delay are performed. Such systems are governed by the so-called " time-lag " kinetic equations [4] and also exhibit a variety of nonlinear oscillations including chaos [81].

References

1) Hao Bai-lin, Chaos, World Scientific, Singapore, 1984

2) Cvitanovic P., Universality in chaos, Adam Hilger, 1984

3) Hao Bai-lin, Chaos II, World Scientific, Singapore, 1990

4) Kawczyński A.L., Chemical Reactions from Equilibrium through Dissipative Structures to Chaos, Wydawnictwa Naukowo-Techiczne, Warsaw, 1990 (in Polish)

5) Field R.J., Györgyi L., Chaos in chemistry and biochemistry, World Scientific, Singapore, 1993

6) Bray W.C., *J. Am. Chem. Soc.*, **43**, 1262 (1921)

7) Lotka A.J., *J. Phys. Chem.*, **14**, 271 (1910)

8) Lotka A.J., *J. Am. Chem. Soc.*, **42**, 1595 (1920)

9) Volterra V., Leçons sur la théorie mathêmatique de la lutte pour la vie, Gautier-Villars, Paris, 1936

10) Belousov B.P., Sbornik referatov po radiacjonnoj medicinie, Medizd. Moscow, 1959 (Russian)

11) Zhabotinskii A.M., Self - oscillating concentrations, Nauka, Moscow, 1974 (in Russian)

12) Glansdorff P., Prigogine I, Thermodynamics of structure, stability and fluctuations, Wiley-Intsci., New York, 1971

13) Nicolis G., Prigogine I., Self-organization in nonequilibrium systems, Wiley-Intsci., New York, 1977

14) Andronov A.A., Vitt A.A., Chaikin S.E., Theory of oscillations, Fizmatgiz., Moscow, 1959

15) Poincaré H., Les médothes nouvelles de la mécanique celeste, Gautier-Villars, Paris, 1899

16) Lorenz E.N., *J. Atmos. Sci.*, **20**, 130 (1963)

17) Ruelle D., Takens F., *Commun. Math. Phys.*, **20**, 167 (1971)

18) Li T.Y., Yorke J.A., *Am. Math. Monthly*, **82**, 985 (1975)

19) Feigenbaum M.J., *J. Stat. Phys.*, **19**, 25 (1978)

20) Rössler O.E., *Z. Naturforsch.*, **31a**, 259 (1976)

21) Olsen L.F., Degn H., *Nature*, **267**, 177 (1977)

22) Schmitz R.A., Graziani K.R., Hudson J.L., *J. Chem. Phys.*, **67**, 3040 (1977)

23) Lichtenberg A.J., Liberman M.A., Regular and stochastic motion, Springer-Verlag, New York, 1983

24) Oseledec V I , *Trans. Moscow Math. Soc.*, **19**, 197 (1968)

25) Schuster H.C., Deterministic chaos, Physik-Verlag, Weinheim, 1984

26) Farmer J.D., Ott E., Yorke J.A., *Physica*, **7D**, 153 (1983)

27) Mandelbroot B.B., The fractal geometry of nature, W.H. Freeman and Co, 1982

28) Grassberger P., Procaccia I., *Phys. Rev. Lett.*, **50**, 346 (1983)

29) Henon M., *Commun. Math. Phys.*, **50**, 69 (1976)

30) Collet P., Eckmann I.P., Itorated maps of the interval as dynamical systems, Birkhäuser, Boston, 1980

31) Swinney H.L., *Physica*, **7D**, 3 (1983)

32) Feigenbaum M.J., *Physica*, **7D**, 16 (1983)

33) Metropolis M., Stein M.L., Stein P.R., *J. Combinatorial Theory (A)*, **15**, 25 (1973)

34) Kawczyński A.L., Misiurewicz M., Leszczyński., *Pol. J. Chem.*, **63**, 239 (1989)

35) Kawczyński A.L., Misiurewicz M., *Z. phys. Chem. (Leipzig)*, **271**, 1037 (1990)

36) Misiurewicz M., Kawczyński A.L., *Commun. Math. Phys.*, **131**, 605 (1990)

37) Misiurewicz M., Kawczyński A.L., *Physica*, **52**, 191 (1991)

38) Pomeau Y., Manneville P., *Commun. Math. Phys.*, **74**, 189 (1980)

39) Glazier J.A., Libchaber A., *IEEE Trans. Circuits and Systems*, **35**, 790 (1988)

40) Glass L., *Chaos*, **1**, 13 (1991)

41) Takens F., Lecture Notes in Mathematics, **898**, Springer, Heidelberg, 1981

42) Hudson J.L., Hart M., Marinko D., *J. Chem. Phys.*, **71**, 1601 (1979)

43) Hudson J.L., Mankin J.C., *J. Chem. Phys.*, **74**, 6171 (1981)

44) Turner J.S., Roux J.C., McCormick W.D., Swinney H.L., *Phys. Lett.*, **85A**, 9 (1981)

45) Simoyi R.H., Wolf A., Swinney H.L., *Phys. Rev. Lett.*, **49**, 245 (1982)

46) Roux J.C., Swinney H.L., in Nonlinear phenomena in chemical dynamics, ed. Vidal C., Pacault A., Springer, Berlin, 1981

47) Roux J.C., *Physica*, **7D**, 57 (1983)

48) Roux J.C., Simoyi R.H., Swinney H.L., *Physica*, **8D**, 257 (1983)

49) Masełko J., Swinney H.L., *J. Chem. Phys.*, **85**, 6430 (1986)

50) Swinney H.L., Masełko J., *Phys. Rev. Lett.*, **55**, 2366 (1985)

51) Coffman K.G., McCormick W.D., Noszticzius Z., Simoyi R.H., Swinney H.L., *J. Chem. Phys.*, **86**, 119 (1987)

52) Noszticzius Z., McCormick W.D., Swinney H.L., *J. Phys. Chem.*, **91**, 5129 (1987)

53) Noszticzius Z., McCormick W.D., Swinney H.L., *J. Phys. Chem.*, **93**, 2796 (1989)

54) Vidal C., Roux J.C., Bachelart S., Rossi A., *Ann. N. Y. Acad. Sci.*, **357**, 377 (1980)

55) Roux J.C., Rossi A., Bachelart S., Vidal C., *Phys. Lett.*, **77A**, 391 (1980)

56) Vidal C., Bachelart S., Rossi A., *J. Phys. (Paris)*, **43**, 7 (1982)

57) Pomeau Y., Roux J.C., Rossi A.,Bachelart S., Vidal C., *j. Phys. Lett.*, **42**, L271 (1981)

58) Argoul F., Roux J.C., *Phys. Lett.*, **108A**, 426 (1985)

59) Argoul F., Arneodo A., Richetti P, Roux J.C., *J. Chem. Phys.*, **86**, 3325 (1987)

60) Epstein I.R., *Physica*, **7D**, 47 (1983)

61) Field R.J., Noyes R.M., *J. Chem. Phys.*, **60**, 1877 (1974)

62) Field R.J., Körös E., Noyes R.M., *J. Am. Chem. Soc.*, **94**, 8649 (1972)

63) Showalter K., Noyes R.M.,Bar Eli K., *J. Chem. Phys.*, **69**, 2514 (1978)

64) Ganapathisubramanian N., Noyes R.M., *J. Chem. Phys.*, **76**, 1770 (1982)

65) Schwartz B., *Phys. Lett.*, **102A**, 25 (1984)

66) Rinzel J., Schwartz B., *J. Chem. Phys.*, **80**, 5610 (1984)

67) Ringland J., Turner J.S., *Phys. Lett.*, **105A**, 93 (1984)

68) Barkley D., Ringland J., Turner J.S., *J. Chem. Phys.*, **87**, 3812 (1987)

69) Linderberg D., Turner J.S., Barkley D., *J. Chem. Phys.*, **92**, 3238 (1990)

70) Richetti P., Roux J.C., Argoul F., Arneodo A., *J. Chem. Phys.*, **86**, 3339 (1987)

71) Field R.J., Försterling H.D., *J. Phys. Chem.*, **90**, 5400 (1986)

72) Györgyi L., Turányi T., Field R.J., *J. Phys. Chem.*, **94**, 7162 (1990)

73) Györgyi L., Rempe S.L., Field R.J., *J. Phys. Chem.*, **95**, 3159 (1991)

74) Györgyi L., Field R.J., *J. Phys. Chem.*, **95**, 6594 (1991)

75) Györgyi L., Field R.J., *Nature*, **355**, 808 (1992)

76) Aicardi F., Kawczyński A.L., *Complex Systems*, **4**, 1 (1990)

77) Aicardi F., Kawczyński A.L., *Complex Systems*, **6**, 95 (1992)

78) Kawczyński A.L., Przasnyski M., Baranowski B., *J. Electroanal.*

Chem., **179**, 285 (1984)

79) Kawczyński A.L., Raczyński W., Baranowski B., Z. *phys. Chem. (Leipzig)*, **269**, 596 (1988)

80) Bak P., *Rep. Prog. Phys.*, **45**, 587 (1982)

81) Schell M., Ross J., *J. Chem. Phys.*, **85**, 6489 (1986)

82) Okniński A., Catastrophe Theory in Chemistry, Państwowe Wydawnictwo Naukowe, Warszawa, 1990 (in Polish)

83) Scott S. K., Chemical chaos, Oxford University Press, 1991

OSCILLATING CHEMICAL REACTION IN OIL/WATER SYSTEMS GENERATION OF MACROSCOPIC OSCILLATORY FORCE

Toshinori Kusumi, Kenichi Yoshikawa[†] and Satoshi Nakata*

Graduate School of Human Informatics, Nagoya University,
Nagoya 464-01, Japan

*Department of Chemistry, Nara University of Education,
Takabatake-cho, Nara 630, Japan

ABSTRACT

We have found that a periodic change of the surface tension is generated at an air/water interface and also at an oil/water interface, when the water phase is the Belousov-Zhabotinsky (BZ) medium. It has been shown that the difference of the surface activity of the iron-catalyst between $[Fe(phen)_3]^{3+}$ and $[Fe(phen)_3]^{2+}$ is the driving force of this rhythmic phenomenon. We also discovered a rhythmic phenomenon for an oil/water system, where the oil phase contains an ester and the water phase is an alkaline solution. These results indicate a new field of research toward the realization of chemo-mechanical coupling based on the idea of nonlinear dynamics.

1. INTRODUCTION

In the present paper, we will show that macroscopic mechanical force is induced by chemical reaction. Living organisms generate mechanical work with direct conversion of chemical energy through the hydrolysis of ATP. In contrast to this, in conventional thermal engines, e.g. gasoline engines, chemical energy is first converted into heat, or thermal energy, and then macroscopic mechanical motion is generated. According to the "Carnot cycle", the temperature difference in the thermal engine is the essential factor that determines the efficiency of the chemo-mechanical energy conversion. It is therefore apparent that the mechanism of chemo-mechanical conversion in living organisms, such as muscles, is quite different from that of the thermal engine. More than four decades ago,

---

†: To whom correspondence should be addressed.

Katchalsky et al. reported a nice idea on the mechano-chemical transduction in an isothermal system utilizing the phase transition of polymer, induced by the concentration difference of salt[1, 2]. However, as far as we know, there has been no study to construct an artificial chemical engine driven by actual "chemical reaction".

2. EXPERIMENTAL

All reagents were analytical grade and used without further purification. The aqueous solution of ferroin, tris(1,10-phenanthroline) iron(II) sulfate, was prepared according to the previous report[3]. The water was distilled and purified by using a Millipore-Q system. The experiments were carried out at 25 °C except for the case mentioned otherwise.

For the experiments with the BZ reaction, the medium was continuously stirred in order to generate the chemical oscillation all-in-phase throughout the whole solution. The redox potential was measured between a platinum and a Ag/AgCl reference electrode, using a salt bridge filled with K_2SO_4 aqueous solution. The change of the surface tension was monitored by the Wilhelmy method[4]. In order to avoid the effect of the stirring on the apparent force to the Wilhelmy plate, cylindrical plate made of a thin platinum sheet (0.1 mm thickness, 15 mm diameter) was used. We have confirmed that stirring effect on the measurement of the surface tension is negligible, using this cylindrical plate in a reference experiment. The measurement of the redox potential and the surface tension in an oil/water system with the BZ medium was carried out in a similar procedure. The measurements of the electrical potential between oil and water phases in the system of hydrolysis reaction were carried out in an apparatus similar to those in the previous reports[5-7].

3. RESULTS AND DISCUSSION

3. 1 Generation of Periodic Force in the BZ Aqueous Medium

Figure 1 shows the time traces of the redox potential (ΔE) and the surface tension ($\Delta \gamma$) for the BZ aqueous medium. In Figure 1, it is clear that the surface tension changes in a rhythmic manner synchronized with the oscillation of the redox potential. When the catalyst is cerium ion without ferroin, the surface tension does not exhibit rhythmic

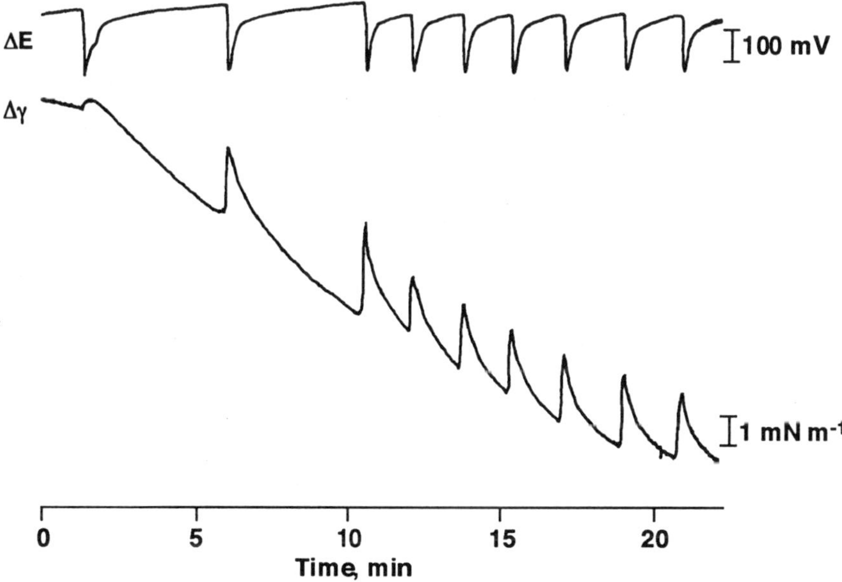

Figure 1. Rhythmic changes of the redox potential (ΔE) in the BZ medium and the surface tension (Δγ) at the air/water interface of the medium. Compositions: [malonic acid] = 0.05 M, [H_2SO_4] = 0.3 M, [$NaBrO_3$] = 0.15 M, [NaBr] = 0.03 M and 6 mM ferroin.

changes. It is therefore plausible that the temporal change between ferric and ferrous ions complexed with 1,10-phenanthroline is the root of the rhythmic change of surface tension. It is expected that the surface activity between Fe(III) and Fe(II) complexes should be different because of the difference in solvation or hydrophilicity. In order to make clear the effect of the iron-complex on the surface tension, we have carried out the measurement of surface tension for the solutions with Fe(III) and Fe(II) complexes in similar conditions as in the BZ medium expect for the absence of the substrate, malonic acid (Table I). It is clear that the surface activity of the Fe(II) complex, is larger than that of the Fe(III) complex, in other words, the surface tension is low when the redox potential is low, that is, when the Fe(II) complex is predominant in the BZ medium. The above mentioned result clearly indicates that the difference of the surface activity of the catalyst with the different redox state is the main driving force of the rhythmic change in the surface tension.

Table I. Surface tension (γ) of iron-1, 10-phenanthroline complexes in the different redox states

concentration c / mM	surface tension γ / mN m^{-1}	
	$[Fe(phen)_3]^{2+}$	$[Fe(phen)_3]^{3+}$
1	64.0	68.2
3.2	63.4	65.7
10	61.3	64.1
32	60.4	61.9

All data was obtained at 25 °C in the presence of 0.15 M NaBrO$_3$. $[Fe(phen)_3]^{3+}$ was prepared from $[Fe(phen)_3]^{2+}$ in the presence of 0.15 M NaBrO$_3$ and 0.3 M H$_2$SO$_4$.

Recently, there have been some reports describing the onset of convective flow accompanying the spatio-temporal pattern formation in the BZ reaction[8-16]. Some of these reports considered the effect of the temperature gradient due to the repetitive redox reaction, because the BZ reaction produces heat only in its oxidative state and generates essentially no heat in the reductive state. The temperature difference between the two states is the order of 0.1 °C.[11] As for the temperature effect on the surface tension, it is well established that, in general, the surface tension becomes smaller with higher temperature. Thus, if the change in the temperature of the BZ medium causes the rhythmic change of the surface tension, the surface tension should be small for the oxidative state. This contradicts with the experimental results. Then, as for the temperature effect on the density of the reacting liquid, rise in the temperature should increase the volume of the solution and this, in turn, should decrease the apparent surface tension. This is opposite with the experimental trend.

Next, as for the possibility of the hydration effect, Pojman et al.[11] speculated that the difference of the degree of hydration in the catalyst ion may induce the buoyancy flow. However this idea can give no explanation on the different effect between the catalysts (see Table I). It is also noted that we have found that oscillatory change of the interfacial tension is enhanced for the interface between benzene and the BZ aqueous medium(Figures 2 and 3). From the above results and discussion, it become apparent that

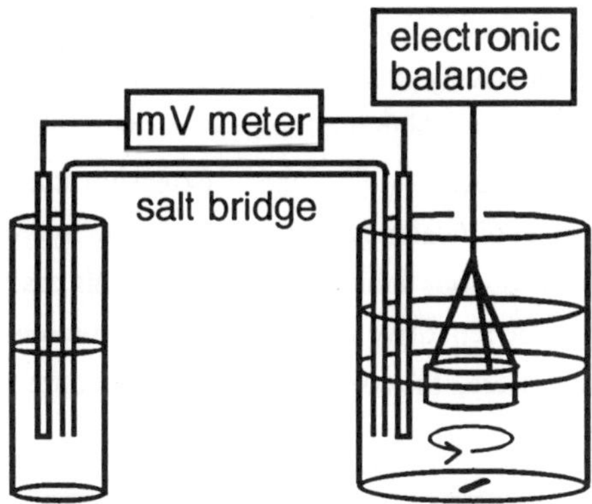

Figure 2. Schematic representation of the experimental apparatus used to monitor simultaneously the periodic changes of surface tension and redox potential in an oil-water system.

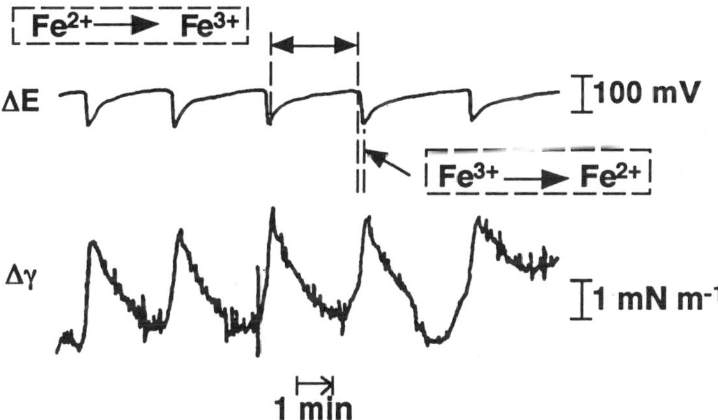

Figure 3. Rhythmic changes of the redox potential (ΔE) and the surface tension ($\Delta\gamma$) in an oil-water system for the experiment schematically shown in Figure 2. Oil phase: benzene and water phase: the BZ medium with [malonic acid] = 0.05 M, [H_2SO_4] = 0.3 M, [$NaBrO_3$] = 0.15 M, [NaBr] = 0.03 M and 6 mM ferroin.

the rhythmic change of the surface tension should be the main cause to induce convective motion in the BZ medium. We also expect that, with the suitable modification of the boundary condition in the experimental system, it will be possible to generate large fluid motion driven by the rhythmic change of the interfacial tension.[17]

92

3. 2 Oscillatory Phenomena in Oil-Water Systems

We have been extensively studying the oscillatory phenomena in oil/water systems which are caused by the high nonlinearity of the transfer process of surfactant molecules through the interface.[5-7, 18] As extension studies, we discovered the oscillatory phenomena in an oil/water system with a simple hydrolysis reaction of ester in the presence of alkaline aqueous solution(Figure 4). In this case, the mechanism of the oscillation is attributed to the repetitive formation and destruction of the soap monolayer at the oil-water interface. Here, the soap molecules are produced successively with the

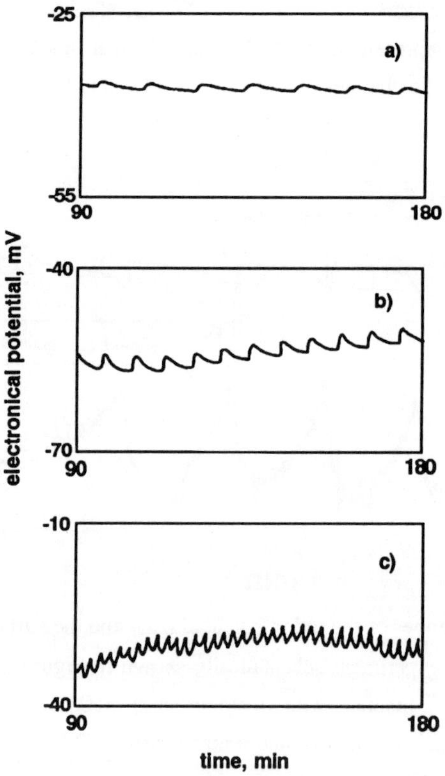

Figure 4. Oscillatory change of the electronical potential in an oil/water interface. Oil phase: 20v/v% n-octyl n-butyrate in nitrobenzene, Water phase:1 M KOH. Temperature: a) 298 K, b) 303 K, c) 308 K. The experimental apparatus was similar to those in the previous studies.[5, 6]

hydrolysis reaction of the ester. As the change of the interfacial tension in the experiment of hydrolysis of the ester seems to be rather small, we have not succeeded yet in observing the synchronized changes between the surface tension and the electronical potential. We hope that it becomes possible to find the experimental condition to observe marked rhythmic change in the surface tension for a similar system in the near future.

According to the Curie-Prigogine theorem,[19, 20] it is impossible to couple scalar and vectorial processes in linear systems with isotropic environment. The violation of either of the conditions, linearity or isotropicity, is enough to accomplish the coupling. However, it may be expected that coupling becomes more preferable with the double negation of the both conditions. In the present study, we have succeeded in verifying the generation of periodic change in the surface tension, or the oscillatory force working at the interface, as a macroscopic vectorial variable, utilizing the nonlinear dynamics in anisotropic environment.

ACKNOWLEDGMENT

The authors would like to express thanks Dr. T. Ishii, Tsurumi University, for useful technical advice and to Ms. M. Ukitsu for the collaboration in the experiment of the hydrolysis of ester. This work was partly supported by a Grant-in-Aid for Scientific Research from the Ministry of Education, Science and Culture of Japan.

REFERENCES

[1] W. Kuhn. B. Hargitay, A. Katchalsky and H. Eisenberg, Nature, <u>165</u>, 514 (1950).

[2] M. V. Sussman and A. Katchalsky, Science, <u>167</u>, 45 (1970).

[3] K. Yoshikawa, T. Kusumi, M. Ukitsu, and S. Nakata, Chem. Phys., Lett. in press.

[4] N. K. Adam, "The physics and chemistry of surfaces 2nd edition", Oxford University press (1938).

[5] K. Yoshikawa and Y. Matsubara, J. Am. Chem. Soc.,<u>106</u>, 4423 (1984).

[6] K. Yoshikawa, M. Shoji, S. Nakata, S. Maeda and H. Kawakami, Langmuir, <u>4</u>, 759 (1988).

[7] K. Yoshikawa, S. Maeda and H. Kawakami, Ferroelectronics, <u>86</u>, 281 (1988).

[8] H. Miike, S. C. Müller and B. Hess, Chem. Phys. Lett., <u>144</u>, 515 (1988) .

[9] H. Miike, S. C. Müller and B. Hess, Phys. Rev. Lett., <u>31</u>, 2109 (1988).

[10] H. Miike, S. C. Müller and B. Hess, Phys. Lett., <u>141A</u>, 25 (1989).

[11] J. A. Pojman and I. R. Epstein, J. Phys. Chem., <u>94</u>, 4966 (1990).

[12] J. A. Pojman, I. R. Epstein, T. J. McManus and K. Showalter, J. Phys. Chem., <u>95</u> 1299 (1991).

[13] J. A. Pojman, I. P. Nagy and I. R. Epstein, J. Phys. Chem., <u>95</u>, 1306 (1991) .

[14] A. Tzalmona, R. L. Armstrong, M. Menzinger, A. Cross and C. Lemaire, Chem. Phys. Lett., <u>174</u>, 199 (1990).

[15] A. Tzalmona, R. L. Armstrong, M. Menzinger, A. Cross and C. Lemaire, Chem. Phys. Lett., <u>188</u>, 475 (1992).

[16] M. Menzinger, A. Tzalmona, R. L. Armstrong, A. Cross and C. Lemaire, J. Phys. Chem., <u>96</u>, 4725 (1992).

[17] K. Yoshikawa, N. Magome, Bull. Chem. Soc. Jpn. in press.

[18] K. Yoshikawa and Y. Matsubara, J. Am. Chem. Soc., <u>105</u>, 5967 (1983).

[19] I. Prigogine, "Introduction to the Thermodynamics of Irriversible Processes", Thomas, Springfield, Illinois (1955).

[20] A. Katchalsky and P. F. Curran, "Nonequilibrium Thermodynamics in Biophysics", Harvard Univ. Press, Cambridge (1965).

INTERACTION OF HOPF AND TURING INSTABILITIES IN CHEMICAL SYSTEMS

G. DEWEL, P. BORCKMANS, A. DE WIT

Service de Chimie-Physique and Center for Nonlinear Phenomena and Complex Systems
CP 231 - Campus Plaine
Université Libre de Bruxelles, 1050 Brussels, Belgium

Abstract: The competition between Turing and Hopf bifurcations may give rise to either localized patterns or complex spatio-temporal behaviour.

1. INTRODUCTION

Chemical species involved in stable nonlinear chemical reactions may organize spontaneously in space if their diffusion coefficients are sufficiently different [1,2]. More precisely in two species systems, the activator which promotes its own production must diffuse more slowly than its antagonist, the inhibitor. For realistic models one finds that the corresponding diffusion coefficients must differ by an order of magnitude. This necessary condition has prevented during forty years the experimental observation of Turing structures in aqueous solutions of small molecules or ions since in such media they all have diffusion coefficients within a factor two of $1{,}5 \ 10^{-5} \ cm^2/s$.

It is now established that the steady periodic concentration (Turing) patterns obtained in the CIMA reaction in gel reactors [3,4] arise because the iodine species form a reversible complex with starch, the color indicator of the reaction, that is immobilized in the gel matrix. The formation of this complex introduces the difference between the effective mobilities of the reactants necessary to shift a competing Hopf bifurcation to larger parameter values so that the Turing instability is the first to appear and is therefore observable [5].

The observation of Turing patterns has sparked off subsequent experimental studies devoted to the determination of bifurcation diagrams, of the role played by the gel and of the dimensionality of the structures. Theoretical work went about discovering the different pattern modes, their selection and the role of the feeding concentration ramps. This work has recently been reviewed [6].

By varying the starch content of the gel it is then possible to monitor the distance in parameter space between these two instabilities. In particular in the region where they interact nontrivial spatio-temporal patterns can be observed [7]. In this note we describe two examples of such behaviors that might have some relevance to the experiments performed by P. De Kepper and coworkers. Their results and the experimental set-up are described in the contribution of B. Rudovich et al. to this volume. For the sake of simplicity we present the theoretical discussion in the framework of the amplitude equations approach [8]. We also mainly focus on one-dimensional systems because it is experimentally possible to restrict the phenomena of interest to a quasi 1D domain by an appropriate choice of the feeding gradients.

2. THE CODIMENSION TWO TURING-HOPF BIFURCATION

In the vicinity of the degenerate bifurcation point, the concentration field C that appears in the reaction-diffusion equations describing the system may be expressed in terms of two complex amplitudes T and H [9]

$$C(x,t) = C_0 + e_T\, T(X,\tau)\, e^{i\,q_c x} + e_H\, H(X,\tau)\, e^{i\,\Omega_c t} + \text{c.c.} \tag{1}$$

C_0 is the uniform reference state e_T and e_H are respectively the critical Turing and Hopf eigenvectors, Ω_c is the critical frequency of the limit cycle while q_c is the critical Turing wavevector.

The competition between the two modes H and T can be described by amplitude equations that are obtained by the use of standard multiscale techniques [8,9] if X and τ are the slow space and time scales then, in one dimension,

$$\frac{\partial T}{\partial \tau} = \mu_T T - g|T|^2 T - \lambda |H|^2 T + D^T \frac{\partial^2 T}{\partial X^2} \tag{2}$$

$$\frac{\partial H}{\partial \tau} = \mu_H H - \left(\beta_r + i\beta_i\right)|H|^2 H - \left(\delta_r + i\delta_i\right)|T|^2 H + \left(D_r^H + i D_i^H\right)\frac{\partial^2 H}{\partial X^2} \tag{3}$$

where μ_M and μ_T are the two unfolding parameters. We will also assume that g and β_r are positive so that both bifurcations are supercritical. Eqs. (3) and (4) possess three non trivial global solutions:

(i) a band of steady Turing structures

$$T = \left\{ (\mu_T - D^T Q^2)/g \right\}^{1/2} e^{\,i\,Q\,X} \quad , \qquad H = 0 \tag{4}$$

(ii) a one parameter family of plane waves

$$T = 0 \quad , \qquad H = \left\{ (\mu_H - D_r^H \kappa^2)/\beta_r \right\}^{1/2} e^{\,i\,(\Omega_\kappa \tau - \kappa X)} \tag{5}$$

with the frequency renormalization $\Omega_\kappa = -\beta_i \left| H_\kappa \right|^2 - D_i^H \kappa^2$ where H_κ is the

preexponential factor in H.

(iii) a two-parameter family of mixed-modes

$$T = \left\{ \frac{\beta_r (\mu_T - D^T Q^2) - \lambda (\mu_H - D_r^H \kappa^2)}{\Delta} \right\}^{1/2} e^{\,i\,QX}$$

$$H = \left\{ \frac{g (\mu_H - D_r^H \kappa^2) - \delta_r (\mu_T - D^T Q^2)}{\Delta} \right\}^{1/2} e^{\,i\,(\Omega_{\kappa Q} - \kappa X)} \tag{6}$$

with $\Delta = \beta_r g - \lambda \delta_r$ and $\Omega_{\kappa Q} = -\beta_i \left| H_{\kappa Q} \right|^2 - \delta_i \left| T_{\kappa Q} \right|^2 - D_i^H \kappa^2$ where $H_{\kappa Q}$ and $T_{\kappa Q}$

are the preexponential factors in Eq. (6). These solutions correspond to pulsating structures.

3.1 Localized structures

When $\Delta < 0$, the mixed mode is always unstable and bistability occurs between the pure Hopf and Turing modes. Besides the global structures given by Eqs. (4) and (5), various localized structures have been characterized in this domain [10,11]. The simplest consists of a stationary front, stabilized by pinning effects, connecting a domain of Turing pattern to a train of plane waves. Turing-Hopf fronts have also been studied in an array of resistively coupled LC oscillators [12]. These fronts may then serve as building blocks for droplets of one state embedded in the other. Simulations have indeed produced such localized objects the core of which is formed by a Turing structure truncated to a few wavelengths and emitting waves on both sides [10]. Similar patches of standing waves surrounded by traveling waves have also been observed in the Rayleigh-Bénard instability in binary mixtures [13].

On varying the control parameter in the direction where the global Hopf state is dominant, single localized Turing spots playing the role of asynchronous wave sources emitting waves on either side in phase opposition may be obtained. They are similar to the 1D spirals (chemical flip-flop) obtained in the CIMA reaction [10]. 2D computer simulations also yield for the same values of the parameters stable spirals with a Turing spot in the core in agreement with the experimental results [14].

3.2 Spatio-temporal Chaos

For other values of the parameters, the interaction between Hopf and Turing instabilities give rise to some type of spatio-temporal complexity. The study of spatio-temporal chaos in driven extended systems has been the focus of a large activity these recent years [15]. One scenario clearly follows from the numerical simulation of the 1D complex Ginzburg-Landau equations [16,17]. In this picture a homogeneous limit cycle first undergoes a Benjamin-Feir instability. It is a diffusion induced instability that drives the system into a state of phase turbulence in which the amplitude is nearly constant but the phase exhibits spatio-temporal irregularities. A further variation of some control parameter then leads the system into a more chaotic state characterized by a proliferation of topological defects in the core of which the amplitude now goes to zero [18]. The term topological turbulence has been coined to characterize this state.

As pointed out already by Kuramoto [16], the two diffusion driven instabilities (Benjamin-Feir, Turing) are mutually exclusive for most reaction-diffusion models. In particular this is the case for the CIMA reaction. Therefore the scenario presented above cannot explain the "chemical turbulence" that may appear in the vicinity of the Turing-Hopf codimension 2 point. We have proposed [19] an alternative mechanism based on the phase instability of the Turing-Hopf mixed mode. When $\Delta<0$ this mixed state is stable towards spatially homogeneous perturbations. This condition can indeed be fulfilled for some range of parameters in reaction-diffusion systems. On increasing the bifurcation parameter μ_H one then typically observes the following sequence of states : Turing structures$\rightarrow$mixed mode$\rightarrow$homogeneous oscillations. In the absence of spatial modulations, eqs. (2) and (3) are invariant under the transformations $T \rightarrow T\, e^{i\theta}$ and $H \rightarrow H\, e^{i\theta}$. As a result, the corresponding linearized matrix about the mixed state has two zero eigenvalues. When spatially inhomogeneous perturbations are taken into account these marginal modes may induce diffusive instabilities of the phases. In particular, the most stable mixed mode ($Q=0$, $\kappa=0$) undergoes such an instability when

$$D = \frac{D_i^H\left(\beta_i\, g - \lambda\, \delta_i\right) + D_r^H\left(\beta_r\, g - \lambda\, \delta_r\right)}{\Delta} < 0 \tag{7}$$

Let us remark that the standard Benjamin-Feir instability criterion of a homogeneous limit cycle is recovered when all the parameters related to the coupling between the two modes are set equal to zero, i.e.,

$$D_i^H\, \beta_i + D_r^H\, \beta_r < 0 \tag{8}$$

It is important to note that the inequality (7) may be satisfied even when (8) is not fulfilled, i.e., when the limit cycle is stable with respect to the modulational instability.

Simulations show that, when $D<0$, the mixed mode is indeed unstable. According to the value of the parameters, the system enters then either a phase-turbulent regime similar to that of the Kuramoto-Sivashinsky equation [20,21] or a defect chaos regime [16,18] characterized by phase defects and large-amplitude fluctuations on both T and H.

4. ACKNOWLEDGEMENTS

We would like to thank G. Nicolis for his interest in this work and P. De Kepper, J. Boissonade, E. Dulos, J-J. Perraud and B. Rudovics for stimulating discussions. P.B. and G.D. are Research Associates with the F.N.R.S. (Belgium). This work was supported by the EC Science Program (Twinning No.. SC1-CT91-0706)

5. REFERENCES

01. A. Turing, Philos. Trans. R. Soc. Lond. B **237**, 37 (1952)

02. G. Nicolis and I. Prigogine, *Self-Organization in Nonequilibrium Systems* (Wiley, New York, 1977)

03. V. Castets, E. Dulos, J. Boissonade and P. De Kepper, Phys. Rev. Lett. **64**, 2953 (1990)

04. Q. Ouyang and H.L. Swinney, Nature **352**, 610 (1991)

05. I. Lengyel and I.R. Epstein, Proc. Natl. Acad. Sci. USA **89**, 3977 (1992)

06. See the contributions by J. Boissonade et al., P. Borckmans et al., I.R. Epstein et al. and Q. Ouyang and H.L. Swinney in *Chemical Waves and Patterns* (R. Kapral and K. Showalter, Eds., Kluwer, Amsterdam, to appear 1994)

07. J-J. Perraud, K. Agladze, E. Dulos and P. De Kepper, Physica A **188**, 1 (1992)

08. P. Manneville, *Dissipative Structures and Weak Turbulence* (Academic Press, San Diego, 1990)

09. H. Kidachi, Progr. Theor. Phys. **63**, 1152 (1980)

10. J-J. Perraud, A. De Wit, E. Dulos, P. De Kepper, G. Dewel and P. Borckmans, Phys. Rev. Lett. **71**, 1272 (1993)

11. O. Jensen, V.O. Pannbacker, G. Dewel and P. Borckmans, Phys. Lett. A **179**, 91 (1993)

12. G. Heidemann, M. Bode and H.G. Purwins, Phys. Lett. A **177**, 225 (1993)

13. P. Kolodner, Phys. Rev. E **48**, R665 (1993)

14. V.O. Pannbacker, O. Jensen, E. Mosekilde, G. Dewel and P. Borckmans, in *Spatio-Temporal Patterns in Nonequilibrium Complex systems* (P. Palffy-Muhoray and P. Cladis, Eds., Kluwer, Amsterdam, to appear 1994)

15. See *Nonlinear Evolution of Spatio-Temporal Structures in Dissipative Continuous Systems* (F. Busse and L. Kramer, Eds., Plenum, New York, 1992)

16. Y. Kuramoto, *Chemical Oscillations, Waves and Turbulence* (Springer, Berlin, 1984)

17. B.I. Shraiman, A. Pumir, W. Van Saarloos, P.C. Hohenberg, H. Chaté and M. Holen, Physica D **57**, 241 (1992)

18. H. Sakaguchi, Progr. Theor. Phys. **84**, 792 (1990)

19. A. De Wit, G. Dewel and P. Borckmans, Phys. Rev. E **48**, R4191 (1993)

20. Y. Kuramoto, Suppl. Progr. Theor. Phys. **64**, 346 (1978)

21. G.I. Sivashinsky, Acta Astron. **6**, 569 (1979)

TURING STRUCTURES AND WAVE PATTERNS
IN THE CIMA REACTION

B. RUDOVICS, J.-J. PERRAUD, P. DE KEPPER and E. DULOS
Centre de Recherche Paul Pascal
Avenue A. Schweitzer
Université de Bordeaux I
F-33600 Pessac, France

1. ABSTRACT

We describe experimental studies on different concentration patterns obtained in a chemical reaction-diffusion system, using the Chlorite-Iodide-Malonic Acid (CIMA) reaction. After a brief presentation of the basic principles of the open spatial reactors used in this work, we show various stationary (Turing) and time-dependent structures and make notice that, due to the parameter ramps, the patterned region is confined. Then, we focus on other spatio-temporal patterns arising from the interaction between the Turing (spatial) and Hopf (temporal) instabilities. This interaction is characterized by the formation of various types of stable and "metastable" localized Turing modes acting as antisymmetric wave sources, and by dynamical behaviors reminiscent of spatio-temporal intermittency. We also present a pattern growth dynamics suggestive of cell replication.

2. INTRODUCTION

In his famous paper published in 1952, Alan Turing [1] suggested a possible mechanism for spontaneous pattern formation from an initially stable and homogeneous steady state. He showed that this spatial instability can arise from the coupling between nonlinear chemical kinetics and pure diffusive transport processes.

Turing's ideas inspired a lot of theoreticians and experimentalists in the fields of chemistry, physics and biology [2-7] and induced a deeper understanding of pattern formation mechanisms with possible application to biological systems. Although many theoretical and experimental works have been published recently in connection with pattern formation phenomena in the CIMA reaction, the investigation on the subject is still in progress. As it turns out from the following, this system still reserves some "unexpected phenomena". In this paper, we first give a short description of our experimental technique (open spatial reactor and chemical reaction). Then, we present several examples of simple and complex stationary structures and waves as well as various intricated spatio-temporal patterns which result from the interaction between the Turing and Hopf instabilities.

3. EXPERIMENTAL TECHNIQUE

3.1 The Disc Reactor

Until recently, most of the experimental studies on spatial dissipative structures were conducted in closed systems (i.e. in Petri dish). The spatial and spatio-temporal patterns observed under these conditions are only transient because the system inexorably relaxes to thermodynamic equilibrium [2,3]. A decisive step in the field of pattern formation is owed to the design and development of open spatial reactors that allow to maintain the system at a controlled distance from thermodynamic equilibrium.

Gel reactors are especially appropriate open spatial reactors since the gel matrix preserves the reactive medium from any parasitic convective motion, so that the only active transport process in the gel is the molecular diffusion of species. A simple reaction-diffusion system can be obtained. In this work, we used a so-called "gel disc reactor" schematically shown in figure 1.

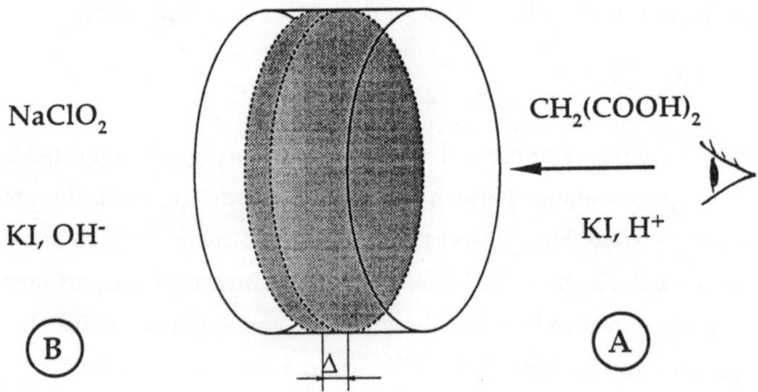

Fig. 1 : Sketch of the Gel Disc Reactor.

Δ is the width of the pattern forming region. Observations are made in a direction perpendicular to the circular feed surfaces.

The main part of this reactor is a disc of hydrogel whose circular faces are in contact with the solutions of reagents contained in two stirred reservoirs A and B. The reagents are permanently renewed by pumps and distributed in the reservoirs in such a way that neither of the solutions A nor B is reactive. The reagents diffuse from the boundaries into the gel where they meet and react.

Due to asymmetric boundary feed compositions, ramps of chemicals establish between the feed surfaces. If no symmetry breaking instability occurs, iso-concentration planes develop parallel to the feed surfaces. The reaction-diffusion instabilities will generally be confined to a region, of width Δ, in the thickness of the gel where appropriate concentrations of major species are met. There only, concentration patterns can develop. During manufacturing, the gel is loaded with a color indicator which is essentially immobile in the gel matrix. As we shall see later, this color indicator can play a crucial role in the pattern development mechanism. The observations are made in a direction perpendicular to the feed surfaces.

3.2 Reaction and Experimental Procedure

The chemical reaction used in this study is the chlorite-iodide-malonic acid (CIMA) reaction [8-10]. This system shows a great variety of transient oscillations in batch conditions as well as sustained oscillations and biotabillty in flow reactors. It also led to the first observation of Turing structures [11] which immediately induced a rejuvenation of research in the field of pattern formation [12-18].

Lengyel and Epstein have shown that besides the initial reagents, chlorine-dioxide and iodine play a key role, both in the oscillatory behavior and in the pattern formation [15]. From the skeletal kinetic mechanism that they proposed, it appears that the difference of diffusity between the activator (iodide) and the inhibitor (chlorite) required for the Turing pattern formation is mediated by the formation of reversible complexes between iodide and some immobile substrate, for instance starch, usually used as a color indicator for iodine species. This complex formation is responsible for a significant decrease in the effective diffusion coefficient of iodide [15]. Note that such a selective decrease of the diffusity of active species due to the formation of immobile complexes was first suggested by Hunding and Sørensen [19]. It was predicted [15] and experimentally verified [12] that, in the CIMA-starch system, a decrease in starch concentration can lead to a transition from stationary (Turing) to time-dependent (oscillations) patterns.

104

Our experiments were performed in agarose gel (2% dry material) containing an iodine color indicator, either polyvinylalcohol or Thiodène from Prolabo. Thiodène contains 7% soluble starch and 93% urea [14] which is washed out before any experiment. These color indicators are macromolecules, the diffusivity of which is much lower than that of all the small reacting species, so that the color indicator can effectively be considered as the immobile substrate.

The reagents were distributed as follows in the feed solutions : iodide was introduced symmetrically in both tanks, malonic acid in acetic acid solution was fed only in tank A and chlorite in basic solution only in tank B. Sodium hydroxide stabilizes chlorite while acetic acid enhances the iodine solubility and acts as a pH buffer. The temperature of the system was maintained constant by a water jacket. Monitoring was provided by a video CCD camera and the gray level contrast was subsequently enhanced by image processing.

4. OBSERVATIONS

4.1 Stationary Patterns

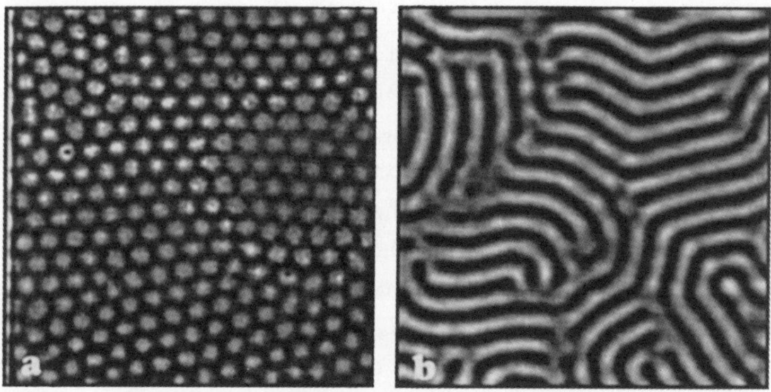

Fig. 2 : Sustained concentration patterns.

Constrast enhanced images of a central part of the reactor. Both images correspond to a view size 3.6 mm x 3.6 mm. Dark and clear regions respectively correspond to reduced and oxidized iodine states.

Experimental conditions : Temperature 6°C, Residence time in reservoirs = 10 min, [Thiodène] = 25 g/liter of gel, $[I^-]^A = [I^-]^B = 2.9 \times 10^{-3}$ M, $[ClO_2^-]^B = 2.0 \times 10^{-2}$ M, $[NaOH]^B = 8.0 \times 10^{-3}$ M, $[CH_3COOH]^A = 2.3$ M,

a - Centered hexagonal array of oxidized iodine state (minimum of iodide concentration) $[CH_2(COOH)_2]^A = 3.5 \times 10^{-3}$ M,

b - Striped pattern $[CH_2(COOH)_2]^A = 5.0 \times 10^{-3}$ M

Figures 2a and 2b show the classic planforms of stationary patterns which are now accepted as Turing structures with one accord. The parameter used for controlling the pattern selection can be the concentration of either malonic acid or iodide or chlorite. Hexagons were observed at low concentration of malonic acid and stripes were obtained at higher value of this control parameter.

Figures 3a presents a stationary organization more complicated than the previous pure hexagons or stripes. This organization could be decomposed into two different patterns by focusing the camera at the limit of detection of the pattern, on both sides. Thus, the pattern seems to result from the superposition of hexagons (Fig 3b) and stripes (Fig 3c) at different depths. This means that two different pattern modes are selected at different depths in the gel. Figures 3a corresponds to a two-dimensional projection of an actually three-dimensional structure.

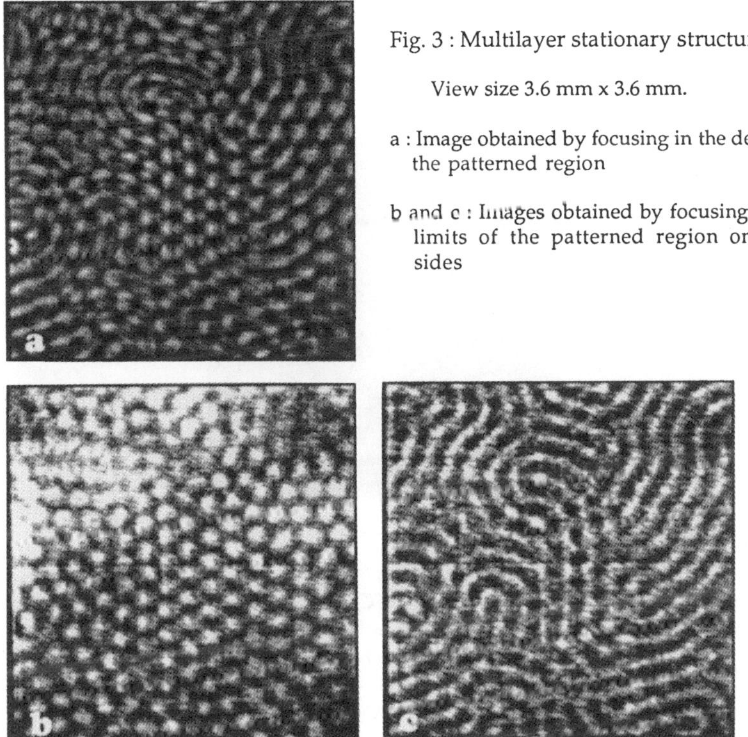

Fig. 3 : Multilayer stationary structure

View size 3.6 mm x 3.6 mm.

a : Image obtained by focusing in the depth of the patterned region

b and c : Images obtained by focusing at the limits of the patterned region on both sides

A pattern develops in the three directions of space if $\Delta>\lambda$, that is if the thickness Δ of the pattern forming region is larger than the wavelength λ of the structure. The pattern can be considered as two-dimensional only if $\Delta<\lambda$. It has been shown that patterns in monolayers embedded in a three-dimensional system exhibit the same sequence of hexagonal and striped planforms as in two-dimensional systems. This is illustrated by the patterns presented in figures 2a and 2b.

4.2 Temporal Patterns

As mentioned above, it is possible to obtain a transition from stationary patterns to time-dependent behaviors (waves) by decreasing the starch concentration in the gel. Because of the necessary manufacturing of a new gel for every starch concentration, this is not a very suitable parameter. The same transition can be observed at constant starch concentration by tuning other parameters, for instance, by an increase in the malonic acid concentration, in accordance with the Lengyel and Epstein's model predictions [15]. In figure 4, we present a snapshot of a time-dependent pattern in the form of traveling waves.

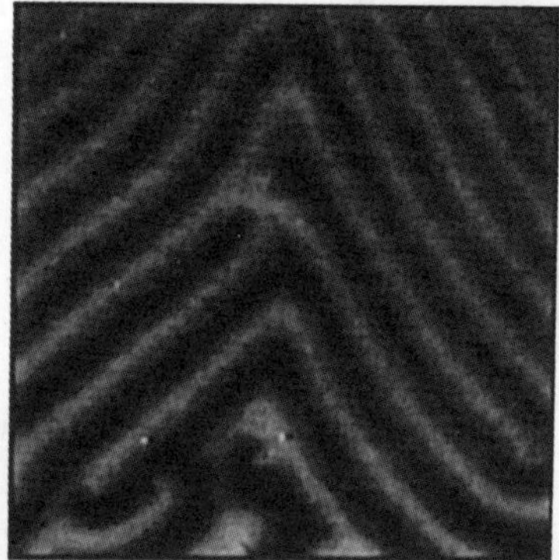

Fig. 4 :   Traveling waves

View size 16 mm x 16 mm.
Experimental conditions : Temperature 6°C, Residence time in reservoirs = 10 min, [polyvinylalcohol] = 5 g/liter of gel, $[I^-]^A = [I^-]^B = 2.8\times10^{-3}$ M, $[ClO_2^-]^B = 2.5\times10^{-2}$ M, $[NaOH]^B = 8.0\times10^{-3}$ M, $[CH_3COOH]^A = 2.3$ M, $[CH_2(COOH)_2]^A = 9.8\times10^{-3}$ M

This temporal behavior results from a Hopf instability of the homogeneous reference state [15,18,20]. The observed spatio-temporal pattern corresponds to a train of phase waves that can either start from the disc-side region where feed conditions are ill-defined, or from defects in the gel (i.e. notches, cracks, dust particles). The wavelength of these waves are about an order of magnitude greater than the wavelength of the stationary patterns of figure 2.

4.3 Interactions of Turing and Hopf Instabilities

In the transition region, a more moderate increase in the malonic acid concentration leads to a large variety of complex spatio-temporal behaviors. In these cases, we observe the coexistence of stationary patterns and temporal oscillations which corresponds to interactions between the Turing and Hopf modes [18].

4.3.1 <u>Superposition of modes</u> : Part of the complexity of the spatio-temporal phenomena comes from the fact that different modes can develop at different positions within the thickness of the gel. For example, though we have no direct optical evidence, observations made in reactors with other geometry and direction of observation convince us that two modes can develop in two neighboring layers in the depth of the gel, with a weak interaction between both modes.

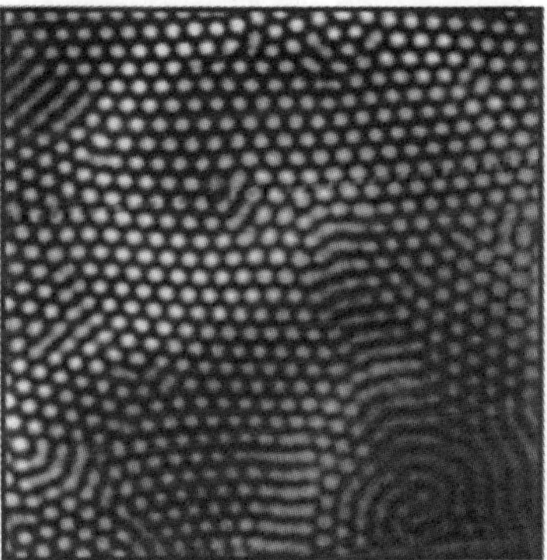

Fig. 5 : Superposition of standing Turing structure and spiral waves

View size : 4.7 x 4.7 mm

Experimental conditions as in figure 2 but, $[I^-]^A = [I^-]^B = 3.0 \times 10^{-3}$ M, $[ClO_2^-]^B = 2.3 \times 10^{-2}$ M, $[NaOH]^B = 8.0 \times 10^{-3}$ M, $[CH_3COOH]^A = 2.3$ M, $[CH_2(COOH)_2]^A = 7.5 \times 10^{-3}$ M

Figure 5 shows an hexagonal array of clear spots on which a wave pattern superimposes. The latter pattern is made of an alternation of dark and clear waves arranged in groups of two counter rotating spirals. The spirals have a rotation period of 29 s and a wavelength approximatively eight times larger than that of the Turing pattern (λ=0,19 mm). Here, the main result of the inetraction is the pinning of the core of the spirals by the neighboring stationary Turing structure.

4.3.2 <u>Spatio-temporal intermittency</u> : Starting from the preceding organization, a completely new behavior was found by further increasing the malonic acid concentration. This spatio-temporal pattern then becomes much more irregular. The Turing and Hopf modes now strongly interact and both modes apparently coexist at the same depth in the gel. In this layer, Turing structures and pure waves compete for space. Inside the irregular Turing-like structures, patches of the standing pattern suddenly collapse and give place to a pure oscillatory mode which, after a while, is slowly reinvaded by the irregular Turing mode. The whole process repetes at different random locations and holds on for several hours, without apparent periodicity in time or even in space.

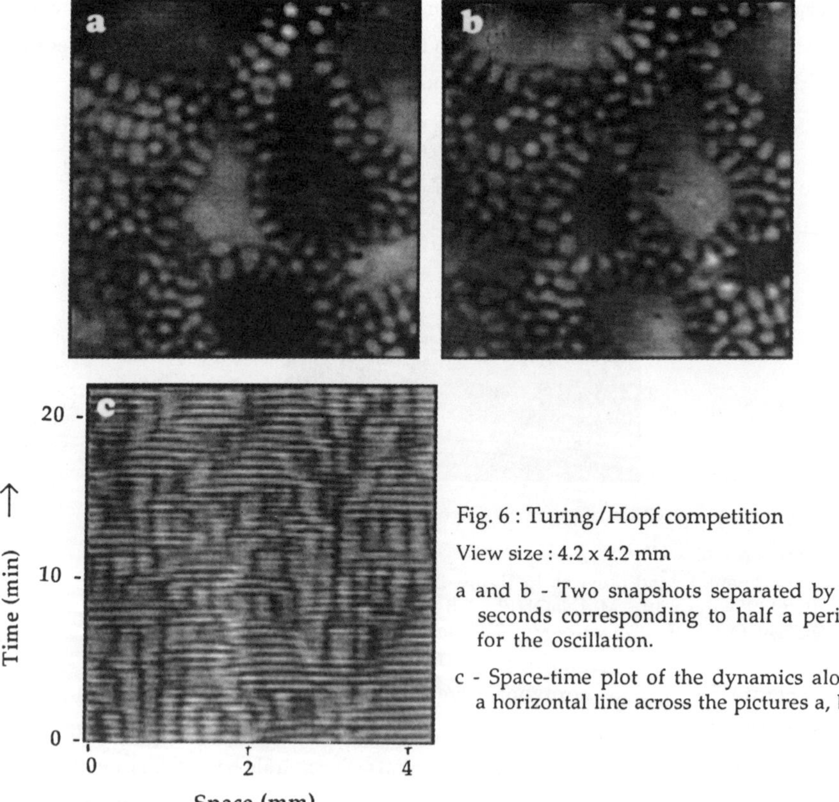

Fig. 6 : Turing/Hopf competition

View size : 4.2 x 4.2 mm

a and b - Two snapshots separated by 15 seconds corresponding to half a period for the oscillation.

c - Space-time plot of the dynamics along a horizontal line across the pictures a, b.

The pictures shown in figures 6a and 6b are taken at 15 s interval, corresponding approximately to half a period of oscillations. Figure 6c presents a space-time plot of the light intensity along an horizontal line through the middle of the previous figure. The branched structures correspond to the Turing patterns and the vertical modulations correspond to the Hopf oscillations. This type of space-time plot is reminiscent of spatio-temporal intermittency [21-24].

4.3.3 <u>Further organizations :</u> The Turing-Hopf competition can also produce other relatively well characterized situations [25,26]. For instance, at high enough malonic acid concentration, the pattern is mainly made of traveling waves that generally arrange around spiral defects as shown in figure 7a. Note that an isolated Turing spot is discernible in the core of the spiral. This spot is more clearly revealed in the time-averaged image of figure 7b where it is surrounded by an uniform gray background.

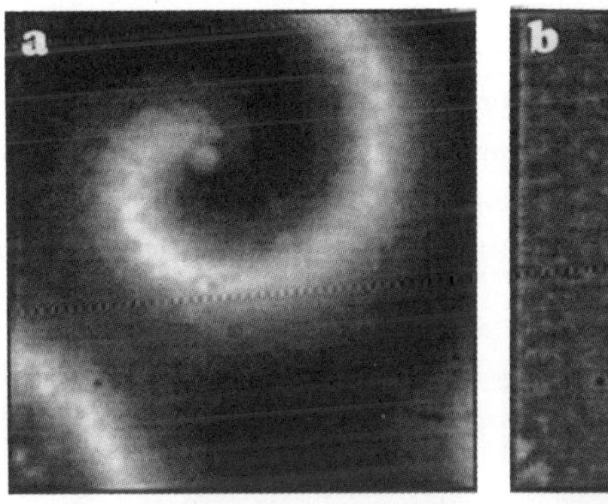
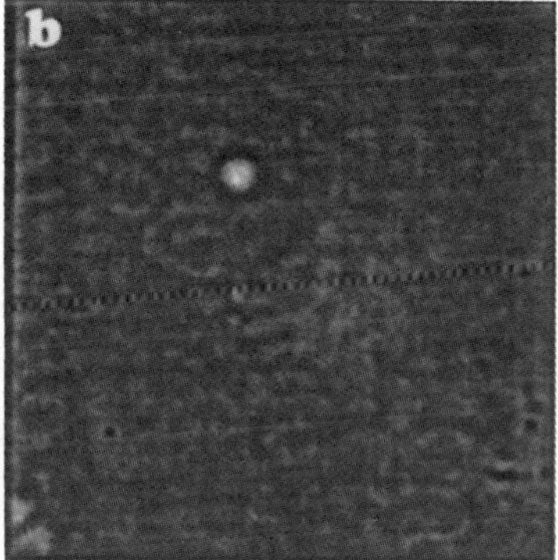

Fig. 7 : Turing/Hopf spiral wave

View size : 3.8 mm x 3.8 mm

Experimental conditions as in Fig. 5 but $[CH_2(COOH)_2]^A$ = 9.5 x 10^{-3} M
a - Snapshot,   b - Time-averaged image

In the core of the spiral, the amplitude of the Hopf mode drops to zero [27] and at that location, the repressed Turing mode can develop, showing that the two modes are still competing [28]. The above description corresponds to the case of a ponctual defect in the wavetrains generated by the Hopf mode. These spirals are the most stable wave defects observed our system when the Hopf mode dominates.

Other less stable wave defect geometries associated with a Turing mode amplitude were also observed. Figures 8a and 8b illustrate the case of isolated Turing

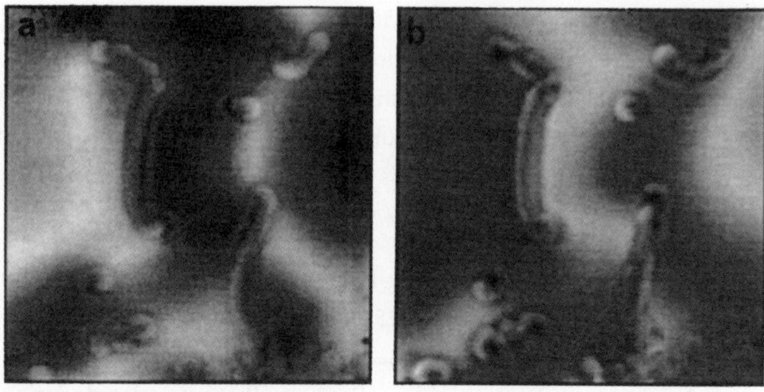

Fig. 8 : Turing/Hopf filament

View size : 3.8 mm x 3.8 mm

Experimental conditions as in Fig. 5 but $[CH_2(COOH)_2]^A = 9.3 \times 10^{-3}$ M
a and b - Two snapshots separated by half a period for the oscillation.

mode filaments acting as the core of elongated spiral waves. The two snapshots separated by half a period of the oscillation show that on opposite sides of the filaments, the oscillatory dynamics exhibits a phase difference of π. The filament geometry holds for more than a dozen oscillations before it eventually decays to a spot.

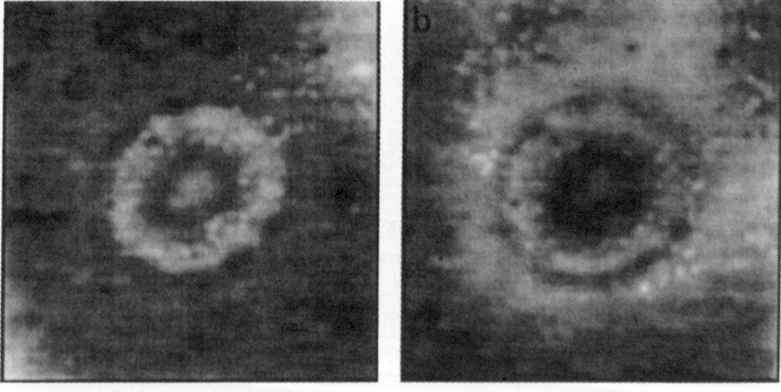

Fig. 9 :   Turing/Hopf ring

View size : 3.8 mm x 3.8 mm

Experimental conditions as in Fig. 8
a and b - Two snapshots separated by half a period for the oscillation.

Another interesting geometry, for the Turing mode embedded in the wave state, is the ring (Fig. 9a and 9b). Here again, the two snapshots are separated by half a period of the oscillation and the oscillations inside and outside of the ring also exhibit a phase difference of π. This is also clearly evidenced by the space-time plot of the amplitude of oscillations (in gray scale) (Fig. 10) along a horizontal line going through the middle of the ring in figure 9. The two essentially time-independent regions correspond to the intersections of the ring with the diameter.

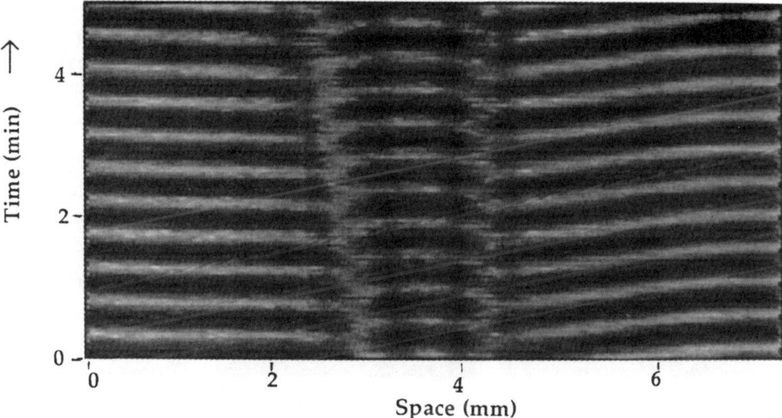

Fig. 10 : Space-time plot of the light intensity along a line across the ring of Fig 9.

5. PATTERN FORMATION BY SPOT DIVISION

Until now, we have studied the asymptotic states of the system, without considering the setting up dynamics of the stationary pattern. In most cases, on crossing the critical parameter value for the onset of Turing patterns, the stable state is nucleated at random locations with the amplitude of the spatial mode slowly growing out from the uniform state. Each individual extremum of the spatial pattern grows at a distance from the already developed elements of the pattern. The patterns are often first nucleated by heterogeneities (i.e. dust particles trapped in the gel) around which the new spatial pattern grows in concentric circles that ultimately organize in an hexagonal array of spots.

However, in a well defined region of the phase diagram, a more surprising pattern growth can be observed in an agarose gel loaded with polyvinylalcohol. Several hours after a small supercritical increase in the malonic acid concentration, growing patches of Turing spots appear in the previously uniform background. They are nucleated, as usual, by heterogeneities of the gel. The already developed spots elongate in the radial and azimuthal directions and, after having reached a critical size, they divide into two spots of quasi-identical size (Fig. 11). The separated daughter spots

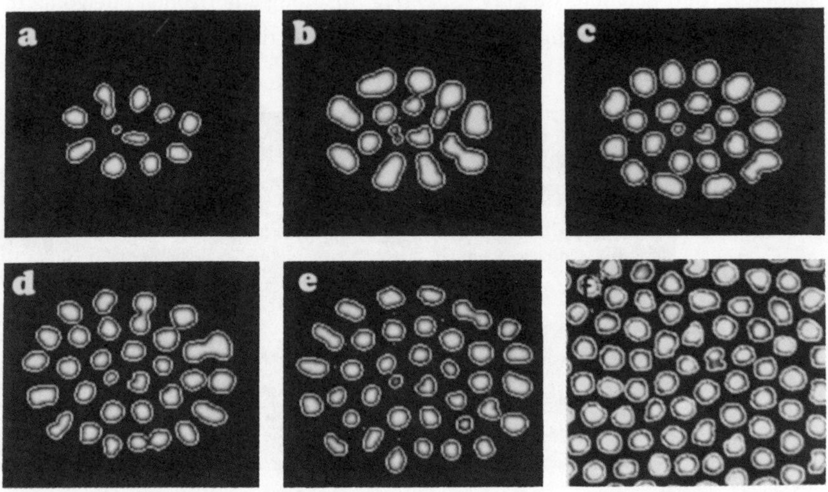

Fig. 11 : Pattern growth dynamics by spot-splitting

Contrast enhanced images - Contour lines outline spots at constant gray level
a to e - Series of snapshots taken at 4 min intervals
f - Hexagonal patterns obtained at the end of the pattern growth ; pentagon-
    heptagon defects are observed around the heterogeneity that triggered the
    initial spots.
Experimental conditions : Temperature 5°C, residence time in reservoirs = 10 min,
[polyvinylalcohol] = 5 g/liter of gel, $[I^-]^A = [I^-]^B = 2.9 \times 10^{-3}$ M, $[ClO_2^-]^B = 2.0 \times 10^{-2}$ M,
$[NaOH]^B = 8.0 \times 10^{-3}$ M, $[CH_3COOH]^A = 2.3$ M, $[CH_2(COOH)_2]^A = 7.0 \times 10^{-3}$ M

located at the border of the growing pattern repete the splitting process. The divisions go on until the whole plane is filled. While the pattern keeps growing at the border, spots in the inner part rearrange into regular hexagons. A similar growth mechanism has recently been found in numerical simulations [29] of a simple reaction-diffusion system exhibiting bistability between two uniform state branches after locally triggering the system from the trivial stable uniform branch of state into the spatially unstable branch.

This pattern formation strongly recalls cell-division observed in living systems and could bring another perspective to replicating systems.

6. ACKNOWLEDGMENTS

We thank J. Boissonade, P. Borckmans, G. Dewel and A. De Wit for many useful and stimulating discussions.

7. REFERENCES

1. A. M. Turing, "The Chemical Basis of Morphogenesis", Phil. Trans. Roy. Soc. London, $\underline{B237}$, 37-72 (1952).

2. I. Prigogine and G. Nicolis, "On Symmetry-Breaking Instabilities in Dissipative Systems", J. Chem. Phys., $\underline{46}$, 3542-3550 (1967).

3. G. Nicolis and I. Prigogine, "Self-Organization in Nonequilibrium Systems", Academic Press, New-York (1977).

4. H. Haken, "Synergetics, an Introduction", Springer, Berlin (1977).

5. H. Meinhardt, "Models of Biological Pattern Formation", Academic Press, New-York (1982).

6. J. D. Murray, "Mathematical Biology", Springer, Berlin (1989).

7. J. Boissonade, "Stationary structure induced along a reaction-diffusion front by a symmetry-breaking instability", J. Phys. France, $\underline{49}$, 541-546 (1988).

8. P. De Kepper, I. R. Epstein, K. Kustin and M. Orbàn, "Batch oscillations and spatial wave patterns in chlorite oscillating systems", J. Phys. Chem., $\underline{86}$, 170 (1988).

9. I. Lengyel, G. Ràbai and I. R. Epstein, "Experimental and modeling study of oscillations in the chlorine dioxide-iodine-malonic acid reaction", J. Am. Chem. Soc., $\underline{112}$, 9104-9110 (1990).

10. I. Lengyel, G. Ràbai and I. R. Epstein, "Batch oscillations in the reaction of ClO_2 with I_2 and AM", J. Am. Chem. Soc., $\underline{112}$, 4606-4607 (1990).

11. V. Castets, E. Dulos, J. Boissonade and P. De Kepper, "Experimental Evidence of a Sustained Standing Turing-Type Nonequilibrium Chemical Pattern", Phys. Rev. Lett., $\underline{64}$, 2953-2956 (1990).

12. K. Agladze, F. Dulos and P. De Kepper, "Turing Patterns in Confined Gel and Gel-Free Media", J. Phys. Chem., $\underline{96}$, 2400-2403 (1992).

13. Q. Ouyang and H. L. Swinney, "Transition from a uniform steady state to hexagonal and striped Turing patterns", Nature, $\underline{352}$, 610-612 (1991).

14. Z. Noszticzius, Q. Ouyang, W. D. McCormick and H. L. Swinney, "Effect of Turing Pattern indicators on CIMA oscillators", J. Chem. Phys., $\underline{96}$, 6302-6307 (1992).

114

15. I. Lengyel and I. R. Epstein, "Modeling of Turing Structures in the Chlorite-Iodide-Malonic Acid-Starch Reaction", Science, $\underline{251}$, 650-652 (1990).

16. , P. Borckmans, G. Dewel and A. De Wit, "Competition in ramped Turing structures", Physica A, $\underline{188}$, 137-157 (1992).

17. V. Dufiet and J. Boissonade, "Numerical simulations of Turing patterns in a two-dimensional system", Physica A, $\underline{188}$, 158-171 (1992).

18. A. De Wit, thesis, Brussels, Belgium (1993).

19. A. Hunding and P. G. Sørensen, "Size Adaptation of Turing Prepatterns", J. Theor. Biol., $\underline{26}$, 27-39 (1988).

20. A. Rovinsky and M. Menzinger, "Interaction of Turing and Hopf Bifurcations in Chemical Systems", Phys. Rev. A, $\underline{46}$, 6315-6322 (1992).

21. H. Chaté and P. Manneville, "Continuous and discontinuous transition to spatio-temporal intermittency in two-dimensional coupled map lattices", Europhys. Lett., $\underline{6}$, 591-595 (1988).

22. F. Daviaud, J. Lega, P. Berger, P. Coullet and M. Dubois, "Spatio-temporal intermittency in a 1D convective pattern : theoretical model and experiment", Physica D, $\underline{55}$, 287-308 (1992).

23. K. Kaneko, "Pattern Dynamics in spatio-temporal chaos", Physica D, $\underline{34}$, 1-41 (1989).

24. E. Bosch and van de Water, "Spatio-temporal intermittency in the Faraday experiment", Phys. Rev. Lett., $\underline{70}$, 3420-3423 (1993).

25. J.-J. Perraud, K. Agladze, E. Dulos and P. De Kepper, "Stationary Turing patterns versus time-dependent structures in the chlorite-iodide malonic acid reaction", Physica A, $\underline{188}$, 1-16 (1992).

26. J.-J. Perraud, A. De Wit, E. Dulos, P. De Kepper, G. Dewel and P. Borckmans, "One-Dimensional 'Spirals' : Novel Asynchronous Chemical Wave Sources", Phys. Rev. Lett., $\underline{71}$, 1272-1275 (1993).

27. A. S. Mikhailov, "Foundations of Synergetics I : Distributive Active Systems", Springer Series in Synergetics, $\underline{51}$, Springer-Verlag, Berlin (1990).

28. S. Ciliberto, P. Coullet, J. Lega, E. Pampaloni and C. Perez-Garcia, "Defects in roll-hexagon competition", Phys. Rev. Lett., $\underline{65}$, 2370-2373 (1990).

29. J. E. Pearson, "Complex patterns in a simple system", Science, $\underline{261}$, 189-191 (1993).

Hyperbolic Reaction-Diffusion Equations and Chemical Oscillations

Byung Chan Eu
Department of Chemistry and Department of Physics, McGill University
Montreal, Quebec H3A 2K6, Canada

Abstract: Hyperbolic reaction-diffusion equations, which appear in the recent theory of irreversible thermodynamics, are discussed for chemical oscillations. The prediction for the behavior of the system is compared with that of the parabolic reaction-diffusion equations conventionally used in connection with oscillating chemical reactions and chemical waves. An approximate dispersion relation is computed for the Brusselator from the hyperbolic reaction-diffusion equations. It has a qualitatively correct wave number dependence whereas the parabolic equations do not yield a dispersion relation when a similar method is used. It is also shown that chaotic solutions can arise in the case of hyperbolic reaction-diffusion equations. Linear stability characteristics can be quite different in the two cases of evolution equations.

I.   Introduction

Chemical oscillations and waves[1] have been theoretically studied by means of reaction-diffusion equations since A. Turing first used them in his path breaking paper on morphogensis[2]. The reaction-diffusion equations, however, are in the class of partial differential equations called parabolic. They are inadequate for describing wave phenomena since they imply that a wave appearing as a response by the system to the disturbance propagates at an infinite speed, but such waves do not propagate at an infinite speed as experimental observations clearly indicate. This inadequacy of parabolic partial differential equations such as the Navier-Stokes equation, Fourier equation, etc. was recognized by Cattaneo[3], Vernotte[4], Morse and Feshbach[5] and others[6], and has been one of the motivations for generalizing the linear theory of irreversible thermodynamics[7].

Maxwell[8] first obtained flux evolution equations which, when combined with the conservation laws of momentum and energy, yield wave equations with a finite wave speed. However, since in the subsequent development of his and Boltzmann's kinetic theory as implemented by Chapman and Enskog the steady-state approximation was taken for the fluxes, it produced the Navier-Stokes, Fourier, and diffusion equations which are parabolic differential equations. It is now well known how to remove such an inadequacy by starting from the kinetic theory of matter[9]. Curiously, in the field of chemical oscillations and chemical waves the aforementioned point has not as yet been taken into consideration since theories of such phenomena are always treated with parabolic reaction-diffusion equations. Our recent work indicates that a system of hyperbolic partial differential equations is thermodynamically better grounded and more adequate for studying aspects of chemical wave phenomena. We review our efforts on the subject in this paper. The limitations of parabolic differential equations will be pointed out in comparison with an alternative hyperbolic system of partial differential equations. Since chemical oscillations are intrinsically dissipative and also typical examples of nonlinear irreversible processes, we first give irreversible thermodynamic foundations for the hyperbolic reaction-diffusion equations. This particular line of discussion is aimed at providing the chemical oscillation and wave phenomena with irreversible thermodynamic basis.

As is well known, in classical physics wave phenomena are described by wave equations, the simplest example of which is

$$\alpha^2 \phi_{tt} - \phi_{xx} = 0 \tag{1}$$

where α is a constant, $\phi_{tt} = \partial^2 \phi / \partial t^2$, and $\phi_{xx} = \partial^2 \phi / \partial x^2$. This equation is hyperbolic and admits traveling wave solutions $\exp[i(kx \pm \omega t)]$, provided that the wave number k and the frequency ω stand in the relation

$$\omega^2 - \alpha^{-2} k^2 = 0 .$$

This means that the dispersion relation is $\omega = k/\alpha$. The phase velocity of the wave is then

$$c = \omega/k = \alpha^{-1}$$

which coincides with the group velocity defined by $C = \partial \omega / \partial k = \alpha^{-1}$. It is a constant in this particular example. Now in the case of a reaction-diffusion equation

$$\phi_t - D\phi_{xx} = f(\phi), \tag{2}$$

it is first-order in time. Here D is the diffusion constant and $f(\phi)$ is the source term attributable to the chemical reaction. For the discussion of the basic aspect of waves the source term is not relevant. Therefore, it is sufficient to consider the equation

$$\phi_t - D\phi_{xx} = 0. \tag{3}$$

This equation is the well-known diffusion equation and parabolic. Consequently, it does not admit a traveling wave solution as does (1). In fact, there is no real dispersion relation for this equation. If there is a traveling wave, the phase velocity of the wave is in fact infinite, implying that *a disturbance at a point in the medium is instantly and simultaneously felt throughout the medium.* This point is easily seen by explicitly solving (3) subject to a suitable initial and boundary conditions since the solution spans the entire space instantly for $t > 0$. It can also be seen in another way as follows. We modify (3) a little bit by adding a term

$$\alpha^2 \phi_{tt} + \phi_t - D\phi_{xx} = 0 \tag{4}$$

where the parameter α will be taken to zero eventually. First, we remove the first time derivative term by transformation

$$\phi(x,t) = \exp(-t/2\alpha^2)g(x,t),$$

which yields the equation

$$\alpha^2 g_{tt} - Dg_{xx} = (4\alpha^2)^{-1}g.$$

It admits traveling wave solutions and a real dispersion relation

$$\omega = (k\sqrt{D}/\alpha)[1 - (4D\alpha^2 k^2)^{-1}]^{1/2},$$

provided that $k < 1/2\alpha\sqrt{D}$. The phase velocity clearly becomes infinite for the wave numbers satisfying the condition $k\alpha\sqrt{D} = $ constant $< 1/2$ in the limit $\alpha \to 0$. Therefore, it is not possible to calculate the phase velocity of a wave if the reaction-diffusion equation is used. This contradicts the fact that there are waves experimentally observed for a given chemical reaction system. This does not mean that the reaction-diffusion equations do not give rise to a stationary oscillatory solution; they do, but there are no traveling waves with a finite phase velocity possible in the case of a parabolic reaction-diffusion equation. This sort of difficulty associated with the parabolic reaction-diffusion equations can be seen quite clearly, for example, in the discussions on traveling waves in the monograph by Gray and Scott[10] who have not been able to definitely characterize traveling waves on the basis of reaction-diffusion equations for most of cases considered. What looks like a wave in their study is, in fact, not a traveling wave in the usual sense but an instantaneous density diffusion whose profile spans the entire space and changes in time. The aforementioned problem may be attributed to the parabolic reaction-diffusion equations used, which are inappropriate for description of traveling waves of finite speeds of propagation. Wave phenomena must be treated with wave equations.

II. Irreversible Thermodynamics of Chemical Oscillations

Since chemical oscillations are macroscopic phenomena and, as such, must be subject to the thermodynamic laws, we first would like to lay the irreversible thermodynamic foundations for phenomena of chemical oscillations. A brief review of irreversible thermodynamics will be given. Its further details are referred to ref. 9. Let us assume that the system consists of $(r + 1)$ components, the first r components of which are involved in k chemical reactions. The last component is regarded as an inert species and serves as the medium (solvent). Furthermore, the solution is assumed to be ideal. This assumption is inessential and removable. The energy density is denoted by $\mathscr{E}$, the specific volume by v, the mass fraction of species l by c_i, and the density of nonconserved variables of species i, suitably arranged, by ϕ_{ai} where the subscript a stands for the ath element of the nonconserved variable set consisting of diffusion fluxes, heat fluxes, stress tensors, polariations, etc. The space spanned by these conserved and nonconserved variables will be called the Gibbs space $\mathscr{G}$ and the variables are called the Gibbs variables. The chemical potentials will be denoted by $\hat{\mu}_i$, the pressure by p, and the temperature by T. Their conjugates to the nonconserved variables will be denoted by X_{ai}. These conjugate variables are dependent on position and time as are the Gibbs variables. The entropy density $\mathscr{S}$ then obeys the equation[9]

118

$$d_t \mathscr{S} = d_t \Psi + d_t \mathscr{B} \tag{5}$$

where d_t is the substantial derivative defined by $d_t = \partial/\partial t + \mathbf{u} \cdot \nabla$ with $\mathbf{u}$ denoting the fluid velocity and

$$d_t \Psi = T^{-1}[d_t \mathscr{E} + p d_t v - \sum_{i=1}^{r+1} \hat{\mu}_i d_t c_i + \sum_{i=1}^{r+1} \sum_{a \geq 1} X_{ai} d_t \phi_{ai}], \tag{6}$$

$$d_t \mathscr{B} = - \hat{\sigma}_L, \tag{7}$$

$$\hat{\sigma}_L = - (\rho T)^{-1} \sum_{i=1}^{r+1} [\Pi_i : \nabla \mathbf{u} + \Delta_i \nabla \cdot \mathbf{u} + Q_i \cdot \nabla lnT + J_i \cdot T \nabla(\hat{\mu}_i^e/T)]. \tag{8}$$

Here Π_i, Δ_i, Q_i, J_i, and $\hat{\mu}_i^e$ denote the traceless symmetric part of the stress tensor, its trace part in excess of pressure, heat flux, diffusion flux, and local equilibrium chemical potential of species i, respectively. The ρ is the mass density, The $d_t \Psi$ is called the compensation differential and $\mathscr{B}$ is the Boltzmann function[9]. The compensation differential is integrable and its integrability condition[11] is

$$\mathscr{E} d_t T^{-1} + v d_t(pT^{-1}) - \sum_{i=1}^{r+1} c_i d_t(\hat{\mu}_i/T) + \sum_{i=1}^{r+1} \sum_{a \geq 1} \phi_{ai} d_t(X_{ai}/T) = 0. \tag{9a}$$

Here, T and p are local quantities depending on position $\mathbf{r}$ and time t, which are, in fact, contact quantities[12] of the elementary volume ar $\mathbf{r}$ and t. These local quantities are related to the local chemical potential by the relation

$$\mathscr{E} d_t T^{-1} + v d_t(pT^{-1}) = \sum_{i=1}^{r+1} c_i d_t(\hat{\mu}_i^e/T). \tag{9b}$$

Since the Boltzmann function $\mathscr{B}$ is dependent on the path of the irreversible process followed by the system in the Gibbs space, the entropy differential is not exact in $\mathfrak{G}$. Since by Carnot's theorem $\hat{\sigma}_L \geq 0$[9] and therefore the Boltzmann function reckoned from equlibrium where it may be set equal to zero is nonpositive, the entropy density is always less than the compensation function (entropy) Ψ which by the integrability condition may be written, apart from the integration constant, as

$$\Psi = T^{-1}[\mathscr{E} + pv - \sum_{i=1}^{r+1} \hat{\mu}_i c_i + \sum_{i=1}^{r+1} \sum_{a \geq 1} X_{ai} \phi_{ai}]. \tag{10}$$

The evolution of the Gibbs variables $\mathfrak{G} = \{\mathscr{E}, v, c_i, \phi_{ai} : i = 1, ..., r+1; a \geq 1\}$ is governed by the conservation laws of energy, mass, and mass fractions and the evolution (constitutive) equations for the nonconserved variables ϕ_{ai}. They are given by the equations:

$$\rho d_t v = \nabla \cdot \mathbf{u}, \tag{11}$$

$$\rho d_t c_i = - \nabla \cdot \mathbf{J}_i + \Lambda_{0_i}, \tag{12}$$

$$\rho d_t \mathbf{u} = - \nabla \cdot \mathbf{P}, \tag{13}$$

$$\rho d_t \mathscr{E} = - \nabla \cdot \mathbf{Q} - \mathbf{P}:\nabla \mathbf{u}, \tag{14}$$

$$\rho d_t \phi_{ai} = Z_{ai} + \Lambda_{ai}, \tag{15}$$

where Z_{ai} and Λ_{ai} are the kinematic and dissipation term, respectively. They depend on ϕ_{ai}, other Gibbs variables, T, and p. Especially, Λ_{0_i} is the reaction rate of species i. The entropy production in the system is then given in terms of the dissipation terms Λ_{ai} in the evolution equations of the nonconserved variables (fluxes):

$$\sigma_{\text{ent}} = T^{-1} \sum_i \sum_{a \geq 0} X_{ai} \Lambda_{ai} \geq 0, \; (X_{0_i} = \hat{\mu}_i - \hat{\mu}_i^e), \tag{16}$$

which must be positive semidefinite and vanishes at equilibrium. This equation implies that dissipative evolutions of nonconserved variables are intimately related to transforming a useful form of energy to a less useful form. Since the entropy production is nonnegative, the dissipative terms of the evolution equations for nonconserved variables must be constructed in such a way that σ_{ent} is nonnegative. The kinematic terms Z_{ai} of the evolution equations for the nonconserved variables are also subject to a condition which may be expressed as follows:

$$\rho d_t (\sum_i \sum_{a \geq 0} \bar{X}_{ai} \phi_{ai}) = - \nabla \cdot (\sum_i \sum_{a \geq 0} \bar{X}_{ai} \psi_{ai}) + \sum_i \sum_{a \geq 0} \bar{X}_{ai} \Lambda_{ai} \tag{17}$$

where $\bar{X}_{ai} = X_{ai}/T$; ψ_{ai} is the flux of ϕ_{ai}; $\phi_{0_i} \equiv c_i$; $X_{0_i} \equiv \hat{\mu}_i - \hat{\mu}_i^e$; and $\psi_{0_i} \equiv \mathbf{J}_i$. The relation of (17) to Z_{ji} can be seen by recasting (17) as follows:

$$\sum_i \sum_{a \geq 1} \bar{X}_{ai} Z_{ai} = - \nabla \cdot [\sum_i \sum_{a \geq 0} \bar{X}_{ai} \psi_{ai} - \sum_i \mathbf{J}_i (\mu_i - \mu_i^e)]$$

$$+ \sum_i \rho d_t [c_i (\mu_i - \mu_i^e)] - \sum_i \sum_{a \geq 1} \phi_{ai} \rho d_t \bar{X}_{ai}. \tag{18}$$

This equation can be derived from statistical models, for example, the Boltzmann equation[9]. If the integrability condition (9b) for the compensation differential $d_t \Psi$ is used in (18), the latter becomes

$$\sum_i \sum_{a \geq 1} \bar{X}_{ai} Z_{ai} = - \nabla \cdot (\sum_i \sum_{a \geq 1} \bar{X}_{ai} \psi_{ai}) + \sum_i \mathbf{J}_i \cdot \nabla (\mu_i - \mu_i^e). \tag{18'}$$

120

We observe that this form can be put into the form

$$\sum_i \sum_{a\geq 1} \bar{X}_{ai} Z_{ai} + \nabla \cdot (\sum_i \sum_{a\geq 1} \bar{X}_{ai} \psi_{ai}) - T^{-1} \sum_i [\Pi{:}\nabla\mathbf{u} + \Delta_i \nabla \cdot \mathbf{u} + \mathbf{Q}_i \cdot \nabla lnT + \mathbf{J}_i \cdot T\nabla\mu_i]$$

$$= \rho\hat{\sigma}_L. \tag{18''}$$

We then recall that the left-hand side is the consistency condition used in the earlier version[13] of the modified moment method which was set equal to zero and used as the conditions for X_{ai}. Since $\sigma_L = \rho\hat{\sigma}_L$ does not vanish if the system is away from equilibrium, the consistency condition was incorrect. When the incorrect assumption is removed, there follows *the conclusion that the entropy differential is not exact*[9,14] In the revised version of the theory, (18) or (18'') imposes conditions on Z_{ai}, namely, on the evolution equations for the nonconserved variables. Finally, we note that the entropy balance equation is given by the form

$$\rho d_t \mathscr{S} = - \nabla \cdot [\sum_i T^{-1}(\mathbf{Q}_i - \hat{\mu}_i \mathbf{J}_i) + \sum_i \sum_{a\geq 1} \bar{X}_{ai} \psi_{ai}] + \sum_i \sum_{a\geq 0} \bar{X}_{ai} \Lambda_{ai}$$

$$\tag{19}$$

which means that the entropy flux is given by the formula

$$\mathbf{J}_s = \sum_i T^{-1}(\mathbf{Q}_i - \hat{\mu}_i \mathbf{J}_i) + T^{-1} \sum_i \sum_{a\geq 1} X_{ai} \psi_{ai}. \tag{20}$$

The set of macroscopic evolution equations for the Gibbs variables (11) - (15) subject to the entropy balance equation or the second law of thermodynamics is called the generalized hydrodynamic equations. The latter together with the entropy balance equation (19) form the mathematical structure of theory of irreversible processes. In this paper, we consider a special case of the generalized hydrodynamic equations in the case of chemically oscillating systems.

III.   Hyperbolic reaction-diffusion equations

The parabolic reaction-diffusion equations conventionally used for chemical oscillations are an approximation to hyperbolic reaction-diffusion equations which are more appropriate from the viewpoint of the theory[9] of irreversible processes in systems removed far from equilibrium. To make the discussion as simple as possible within the context of chemical oscillations and waves, we will assume that the system is homogeneous in temperature and there is no convection flow and thus no stress present in the system. Therefore, the medium is stationary. The fluid is also assumed to be incompressible. Under these assumptions it is sufficient to take into account the changes in concentrations and the accompanying diffusion flow. Then their evolution equations may be written in the forms[9]

$$\partial_t c_i = - \nabla \cdot \mathbf{J}_i + \Lambda_i(c), \tag{21a}$$

$$\partial_t \mathbf{J}_i = - (\rho k_B T/m_i)\nabla c_i - \sum_j L_{ij} \mathbf{J}_j, \tag{21b}$$

if we further assume that the evolution of diffusion fluxes follows a linear law. This latter assumption is not necessary, but makes the equations simpler than otherwise. In (21), $\Lambda_i(c)$ is the reaction rate term which is generally nonlinear unless the chemical reaction is unimolecular, m_i is the mass of species i, and L_{ij} are phenomenological coefficients. They obey the Onsager reciprocal relations. It is convenient to introduce a column vector $\mathbf{w}$ defined by

$$\mathbf{w} = (c_1, c_2, ..., c_r, \mathbf{J}_1, \mathbf{J}_2, ..., \mathbf{J}_r) \tag{22}$$

and a $2r{\times}2r$ matrix $\mathbf{K}$ defined by

$$\mathbf{K} = \begin{bmatrix} 0 & v\mathbf{I} \\ v\mathbf{p} & 0 \end{bmatrix}, \tag{23}$$

where $\mathbf{I}$ is the r-dimensional unit matrix, $\mathbf{p}$ is an r-dimensional diagonal matrix defined by $\mathbf{p} = (\rho k_B T/m_i) \equiv (p_i)$, and v is a unit three-dimensional vector. Here, it is assumed that the diffusion fluxes $\mathbf{J}_1, ..., \mathbf{J}_r$ are linearly independent. Then, for an $(r + 1)$ component system the $(r + 1)$th diffusion flux is determined by the independent fluxes since $\sum_{i=1}^{r+1} \mathbf{J}_i = 0$ by the definition of diffusion fluxes.

System (21) is called hyperbolic[15] if the eigenvalue problem associated with (21)

$$(- \lambda \mathbf{I} + \mathbf{K}) \cdot \mathbf{f} = 0, \tag{24}$$

where $\mathbf{I}$ is the $2r$-dimensional unit matrix, $\mathbf{f}$ is a $2r$-dimensional vector, has $2r$ real eigenvalues, not necessarily distinct. Indeed, the eigenvalues are all real for the system (21): $\lambda_i^{\pm} = \pm v\sqrt{p_i}$, $i = 1, 2, ..., r$, and hence the system is hyperbolic. It must be noted that the inhomogeneous terms in the differential equations in (21) do not play a role in determining the hyperbolicity. The characteristic equations are then given by

$$\frac{d\mathbf{r}}{dt} = \pm v\sqrt{p_i}. \tag{25}$$

The conventional parabolic reaction-diffusion equations arise from (21) if the diffusion fluxes do not change in time. In this case, (21b) becomes a set of linear thermodynamic force-flux relations well known in linear irreversible thermodynamics[16]:

$$(\rho k_B T/m_i)\nabla c_i + \sum_j L_{ij}\mathbf{J}_j = 0 \tag{26}$$

which may be solved to obtain

$$\mathbf{J}_i = - \sum_j D_{ij}\nabla c_j. \tag{27}$$

Here the diffusion coefficients D_{ij} are defined by

$$D_{ij} = (\rho k_B T/m_i)(L^{-1})_{ij} \tag{28}$$

with $\mathbf{L}$ denoting the matrix consisting of elements L_{ij}: $\mathbf{L} = (L_{ij})$.

On substitution of (27) into (21a), we obtain the conventional reaction-diffusion equations. These reaction-diffusion equations may be considered from another viewpoint using two different time scales for slow conserved and fast nonconserved variables. The details about this argument is referred to ref. 9. Eq.(21b) is a linearized form for the evolution equations for diffusion fluxes, which are nonlinear in general. Some examples for such nonlinear evolution equations for fluxes can be found in ref. 9.

The evolution of the variables $(c_i, \mathbf{J}_i: i = 1, 2, ..., r)$ in (21) is subject to the restriction by the second law of thermodynamics as mentioned earlier. To elaborate on this aspect, we first rewrite the reaction rate term as follows[9]:

$$\Lambda_{0_i} = \sum_\ell [\Lambda_\ell^{(+)} - \Lambda_\ell^{(-)}]_i \tag{29}$$

if the species i is involved in chemical reaction ℓ. The term $\Lambda_\ell^{(+)}$ represents the forward reaction rate whereas $\Lambda_\ell^{(-)}$ represents the reverse reaction rate in reaction ℓ. If it is assumed that the solution in which the reactions occur is ideal to an approximation, then the entropy production for the reacting system can be written as

$$\sigma_{\text{chem}} = k_B \sum_\ell (\Lambda_\ell^{(+)} - \Lambda_\ell^{(-)}) ln(\Lambda_\ell^{(+)}/\Lambda_\ell^{(-)}) \tag{30}$$

which is always positive and becomes equal to zero only at equilibrium where $\Lambda_\ell^{(+)} = \Lambda_\ell^{(-)}$ for all ℓ. There is an additional contribution to the entropy production in the system which arises from diffusion. In the case of the linear model implied by (21b), it is given by

$$\sigma_{\text{dif}} = T^{-1} \sum_i \sum_j L_{ij} \mathbf{J}_i \cdot \mathbf{J}_j \tag{31}$$

which is also positive semidefinite if the symmetric phenomenological coefficient matrix $\mathbf{L}$ is positive. It is assumed that this positivity condition is satisfied. Therefore, the total positive entropy production is given by the sum of the two contributions:

$$\sigma_{\text{ent}} = k_B[\sum_\ell (\Lambda_\ell^{(+)} - \Lambda_\ell^{(-)}) ln(\Lambda_\ell^{(+)}/\Lambda_\ell^{(-)}) + \beta \sum_i \sum_j L_{ij} \mathbf{J}_i \cdot \mathbf{J}_j] \tag{32}$$

where $\beta = 1/k_B T$.

III. The Brusselator

Let us take an example with the Brusselator[17] which has served as a model for the Belousov-Zhabotinskii reaction. It is somewhat limited as a model for the Belousov-Zhabotinskii reaction, but since there are only two species to consider, it is one of the simplest among the models proposed for the reaction. The reaction schemes for the model are as follows:

$$A \rightleftharpoons X \tag{R1}$$

$$B + X \rightleftharpoons Y + D \tag{R2}$$

$$2X + Y \rightleftharpoons 3X \tag{R3}$$

$$X \rightleftharpoons E \tag{R4}$$

for which, in the order of reactions listed above, the forward and reverse rate coefficients are denoted by k_ℓ and $k_{-\ell}$, $\ell = 1, ..., 4$, respectively. In the model it is assumed that species A and B are kept at constant nonvanishing values whereas the concentrations of species D and E are put equal to zero, namely, species D and E are removed from the system as they are produced. The assumption is equivalent to $k_{-2} = k_{-4} = 0$. In addition, it is also assumed that $k_{-1} = k_{-3} = 0$[17]. These additional assumptions render the reactions completely irreversible. The reaction rate terms are then as follows:

$$\Lambda_x = k_1\tilde{A} - (k_2\tilde{B} + k_4)\tilde{X} + k_3\tilde{X}^2\tilde{Y}, \tag{33a}$$

$$\Lambda_y = k_2\tilde{B}\tilde{X} - k_3\tilde{X}^2\tilde{Y}, \tag{33b}$$

where $\tilde{X}$, etc. stand for the concentrations. To make the equations a little simpler, it may be assumed that $L_{xy} = L_{yx} = 0$. Then the evolution equations (21) for the system take the forms

$$\partial_t\tilde{X} = - \nabla\cdot\mathbf{J}_x + k_1\tilde{A} - (k_2\tilde{B} + k_4)\tilde{X} + k_3\tilde{X}^2\tilde{Y}, \tag{34a}$$

$$\partial_t\tilde{Y} = - \nabla\cdot\mathbf{J}_y + k_2\tilde{B}\tilde{X} - k_3\tilde{X}^2\tilde{Y}, \tag{34b}$$

$$\partial_t\mathbf{J}_x = - (k_BT/m_x)\nabla\tilde{X} - L_{xx}\mathbf{J}_x, \tag{34c}$$

$$\partial_t\mathbf{J}_y = - (k_BT/m_y)\nabla\tilde{Y} - L_{yy}\mathbf{J}_y. \tag{34d}$$

Here, we regard $\tilde{X}$, etc. as mass densities and have adjusted the coefficients in (34c) and (34d) accordingly. The variables are scaled as follows.

$$\tau = k_4t; \quad \xi = r/L; \quad X = (k_3/k_4)^{1/2}\tilde{X}; \quad Y = (k_3/k_4)^{1/2}\tilde{Y};$$

$$A = (k_1/k_4)(k_3/k_4)^{1/2}\tilde{A}; \quad B = (k_2/k_4)\tilde{B}; \quad \mathbf{u} = (Lk_4)^{-1}(k_3/k_4)^{1/2}\mathbf{J}_x;$$

$$\mathbf{v} = (Lk_4)^{-1}(k_3/k_4)^{1/2}\mathbf{J}_y$$

and the dimensionless parameter N_{rd} is defined by

$$N_{rd} = [k_B T/(m_x m_y)^{1/2}]/k_4 (D_x D_y)^{1/2} \tag{35}$$

where D_i is the diffusion coefficient of species i defined by $D_i = (k_B T/m_i)/L_{ii}$, $i =$ x or y. The linear dimension of the system is denoted by L. The dimensionless number N_{rd} will be called *the reaction-diffusion number.* Roughly, $[k_B T/(m_x m_y)^{1/2}]^{1/2}$ is the sound wave speed whereas $[(D_x D_y)^{1/2}k_4]^{1/2}$ is a measure of mean diffusion of species X and Y over the time scale on which the species X is replenished by chemical reaction R1. Therefore, *the reaction-diffusion number is a measure of how fast particles diffuse compared to the sound wave on the time scale of species being replenished by chemical reactions.* If this number is large, then the evolution of $\mathbf{u}$ and $\mathbf{v}$ approximately follow the linear thermodynamic force-flux law, but if it is small, then $\mathbf{u}$ and $\mathbf{v}$ take a long time to reach a steady state value and hence their evolution time scale becomes comparable to that of the variables X and Y. It must be pointed out that if the evolution equations (34c) and (34d) were nonlinear in $\mathbf{J}_x$ and $\mathbf{J}_y$, then nonlinear thermodynamic force-flux relations would arise in the limit of N_{rd} large. Now, with the reduced variables so defined, the equations in (34) can be cast into the forms:

$$\partial_\tau X = -\partial_\xi \cdot \mathbf{u} + A - (B+1)X + X^2 Y, \tag{36a}$$

$$\partial_\tau Y = -\partial_\xi \cdot \mathbf{v} + BX - X^2 Y, \tag{36b}$$

$$\partial_\tau \mathbf{u} = -N_{rd}(\mathscr{D}_x \partial_\xi X + f\mathbf{u}), \tag{36c}$$

$$\partial_\tau \mathbf{v} = -N_{rd}(\mathscr{D}_y \partial_\xi Y + \mathbf{v}/f), \tag{36d}$$

where

$$f = (m_y D_y/m_x D_x)^{1/2}, \tag{37}$$

$$\mathscr{D}_x = fD_x/k_4 L^2 \equiv f\hat{\mathscr{D}}_x, \tag{38a}$$

$$\mathscr{D}_y = D_y/fk_4 L^2 \equiv \hat{\mathscr{D}}_y/f. \tag{38b}$$

As the reaction-diffusion number N_{rd} gets large, the time derivative terms in (36c) and (36d) become relatively negligible and the approximation

$$\mathscr{D}_x \partial_\xi X + fu = 0, \tag{39a}$$

$$\mathscr{D}_y \partial_\xi Y + v/f = 0 \tag{39b}$$

becomes good. In the limit of $N_{rd} \to \infty$, where (39a) and (39b) hold, the reaction-diffusion equations arise on substitution of (39a) and (39b) into (36a) and (36b):

$$\partial_\tau X = \hat{\mathscr{D}}_x \partial_\xi^2 X + A - (B + 1)X + X^2 Y, \tag{40a}$$

$$\partial_\tau Y = \hat{\mathscr{D}}_y \partial_\xi^2 Y + BX - X^2 Y. \tag{40b}$$

These parabolic differential equations are the basis for the usual analysis for the Brusselator in the literature[10,17].

The role of chemical reactions can be understood if the following hyperbolic system is considered which arises upon neglecting the reaction rate parts in (36a) and (36b):

$$\partial_\tau X = - \partial_\xi \cdot u , \tag{41a}$$

$$\partial_\tau Y = - \partial_\xi \cdot v , \tag{41b}$$

$$\partial_\tau u = - N_{rd}(\mathscr{D}_x \partial_\xi X + fu), \tag{41c}$$

$$\partial_\tau v = - N_{rd}(\mathscr{D}_y \partial_\xi Y + v/f), \tag{41d}$$

We look for the solution to these equations in the form

$$X = X_o \exp[i(k \cdot \xi - \omega\tau)], \text{ etc.} \tag{42}$$

Then the frequencies are given by

$$\omega = \tfrac{1}{2} l_x [-i \pm (4\varepsilon_x k^2 / l_x^2 - 1)^{1/2}], \tag{43a}$$

$$\omega = \tfrac{1}{2} l_y [-i \pm (4\varepsilon_y k^2 / l_y^2 - 1)^{1/2}], \tag{43b}$$

where

$$\varepsilon_x = N_{rd} \mathscr{D}_x, \qquad \varepsilon_y = N_{rd} \mathscr{D}_y, \qquad l_x = N_{rd} f, \qquad l_y = N_{rd}/f. \tag{44}$$

From these roots for the secular determinant for the system (41) it can be concluded that $\text{Im}\,\omega < 0$ for all values of k. This means that the amplitudes of the traveling waves damp out in time, namely, the steady state is stable. When the diffusion is neglected in this example, there are traveling waves, but the presence of diffusion damps out the traveling waves in time. The effect of chemical reactions on such waves now may be examined together with the stability of the linearized system (36).

III.1 Linear stability

System (36) has a homogeneous steady state: $(X,Y,u,v) = (A,B/A,0,0)$ if $A \neq 0$. If $A = 0$, a homogeneous steady state occurs also at $(X,Y,u,v) = (0,0,0,0)$ which is the equilibrium state. Notice that although the number of steady states is the same as for the parabolic reaction-diffusion equations, they are defined in a four-dimensional space whereas in the case of the latter the dimension of the space is 2. We examine under what condition traveling wave solutions are possible for (36). If (36a)-(36d) are linearized around the steady state $(0,0,0,0)$ in the case of $A = 0$, then (41a)-(41d) are obtained. It was seen that the steady state is stable and there are no sustained traveling wave solutions in this case. If system (36) is linearized around the steady state $(A,B/A,0,0)$ in the case of $A \neq 0$, then the following set is obtained:

$$\partial_\tau x = - \partial_\xi \cdot u + (B - 1)x + A^2 y, \tag{45a}$$

$$\partial_\tau y = - \partial_\xi \cdot v - Bx - A^2 y, \tag{45b}$$

$$\partial_\tau u = - \varepsilon_x \partial_\xi x - l_x u, \tag{45c}$$

$$\partial_\tau v = - \varepsilon_y \partial_\xi y - l_y v. \tag{45d}$$

Here

$$x = X - A \text{ and } y = Y - B/A. \tag{46}$$

The differential equation system (36) may be compactly summarized in matrix form

$$\partial_\tau W + M_0 \partial_\xi W + M_1 W = \Lambda(x,y) \tag{47}$$

where

$$M_0 = \begin{bmatrix} 0 & 0 & 1 & 0 \\ 0 & 0 & 0 & 1 \\ \varepsilon_x & 0 & 0 & 0 \\ 0 & \varepsilon_y & 0 & 0 \end{bmatrix}, \quad M_1 = \begin{bmatrix} 1 - B & - A^2 & 0 & 0 \\ B & A^2 & 0 & 0 \\ 0 & 0 & l_x & 0 \\ 0 & 0 & 0 & l_y \end{bmatrix}, \tag{48a}$$

and $\Lambda(x,y)$ is a column vector defined by

$$\Lambda(x,y) = [(B/A)x^2 + 2Axy + x^2 y](1, -1, 0, 0). \tag{48b}$$

Then, the linearized set may be written as

$$\partial_\tau \mathbf{W}_0 + \mathbf{M}_0 \partial_\xi \mathbf{W}_0 + \mathbf{M}_1 \mathbf{W}_0 = 0 \tag{49}$$

where $\mathbf{W}_0$ is a column vector. The boundary conditions on $\mathbf{W}_0$ are: $x = y = 0$ at $\xi = 0$ and 1.

Under these conditions we may look for the solution in the form

$$\mathbf{W}_0 = \sum_m \exp(\omega_m \tau)(p_m \sin k\xi, \, q_m \sin k\xi, \, r_m \cos k\xi, \, s_m \cos k\xi), \tag{50}$$

where $k = m\pi/L$ and p_m, q_m, etc. are the amplitudes. The secular determinant of (49) is given by

$$\begin{vmatrix} \omega + 1 - B & -k & -A^2 & 0 \\ \varepsilon_x k & \omega + l_x & 0 & 0 \\ B & 0 & \omega + A^2 & -k \\ 0 & 0 & \varepsilon_y k & \omega + l_y \end{vmatrix} = 0 \tag{51}$$

where the subscript m to ω_m is dropped. Expanding this, we obtain the quartic polynomial

$$\omega^4 + P\omega^3 + Q\omega^2 + R\omega + S = 0 \tag{52}$$

where

$$P = 1 + A^2 - B + l_x + l_y, \tag{53a}$$

$$Q = A^2 + (1 + A^2 - B)(l_x + l_y) + l_x l_y + (\varepsilon_x + \varepsilon_y)k^2, \tag{53b}$$

$$R = A^2(l_x + l_y) + (1 + A^2 - B)l_x l_y + \varepsilon_x(l_y + A^2)k^2 + \varepsilon_y(1 + l_x - B)k^2, \tag{53c}$$

$$S = A^2 l_x l_y + A^2 \varepsilon_x l_y k^2 + (1 - B)\varepsilon_y l_x k^2 + \varepsilon_x \varepsilon_y k^4. \tag{53d}$$

The quartic polynomial (51) is solvable analytically, but the result is rather uninformative with regard to whether the roots are real or complex and under what condition. To extract the desired information, we may take advantage of the Hurwitz conditions[18]. Since these conditions are complicated, the related but simpler Lienard-Chipart criteria[18] may be used which in the case of (51) become as follows:

$$P > 0, \tag{54a}$$

$$Q > 0, \tag{54b}$$

$$S > 0, \tag{54c}$$

$$PQR - R^2 - P^2S > 0. \tag{54d}$$

The first three conditions are simple and give the conditions which the parameter B must satisfy for the real parts of the roots to be negative, but the last condition is very complicated. To extract some meaningful information out of this condition, we consider the asymptotic limit of k and find

$$PQR - R^2 - P^2S \sim (\varepsilon_x - \varepsilon_y)^2(A^2 + l_y)(1 - B + l_x)k^4. \tag{55}$$

The first and the second condition, (54a) and (54b), mean respectively

$$B < 1 + A^2 + l_x + l_y, \tag{54a'}$$

$$B < 1 + A^2 + \frac{A^2}{l_x + l_y} + \frac{\varepsilon_x + \varepsilon_y}{l_x + l_y}k^2 + \frac{l_x l_y}{l_x + l_y}. \tag{54b'}$$

Condition (54c) is the same as for linear stability in the parabolic case:

$$B < 1 + \frac{\varepsilon_x l_y}{\varepsilon_y l_x}A^2 + \frac{l_y}{\varepsilon_y}A^2 k^{-2} + \frac{\varepsilon_x}{l_x}k^2. \tag{54c'}$$

Note that $\varepsilon_x l_y / \varepsilon_y l_x = D_x / D_y$. If the asymptotic form (55) of condition (54d) is used, there follows the condition

$$B < 1 + l_x. \tag{54d'}$$

The intersection of these conditions (54a')-(54d') implies that the real parts of the characteristic roots are negative only in the domain where (54a')-(54d') are satisfied. See Fig. 1 in ref. 19. In the case of the corresponding parabolic equations, the stability phase diagram involves only two curves corresponding to $S = 0$ and $P = 0$ which give rise to the parabola and the horizontal line, respectively. However, in the case of hyperbolic equations the situation is qualitatively altered. For example, depending on the value of N_{rd}, the B-curve corresponding to (54d) crosses the m (or k) axis, giving rise to a critical value of B beyond which the oscillatory solution become unstable. This presence of the critical B is useful to understand the behavior of the numerical solutions.

Stability analysis of (49) may be treated approximately by another way if A is treated as a small parameter. In this case, to the order of approximation neglecting A^2, we find

$$[(\omega + 1 - B)(\omega + l_x) + \varepsilon_x k^2][\omega(\omega + l_y) + \varepsilon_y k^2] = 0 \tag{56}$$

which yields

$$\omega_1^{(\pm)} = -\frac{1}{2}(1 - B + l_x) \pm \frac{1}{2}i[4\varepsilon_x k^2 - (1 - B + l_x)^2]^{1/2}, \tag{57a}$$

$$\omega_2^{(\pm)} = -\frac{1}{2}l_y \pm \frac{1}{2}(l_y^2 - 4\varepsilon_y k^2)^{1/2}, \tag{57b}$$

for approximate characteristic roots. Comparing these roots with those in (43) after multiplication of $i = \sqrt{-1}$, we see that the $\omega_1^{(\pm)}$ modes are now altered by the chemical reactions. We recall that in the absence of chemical reactions, all four modes were stable, but in the presence of chemical reactions there arises a possibility of oscillatory solutions. Thus, if $B = 1 + l_x$, then $\omega_1^{(\pm)}$ are imaginary if $\varepsilon_x k^2 > l_x^2$ whereas $\mathrm{Re}\,\omega_2^{(\pm)} < 0$ always. Therefore, $\omega_2^{(\pm)}$ represent stable modes for all values of k whereas $\omega_1^{(\pm)}$ give rise to oscillatory modes in the case considered.

III.2 Dispersion relation and phase velocity

By using the linearized equations (49), we can examine an approximate dispersion relation for the traveling waves emerging from the stable steady state under some conditions on the parameters. We resort to an approximate analysis because of the complicated relation implied by the exact solutions of the quartic polynomial (52). In the case of A being small in magnitude, we have shown that (40) holds approximately, and have thereby obtained (57) for the characteristic roots. As already noted, the $\omega_1^{(\pm)}$ modes give rise to an oscillatory, in fact, a traveling, wave if $B = 1 + l_x$. In this case, the frequency of the oscillating traveling wave is given by

$$\omega_1^{(\pm)} = \pm i\omega_0, \tag{58a}$$

$$\omega_0 = k\sqrt{\varepsilon_x}[1 - l_x^2/\varepsilon_x k^2]^{1/2}, \tag{58b}$$

if $l_x^2/\varepsilon_x k^2 < 1$. If this condition is not met, there is no oscillating traveling wave. Therefore, the phase speed of the wave can be defined by

$$c = \frac{\omega_0}{k} = \sqrt{\varepsilon_x}[1 - l_x^2/\varepsilon_x k^2]^{1/2} \tag{59}$$

according to the usual definition of phase speed[20]. There are some experimental data for phase speed available in the case of the Belousov-Zhabotinskii reaction[21]. According to the data the frequency of the traveling wave is approximately proportional to the wave number[20]; put in another way, the phase speed is approximately constant with respect to the wave number. The result in (59) is in qualitative agreement with the experimental finding[21b] in the regime where $l_x^2/\varepsilon_x k^2 < 1$. It will be worthwhile to study dispersion relations with the nonlinear evolution equations (47).

In contrast to this result for the hyperbolic system presented above, if the parabolic reaction-diffusion equations were used, then the characteristic polynomial would be given by

$$\begin{vmatrix} \omega + 1 - B + (\varepsilon_x/l_x)k^2 & - A^2 \\ B & \omega + A^2 + (\varepsilon_y/l_y)k^2 \end{vmatrix} = 0$$

(60)

which, in the limit of negligibly small A, yields

$$\omega^{(\pm)} = \pm k^2 (\varepsilon_y/l_y)$$

(61)

if there holds the condition $B = 1 + A^2 + (\varepsilon_x/l_x + \varepsilon_y/l_y)k^2$. This result is in the same approximation as for (58) and hence the parabolic counterpart to the hyperbolic frequency given in (58). The frequency given by (61) is real and therefore implies that *there is no traveling waves and, consequently, there is no question of phase speed.* This is a qualitative distinction between the linearized hyperbolic equations (49) and their parabolic counterparts (40). This distinction basically stems from the fact that the parabolic reaction-diffusion equations do not sustain traveling waves whereas the corresponding hyperbolic equations do admit traveling waves. Eq. (57b) is an approximate dispersion relation for the system in question. The dispersion relation for reaction-difusion systems has been calculated by means of an asymptotic method, namely, a singular perturbation method[22] which somehow puts the reaction-diffusion equations into forms admitting traveling waves by means of an approximation, since the parabolic reaction-diffusion equations do not admit a dispersion relation in principle.

IV. Numerical results

The hyperbolic reaction-diffusion equations (36a)-(36d) have been numerically solved[19] subject to the initial and boundary conditions:

$$X(\xi,0) = Y(\xi,0) = u(\xi,0) = v(\xi,0) = 0,$$

$$X(0,\tau) = X(1,\tau) = A, \quad Y(0,\tau) = Y(1,\tau) = B/A.$$

(62)

The values of the various parameters taken are as follows:

$$A = 2, \quad B = 4.17, \quad \hat{\mathcal{D}}_x = 0.0016, \quad \hat{\mathcal{D}}_y = 0.006,$$

and the value of the reaction diffusion number is varied to see its effect. The numerical results show that under the boundary conditions in (62), the waves propagate from the boundaries toward the center of the interval $\xi = [0,1]$.

The numerical results[19] show that if $N_{rd} = 0.1$, sharp wave fronts propagate toward the midpoint, whereas if N_{rd} is increased to 2.5, the sharp wave fronts disappear and in their place a diffusive motion takes over. Thus, there is a transition point somewhere below $N_{rd} = 2.5$ at which point the hyperbolic system effectively exhibits a behavior characteristic of a parabolic system. This kind of behavior has been found[23] in the case of heat conduction in the Lennard-Jones fluid, and we see that such a transition in behavior is also present in the case of diffusion.

As the value of N_{rd} is further increased and when the value of B is within the first stable domain (i.e., the lower k domain), the long-time solutions exhibit features similar to those by the parabolic equations for the Brusselator. In this domain of B, seven or eight node oscillatory solutions appear for X and Y, which are stable with the trajectories tending toward the stable steady state.

However, as the value of N_{rd} is chosen such that the system moves into the unstable domain bounded from below by (54d), the system that has behaved as if parabolic begins to exhibit a chaotic behavior after some lapse of time. Trajectories initially move toward the steady state in a regular fashion, but as they approach the latter, they get repelled away from the steady state and start showing a rather chaotic behavior. This behavior can be understood from the stability phase diagram. In the case of parabolic reaction-diffusion equations chaotic trajectories are not possible to arise since there are only two variables, but in the case of the corresponding hyperbolic reaction-diffusion equations, since there are four dependent variables, namely, since the dimension of the variable space is 4, chaotic trajectories cannot be ruled out. This situation would appear generally if hyperbolic systems were taken, since then the number of dependent variables would be larger for hyperbolic systems than for the corresponding parabolic systems.   Therefore, in the case of hyperbolic reaction-diffusion equations chaotic trajectories are possible even for models which do not allow emergence of chaotic motions if parabolic reaction-diffusion equations are used; the Brusselator is a typical example.

The examples of the numerical results mentioned[19] here clearly exhibit the qualitative differences in the behaviors of the solutions for the hyperbolic and parabolic reaction-diffusion equations. The hyperbolic system is, of course, more general than the parabolic system since the latter is contained in the former as a special case as shown previously and the numerical results bear this prediction out.

V. Discussion and Conclusion

The irreversible Brusselator used in ref. 19 is not appropriate for irreversible thermodynamic analysis of the entropy production associated with chemical oscillations. For such a purpose reversible chemical reactions are better suited. The Sel'kov model[24,25] provides a set of evolution equations which appear to be the simplest of the kind. The Sel'kov model assumes the chemical reactions

$$A \rightleftharpoons X$$

$$X + 2Y \rightleftharpoons 3Y$$

$$Y \rightleftharpoons B$$

where A and B are controllable bath concentrations and X and Y are freely responding internal concentrations. A study is in progress by using the Sel'kov model which gives rise to the following reduced equations

$$\partial_\tau X = - \nabla_\xi \cdot \mathbf{u} + B - X + X^2Y - KX^3, \tag{63a}$$

$$\partial_\tau Y = - \nabla_\xi \cdot \mathbf{v} + A - RY - X^2Y + KX^3, \tag{63b}$$

$$\partial_\tau \mathbf{u} = - N_{rd}(\hat{D}_X f \nabla_\xi X + f\mathbf{u}), \tag{63c}$$

132

$$\partial_\tau \mathbf{v} = - N_{rd}(\hat{D}_y f^{-1} \nabla_\xi Y + f^{-1} \mathbf{v}), \tag{63d}$$

where R and K are parameters consisting of rate constants and the reduction of the equations has been achieved suitably scaling the variables in a manner similar to the evolution equation used for the Brusselator. These equations give rise to a bistable system which oscillates between two stable steady states in some domain of the parameter set. The phase diagrams of bifurcation are much more complicated than those of the Brusselator, suggesting more complicated dynamics. The results of this study will be reported elsewhere in the near future[26].

In this paper we have demonstrated by linear analysis and numerical solutions the characteristic features of hyperbolic reaction-diffusion equations for the Brusselator. It is shown that there are regions of the parameter space where the hyperbolic system is qualitatively different from the parabolic reaction-diffusion system. The hyperbolic system admits waves of finite speeds of propagation whereas the parabolic system gives rise to waves of an infinite speed. Furthermore, the former reduces to the latter system as the reaction-diffusion number becomes large. It therefore is more general and flexible than the parabolic system in treating reaction-diffusion systems and traveling chemical waves and oscillations In the case of the Brusselator studied, the hyperbolic reaction-diffusion equations can give rise to chaotic solutions. We hope to report on further studies of hyperbolic reaction-diffusion equations in the future.

References and Footnotes

1. R. J. Field and M. Burger, eds., **Oscillations and Traveling Waves in Chemical Systems** (Wiley, New York, 1985);
    F. Baras and D. Walgraef, eds., **Chemical Dynamics: Expreiments to Microscopic Simulations** (Physica A **188**, Nos. 1-3 (1992)).
2. A. Turing, Phil. Trans. Roy. Soc. London **B237**, 37(1952).
3. C. Cattaneo, Atti. Semin. Mat. Fis. Univ. Modena 3, 3(1948); C. R. Acad. Sci.(Paris) **246**, 31(1958).
4. P. Vernotte, C. R. Acad. Sci.(Paris) **246**, 3154(1958); **247**, 2103(1958).
5. P. M. Morse and H. Feshbach, **Methods of Mathematical Physics** (McGraw-Hill, New York, 1953), p. 865.
6. (a)I. Mullër, Z. Phys. **198**, 329(1967).
    (b) H. D. Weynman, Amer. J. Phys. **35**, 488(1967).
    There is an extensive literature on the subject in connection with thermal conduction. See, for example, (c)G. F. Carey and M. Tsai, Num. Heat Transfer **5**, 309(1982) and (d)H. Q. Yang, Num. Heat Transfer **18**, 221(1990) and references cited therein.
7. See, for example, various papers in Springer Lecture Notes on Physics, vol. 199 (Springer, Berlin, 1984). It is important to remark that although the motivation to use hyperbolic evolution equations in the papers in this volume is correct, the basic premise on the nonequilibrium entropy requires a correction since the nonequilibrium entropy does not possess an exact differential in an extended Gibbs space that includes fluxes as variables. See ref. 9 below for the correct use of the nonequilibrium entropy.
8. J. C. Maxwell, Phil. Trans. Roy. Soc. London A157, 49(1867).
9. B. C. Eu, **Kinetic Theory and Irreversible Thermodynamics** (Wiley, New York, 1992).

10. P. Gray and S. K. Scott, **Chemical Waves and Instabilities** (Clarendon, Oxford, 1990).

11. M. Chen and B. C. Eu, J. Math. Phys. **34**, 3012(1993).

12. W. Muschik, Arch. Rat. Mech. Anal. **66**, 379(1977).

13. B. C. Eu, J. Chem. Phys. **73**, 2958(1980).

14. B. C. Eu, Physica A **171**, 285(1991).

15. C. M. Dafermos in: **Nonlinear Waves**, eds. S. Leibovich and A. R. Seebass (Cornell, Ithaca, 1974).

16. S. R. de Groot and P. Mazur, **Nonequilibrium Thermodynamics** (North-Holland, Amsterdam, 1962).

17. G. Nicolis and I. Prigogine, **Self-Organization in Nonequilibrium Systems** (Wiley, New York, 1977).

18. F. R. Gantmacher, **Théorie des Matrices** (Dunod, Paris, 1966), vol 2.

19. U. I. Cho and B. C. Eu, Physica D (in press).

20. S. Leibovich and A. R. Seebass, **Nonlinear Waves** (Cornell, Ithaca, 1974).

21. (a)P. M. Wood and J. Ross, J. Chem. Phys. **82**, 1924(1985); (b)C. Vidal, J. Stat. Phys. **48**, 1017(1987).

22. J. D. Dockery, J. P. Keener, and J. J. Tyson, Physica D **30**, 177(1988).

23. R. E. Khayat and B. C. Eu, Can. J. Phys. **70**, 62(1992).

24. E. E. Sel'kov, Eur. J. Biochem. **4**, 79(1968).

25. P. H. Richter, P. Rehmus, and J. Ross, Prog. Theoret. Phys. **66**, 385(1981).

26. M. Al-Ghoul and B. C. Eu (in preparation).

134

ON THE MECHANISM OF CONTROL OF VORTEX DRIFT

IN NONEQUILIBRIUM CHEMICAL SYSTEMS

A.P. MUÑUZURI, V. PÉREZ-MUÑUZURI, M. GÓMEZ-GESTEIRA, V.I. KRINSKY[1]

and V. PÉREZ-VILLAR

Group of Nonlinear Physics. Faculty of Physics
University of Santiago de Compostela
15706 Santiago de Compostela, Spain.

1. ABSTRACT

The mechanism of electric field induced vortex drift is analyzed for a nonequilibrium chemical system (Belousov-Zhabotinsty reaction),experimentally and numerically. The electric current has been proven to be a very helpful tool in order to control the spiral dynamics (change of its period, wavelength and core size). Constant parallel electric field induces drift of spirals along a direction determined by the direction of the electric field and the initial position of the spiral. Using alternate electric field, spiral drift was observed at parametric resonant frequency of the current (twice the frequency of the spiral). In this case it is possible to choose the drift direction by introducing a certain phase between the current frequency and the initial position of the spiral.

2. INTRODUCTION

The tendency of excitable media to organize themselves into highly structured periodic waves or spirals has been the object of intense interest. Rotating spiral waves (or vortices) have been observed in various excitable media, including cardiac muscle [1], (their formation is one of the fundamental mechanism of dangerous

---

[1] On leave from ITEB, Pushchino, 142292 Moscow Region, Russia.

arrhythmias which often leads to sudden death), retinae [2], (their appearance is a manifestation of some pathology such as Leao's spreading depression), cultures of the slime mould *Dyctiostelium discoideum* [3, 4] (here, spirals play a constructive role in a morphogenetic process), two dimensional array of coupled electronic circuits [5], chemical waves on the surface of platinum catalyzers [6] and chemical oscillators as the Belousov-Zhabotinsky (BZ) reaction [7, 8].

These waves can move in homogeneous medium or may be pinned by inhomogeneities. Different approaches to control spiral movements have been done. Spiral drift has been induced by high- frequency concentrational waves propagating in the medium [9]. Movement towards the anode has been observed under the influence with a constant electric current created by two parallel electrodes [10]. Light gradient can be used to control spiral waves, spirals drift toward regions of higher light intensity, phenomenon called phototaxis [11].

This paper is devoted to the effect of the electric field on the spiral wave dynamics. Different effects can be expected from different geometries of the electric field. In this way, when a radial geometry is considered, it is possible to modify the characteristic parameters of the spiral (period and core size) [12,13]. On the other hand, in a chemical active medium (BZ reaction), electric field interacts with ionic species so that waves are accelerated when propagating to a positive electrode and decelerated when propagating to a negative one [14, 15]. This gives rise to the drift of a vortex to the positive electrode [10].

In alternate parallel electric field, one might expect no displacement of a vortex. And, actually, no displacement appears when the frequency of the electric field is equal to the frequency of the spiral ($f = f_e$, resonant frequency), but when the frequency of the electric field is twice the frequency of the spiral ($f = 2 f_e$, parametric resonant frequency), the spiral drifts and this phenomenon is called parametric resonance of vortices [16] This can be explained by detailed analysis of normal and tangential wave velocities in a rotating vortex [17].

3. EXPERIMENTAL METHOD

The experimental study of these phenomena requires high sensitivity and long time observations without significant change in the chemical composition of the medium, as provided by a CSTR type reactors.

The experiments were performed in the Belousov-Zhabotinsky reaction. To prevent convection, the catalyst (ferroin, 0.008 M) was immobilized in silica gel (0.9 mm thick) by mixing solutions of sodium silicate (approx. 27% SiO_2 in 3.5 M NaOH), ferroin, water and H_2SO_4 with the ratios 2:4:3:2 as reported in [8,12,18]. The reactor was then filled with the following solution: 0.15 M $NaBrO_3$, 0.15 M $CH_2(COOH)_2$ and 0.30 M H_2SO_4. Experiments were carried out at 20 ± 1 $^{\circ}$C.

Spiral waves were produced in this medium by the following procedure: first, a circular wave was induced by touching the gel layer with a silver wire and then the resulting wave front was broken with an iron wire to produce a pair of spiral waves.

A CCD camera and a video- tape system were used to record the experiments. Digital image processing was carried out before printing the pictures shown in this paper.

3.1. Radial Electrochemical Reactor

When the radial electric field is considered, the electrochemical reactor used consists, mainly, in a Petri dish (9 cm in diameter), where the silica gel is placed at its bottom. The dish is fulfilled with the recipe described above. To prevent air from interfering with the BZ reaction, the depth of the liquid layer was between 6 and 8 mm.

The spiral wave dynamics was perturbed by a radial electric field, created by a positively charged needle- shaped electrode that was located at the core of a spiral wave, and a negative circular electrode, located at the border of the Petri dish. The results showed below were obtained with platinum electrodes.

3.2. Parallel Electrochemical Reactor

The reactor used for these experiments (with parallel electric field) is analogous to that in Ref. [19]. The main part of the open reactor was a small plate (20 mm long, 20 mm wide) between two continuously stirred flow tanks with identical fixed feed concentration (given above). A 1 mm thick silica gel layer was prepared in a Petri dish as described above. After gelled, it was cut and put over the central plate. During the experiments this layer was covered with the feeding solution (1 mm) continuously flowing from the tanks. Platinum electrodes were disposed into the tank reactors in a parallel geometry.

4. NUMERICAL SIMULATION

The two- variable version of the Oregonator model [20-22] was used for simulations. It has been modified in order to take into account the effects of an electric field [13]. Its equations are,

$$\partial u/\partial t = \{u - u^2 - f\, v\, (u-q)/(u+q)\}/\varepsilon + \vec{\nabla}\, \vec{J}_u$$

$$\partial v/\partial t = \{u - v\} + \vec{\nabla}\, \vec{J}_v$$

$$(1)$$

where u is the 'propagator' variable representing the dimensionless concentration $[HBrO_2]$ and v, the 'recovery' variable, is the dimensionless concentration [ferriin]. The parameters q, f and ε are function of concentrations and rate constants. D_u and D_v will be the dimensionless diffusion coefficients of u and v and their ratio is important in controlling wave dynamics in general and wave tip meandering in particular.

$\vec{J}_u$ and $\vec{J}_v$ are the density currents for both variables. For the case of an electric field applied to a medium in which only the propagator variable diffuses, the divergence of these density currents can be written as [13]

$$\vec{\nabla}\vec{J}_u = \vec{\nabla} \{ D_u \vec{\nabla}u + D_u M u \vec{E} \}$$

$$\vec{\nabla}\vec{J}_v = \vec{\nabla} \{ D_v \vec{\nabla}v \} = D_v \nabla^2 v$$

$$(2)$$

where M is the mobility of the relevant variable and $\vec{E}$ is the applied electric field.

4.1. Radial Electric Field

For the case of a radial electric field applied to the medium the divergence of these density currents can be written as [13]:

$$\vec{\nabla}\vec{J}_u = D_u \nabla^2 u + D_u M \left\{ u \left(\frac{E_r}{r} + \frac{dE_r}{dr} \right) + \frac{du}{dr} E_r \right\}$$

$$\vec{\nabla}\vec{J}_v = \vec{\nabla} \{ D_v \vec{\nabla}v \} = D_v \nabla^2 v$$

$$(3)$$

where the applied field has only the radial component E_r.

4.2. Parallel Electric Field

For the case of a parallel electric field applied to the medium the divergence of these density currents can be written as [16,23]:

$$\vec{\nabla}\vec{J}_u = D_u \nabla^2 u + E \, \partial u / \partial x$$

$$\vec{\nabla}\vec{J}_v = \vec{\nabla} \{ D_v \vec{\nabla}v \} = D_v \nabla^2 v$$

$$(4)$$

where now E is the absolute value of the electric field.

The set of equations (1), with the appropriate values of the electric field, were solved using the explicit Euler method of integration and no flux boundary conditions. The calculations were performed for a two- dimensional medium, on a rectangular 200 x 200 grid.

5. RADIAL ELECTRIC FIELD: CONTROL OF SPIRAL PARAMETERS

We have investigated experimentally [8,12] the changes in spiral wave behavior under radial electric current, using the setup described above. When positive constant voltage is applied to the central electrode, period, wavelength, as well as core size of the spiral waves increase by a factor up to three. A typical picture observed during the experiments is presented in Fig. 1. It can be seen that the wavelength, as well as the period of the tested spiral (at the right of the figure) is about two times greater than that of the spiral at the left, which has not been controlled by electric current. When the intensity of the applied current is increased, the period of the spiral is increased too, until it reaches a stationary value.

Figure 1: Effect of the radial electric field on spirals in BZ reaction. The right one, being under electric current stimulation, it has large wavelength in comparison with the nonstimulated spiral in the left part of the figure.

140

This phenomenon was studied numerically [13] using the set of equations described above. We investigated the effects of a constant positive electric field on a fully developed spiral (Fig 2).

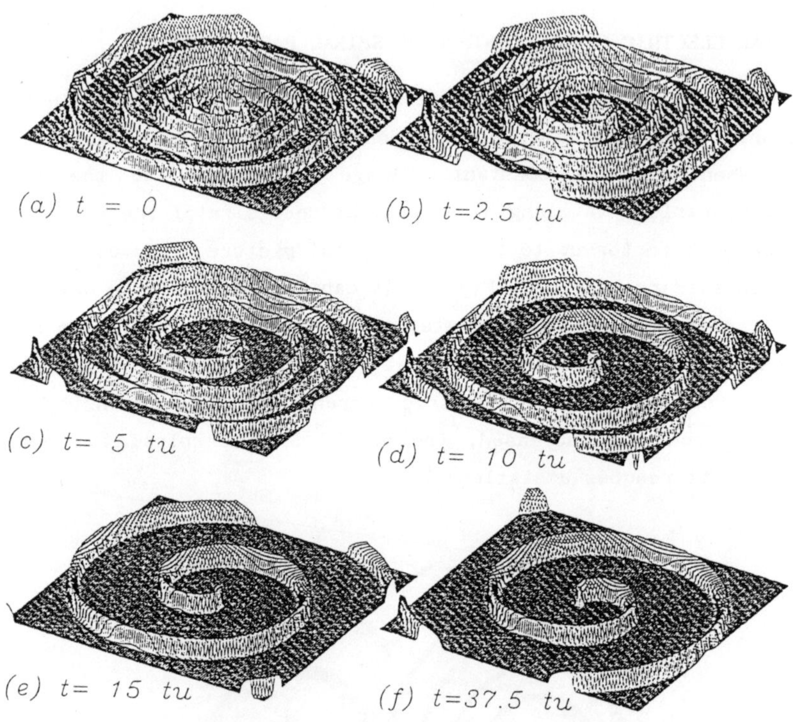

Figure 2: Unfurling of a spiral wave under the influence of an electric field of intensity E = 10 except for the region r < 1.5 s.u. where E = 0.

The spiral shown on Fig. 2a was generated in the absence of an electric field. A total of four rings can be observed with a typical front width of twelve gridpoints (the total grid is 500 X 500). The electric field is then turned on. The evolution of the spiral after 2.5 time units is shown in Fig. 2b. It can be clearly observed that the spiral center is reducing its front curvature and, at the same time, the front becomes wider, the sections near the spiral tip having now a width of approximately sixteen gridpoints. At the same time it can be observed that the affected section of the spiral alters its rotation rate as it slows down. The process can be observed to continue in figures 2c to 2e. The fronts increase to an average width of 28 gridpoints and the spreading out process has reduced the spiral to only one and a half turns. The configuration shown in 2e has reached equilibrium and remains unchanged a further 22.5 time units later, its rotation rate also unchanged. These results reproduce the experimental observations in BZ as can be observed from Fig. 1.

6. PARALLEL ELECTRIC FIELD: DRIFT OF SPIRALS

In this section, the influence of the electric current on the characteristic parameters of the spiral is going to be used to induce the drift of the spiral.

When a constant electric field is applied, it has been seen that the spirals move towards positive electrode [10]. If the electric current used is alternate, no effect is expected. Actually, for a frequency of the current equal to the frequency of the spiral ($f = f_e$, resonant frequency), the spiral does not drift. If the frequency of the current is twice the frequency of the spiral ($f = 2 f_e$, parametric resonant frequency), unexpectedly, the spiral drifts [16].

6.1. Experimental Results

The experiments were done in the electrochemical reactor described in section 3, for a parallel geometry of the electric current.

The resonance behavior of the vortex was induced by electric pulses of alternating polarity, controlled by PC computer. Positive pulses were sent when the spiral tip was oriented towards North and South, and negative pulses of the same amplitude and duration when

142

oriented towards West and East. The duration of the pulses and the current density in the gel were kept constant (5.5 sec. and 3 mA/mm^2 respectively). The vortex dynamics was followed by a CCD camera. For computer analysis, images were taken with intervals 5-7 s, and the position of the tip was calculated with an error not greater than 1/20 mm.

There were no effects at main resonant frequency ($f = f_e$, f_e is the frequency of a single vortex and was 1/37 s^{-1}) and at nonresonant frequencies as well. But at parametric resonant frequency, $f = 2 f_e$, the vortex started drifting. Fig. 3 shows the spiral 40 min. after the electric current was connected and the arrow shows the displacement followed by the spiral. The drift velocity calculated is about 1/20 of the velocity of free wave propagation (without electric field), and 1/5 of the drift velocity induced by a constant electric field of the same intensity.

Figure 3: Behavior of a vortex in an alternate electric current oscillating with parametric resonance frequency. The arrow shows the path followed by the spiral tip during these 40 min.

After 40 min. of the start of the experiment, the phase of the alternate electric field was increased by 90° (positive pulses were now sent when the tip was oriented towards East and West, and negatives ones at North and South orientation). In this case, the direction of the drift has changed to the opposite one, the drift velocity being the same.

The X position of the tip of the vortex in this experiment is shown in Fig 4. Character A shows the initial position of the spiral, B the position 40 min. later (corresponds to picture in Fig. 3) and C is the final position 40 min. after the polarity of the electric current was changed.

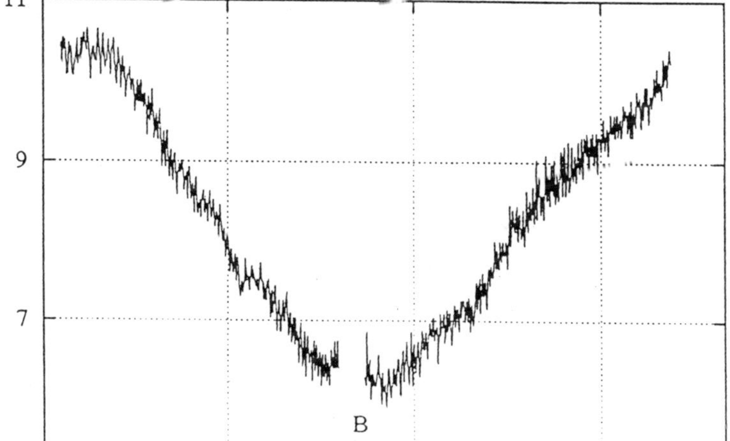

Figure 4: Records of the X position of the vortex tip experimentally obtained. At t= 40 min., phase of the electric field was increased by 90° (character B on the graph), and the drift direction has changed to the opposite one.

Small oscillations of the position of the tip during its global linear displacement correspond to the rotation of the tip around the core. This provides an estimation of the core diameter, about 0.3 mm. The X component of the drift velocity, calculated as the slope of the straight line, is 0.002 mm/s.

6.2. Numerical Results

The computer results (Fig. 5) obtained solving numerically the Oregonator's model [16,23] fit well to our experiment (Fig. 4).

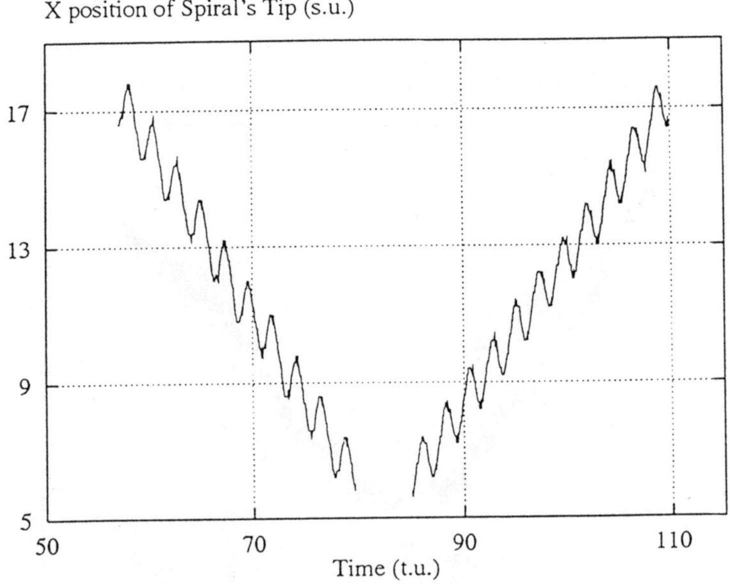

Figure 5: Records of the X position of the vortex tip numerically obtained. At t = 80.0, the phase of the electric field was increased by 90° resulting in the change of the drift direction (compare with Fig. 4).

6.2.1. Resonant curve

Several frequencies of the alternate electric field were considered and two second order resonant frequencies were found at $f = 4\ f_e$ and $f = 0.8\ f_e$ (in these cases the spiral drifts following a straight line, Fig. 6a). At frequencies close to the parametric resonant frequency ($f = 2\ f_e$), the motion of the spiral tip around the core exhibits a more complex behavior, it meanders following a flowerlet pattern (Fig. 6b and 6c).

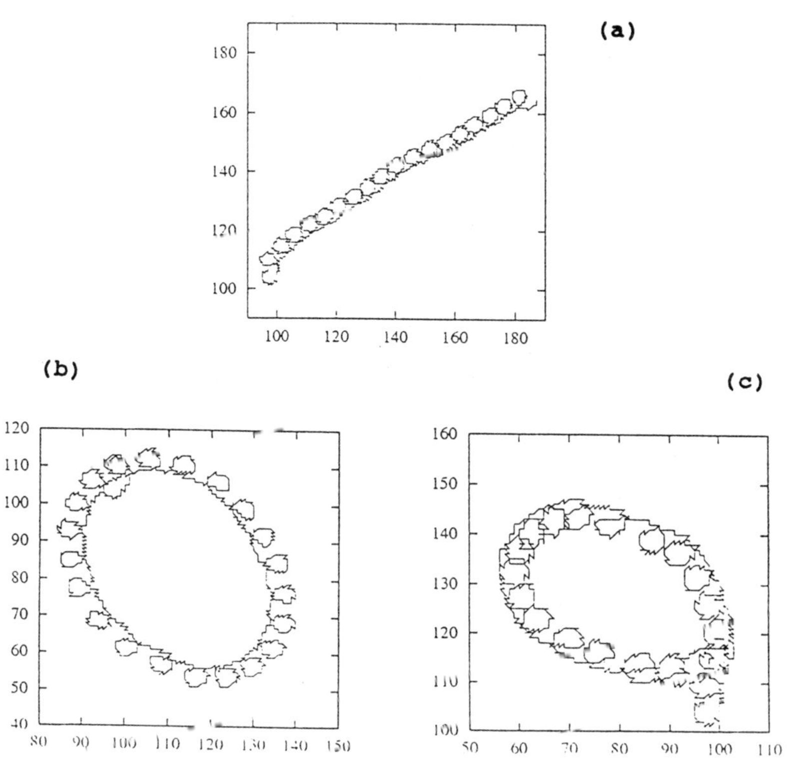

Figure 6: Different patterns, numerically calculated, followed by the tip of the spiral when alternate current is considered. (a) $f = 2\ f_e$, parametric resonance frequency, (b) $f = 2.05\ f_e$ and (c) $f = 1.95\ f_e$.

If the frequency of the electric field is very close to the resonant one, a coupling between this circular movement and a linear drift is observed.

In Fig. 7 the resonant curve is shown, the velocity of the spiral drift is plotted as a function of the frequency of the applied electric field. It is possible to observe a pick at the parametric resonance frequency and two other second order picks. Notice that slight differences in frequency lead to big changes in the drift velocity.

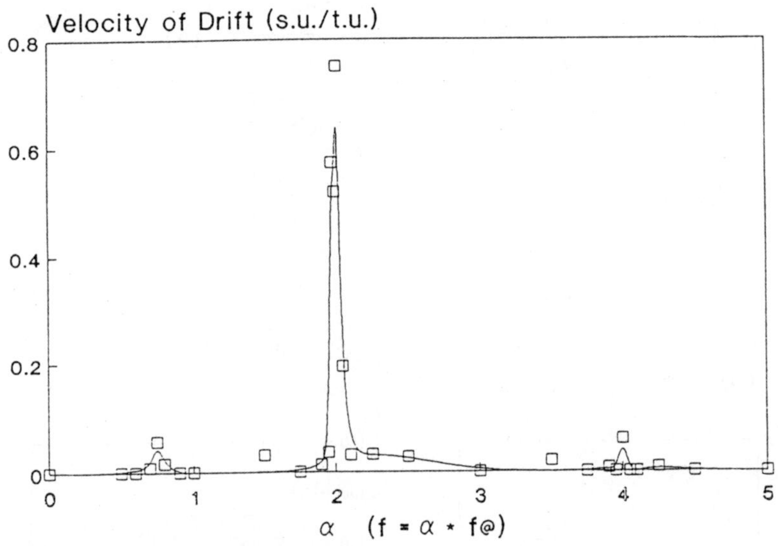

Figure 7: Resonant curve for the drift velocity. Dimensionless frequency of electric field, α, $(f = \alpha f_@)$ is plotted in the abscissa versus the drift velocity.

6.2.2. Effect of the phase of the alternate current

The resonance in electric field is interesting also because a different type of symmetry is involved here. A vortex always drifts in the same direction selected by the symmetry break in the medium, both in constant electric field [10] (to the anode) and heterogeneous light field [11] (illumination) (towards regions of higher light intensity). When illumination is homogeneous in space and alternating in time there is no selected direction in the medium; there is only one direction determined by the position of the rotating vortex's tip, and that position at some selected phase of light oscillation, determines the direction of vortex drift.

In oscillating electric field, there is a selected direction but there is no selected orientation. And besides, at every time, there is another selected direction determined by the rotating vortex tip. So much more interesting physics is observed here.

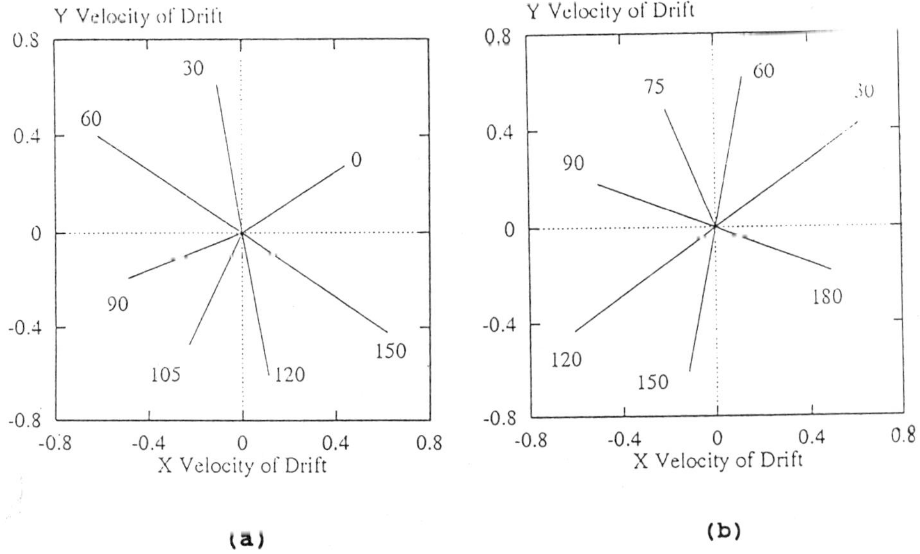

Figure 8: Hodograph of the parametric resonance drift velocities for vortices of different chirality. (a) Anti-clockwise rotating vortex. (b) Clockwise rotating vortex. The vector of the drift velocity (the modulus and the angle) for different phase shifts (indicated near each line) is shown. The phase shift was measured as the angle between the northern position of the vortex tip and its position at the moment when electric field was switched on.

148

It has been found numerically that the direction of the drift can be controlled by considering a different initial position of the vortex tip (an initial shift is introduced between the initial position of the spiral and the frequency of the electric current). This assert is illustrated in Fig. 8a, where drift velocities are plotted for different initial rotation angles of the tip.

The shift in the drift velocity introduced by modifying the initial rotating angle of the tip vortex is about twice the initial angle introduced (the direction of the vortex drift is changing continuously between $0°$ and $360°$ when the phase shift between electric field alterations and vortex rotation is changed between $0°$ and $180°$). In this way direction of the drift can be controlled by introducing an initial shift between the position of the spiral and the frequency of the electric field.

6.2.3. Effect of the spiral chirality

Changing the chirality of a vortex affects its behavior. The simplest way to understand the role of chirality is to find a coordinate transformation which changes the chirality of a vortex without changing the electric field. For an electric field oriented along X axis, this transformation is $X \to X$, $Y \to - Y$. The phase angle of rotation is changed by this transformation as $\varphi \to 180° - \varphi$. It is easily seen now that, really, the behavior shown in Fig. 8b can be obtained from that shown in Fig. 8a by this transformation.

7. DISCUSSION

In conservative systems, parametric resonance is governed by the equation [24],

$$x'' + \omega_0^2 F x = 0 \tag{5}$$

and forced resonance by the equation,

$$x'' + \omega_0^2 x + F = 0 \qquad (6)$$

where F describes parametric pumping of energy in Eq. (5) ($F = 1 + h \cos\omega t$) (the characteristic parameter of the oscillator, frequency, is modulated) and external force in Eq. (6) ($F = h * \cos\omega t$). Different behavior in both cases is determined by the time symmetry breaking of the equations, the external influence F; it is multiplicative in Eq. (5) and additive in Eq. (6). This results in resonance frequency ω_0 in forced oscillations and $2\omega_0$ in parametrically driven oscillations.

Waves in excitable media in the presence of electric field are described by the equation [14],

$$\partial u/\partial t = f(u) + D_u \nabla^2 u + E \, \partial u/\partial x \qquad (7)$$

The effect of electric field E is described by the multiplicative term $E \, \partial u/\partial x$. This results in the same type of symmetry breaking as for parametric resonance in conservative systems, and $2\omega_0$ resonance frequency. Experimentally, the effect of the electric field is to modify the characteristic parameters of the spiral (period), and this corresponds with the parametric resonance case in conservative systems (where the frequency of the oscillator is being modulated by the external force).

It is important to note the difference in parametric resonance behavior in conservative systems and in excitable active media. Energy is not conserved during wave propagation in an excitable medium; a wave propagates due to (chemical) energy stored in the medium. Their amplitude and shape remains constant during propagation, which is in general not the case for waves in conservative systems of arbitrary dimensionality. For vortex parametric resonance in an active medium, no pumping of energy is involved. But the type of symmetry of the equations is the same as for parametric resonance in conservative systems.

150

8. ACKNOWLEDGEMENTS

We want to thank Prof. J.L.F. Porteiro for some of the computer simulations here shown. This work is supported in part by the *"Comisión Interministerial de Ciencia y Tecnología"* (Spain) under project DGICYT-PB91-0660.

9. REFERENCES

[1] M.A. Allesie, F.I.M. Bonke and T.Y.G. Schopman. *"Circus Movement in Rabbit Atrial Muscle as a Mechanism in Tachycardia"*. Circulation Res. 33, 54-62, (1973).

[2] J. Bures, V.I. Koroleva and N.A. Gorelova. *"Leao's Spreading Depression, an Example of Diffusion-Mediated Propagation of Excitation in the Central Nervous System"* in *Autowaves and structures far from equilibrium*, Ed. V.I. Krinsky, Springer-Verlag, 180-183, (1984).

[3] M.H. Cohen and A. Robertson. *"Chemotaxis and Early Stages of Aggregation in Cellular Slime Molds"*. J. Theor. Biol. 31, 119-130, (1971).

[4] P.N. Devreotes, M.J. Potel and S.A. MacKay. *"Quantitative Analysis of Cyclic AMP Waves Mediating Aggregation in Dyctiostelium Discoideum"*. Devl. Biol. 96, 405-415, (1983).

[5] A.P. Muñuzuri, V. Pérez-Muñuzuri, V. Pérez-Villar and L.O. Chua. *"Spiral Waves on a Two-Dimensional Array of Nonlinear Circuits"*. IEEE Trans. on Circ. and Systems, Special Issue. 40, 10, (Dec 1993).

[6] S. Jakubith, H. Rotermund, et al. *"Spatiotemporal Concentration Patterns in a Surface Reaction: Propagating and Standing Waves, Rotating Spirals, and Turbulence"*. Phys. Rev. Lett. 65, 3013-3016, (1990).

[7] S.C. Müller, T. Plesser and B. Hess. *"Two-Dimensional Spectrophotometry of Spiral Wave Propagation in the Belousov-Zhabotinskii Reaction I. Experiments and Digital Representation. II. Geometric and Kinematic Parameters"*. Physica 24D, 71-96, (1987).

[8] V. Pérez-Muñuzuri, R. Aliev, B. Vasiev, V. Pérez-Villar and V.I. Krinsky. *"Super-Spiral Structures in an Excitable Medium"*. Nature 353, 740-742, (1991).

[9] V.I. Krinsky and K.I. Agladze. Physica 8D (1983) 50-56

[10] O. Steinbock, J. Schütze and S.C. Müller. *"Electric Field Induced Drift and Deformation of Spiral Waves in an Excitable Medium"*. Phys. Rev. Lett. 68, 248-251 (1992).

[11] M. Markus, Zs. Nagy- Ungvaray and B. Hess. *"Phototaxis of Spiral Waves"*. Science 257, 225-227 (1992).

[12] V. Pérez-Muñuzuri, R. Aliev, B. Vasiev and V.I. Krinsky. *"Electric Current Control of Spiral Wave Dynamics"*. Physica 56D 229-234 (1992)

[13] J.L.F. Porteiro, V. Pérez-Muñuzuri, A.P. Muñuzuri and V. Pérez-Villar. *"A Numerical Model for the Influence of a Radial Electric Field on Spiral Wave Dynamics"*. To appear in Physica D (1993).

[14] P.J. Ortoleva, *"Nonlinear Chemical Waves"*. John Wiley & Sons (1991).
[15] H. Sevcikova and M. Marek. Physica **9D** 140 (1983).
[16] A.P. Muñuzuri, M. Gómez-Gesteira, V. Pérez-Muñuzuri, V.I. Krinsky and V. Pérez-Villar. *"Parametric Resonance of a Vortex in an Active Medium"*. Submitted to Phys. Rev. Lett. (1993).
[17] A.P. Muñuzuri et al. In preparation.
[18] T. Yamaguchi, L. Kunhert, Zs. Nagy-Unguaray, S.C. Müller and B. Hess. *"Gel Systems for the Belousov-Zhabotinsky Reaction"*. J. Phys. Chem. **95**, 5831 (1991).
[19] K.I. Agladze and P. DeKepper. *"Influence of Electric Field on Rotating Spiral Waves in the BZ-Reaction"*. J. Phys. Chem. **96**, 5239-5242 (1992).
[20] W. Jahnke, W.E. Skaggs and A.T. Winfree. *"Chemical Vortex Dynamics in the Belousov-Zhabotinsky Reaction and in the Two-Variable Oregonator Model"*. J. Phys. Chem. **93** (1989) 740-749.
[21] W. Jahnke and A.T. Winfree. *"A Survey of Spiral-Wave Behaviors in the Oregonator Model"*. J. Bif. and Chaos. **1** 445-466 (1991).
[22] A.T. Winfree. *"The Varieties of Spiral Wave Behaviors in Excitable Media"*. Chaos. **1** (1991) 303.
[23] A.P. Muñuzuri, M. Gómez-Gesteira, V. Pérez-Muñuzuri, V. Pérez-Villar and V.I. Krinsky. *"Mechanism of the Electric Field Induced Vortex Drift in Excitable Media"*. Submitted to Phys. Rev. Lett. (Dec. 1992).
[24] Landau, L. & Lifchitz, E. *"Mecanique"*. Mir, Moscow, (1960).

AUTOWAVE MODEL
OF THE AMOEBOID MOTILITY IN THE PRESENCE
OF THE EXTERNAL LIGHT STIMULI

D. A. Pavlov [a], M. M. Potapov [a] and Yu. M. Romanovsky [b]

[a] Department of Computational Mathematics and
Cybernetics,
[b] Department of Physics, Moscow State University,
119899, Moscow, Russia.

ABSTRACT

To simulate the responses of the amoeboid type cells
to external light stimuli the base distributed nonlinear mathematical model is complemented by ordinary
differential equations. By these equations kinetic coefficients of the distributed model are connected with
intra-cellular concentrations of the inhibitor and activator which depend on the photoreceptors activity. Bio-
physical experimental results are compared with results of a computer simulation.

1. INTRODUCTION

The ability of cells to react with active motions to internal or external stimuli is without any doubt one of the most striking characteristics of living organisms. The mathematical models of active motility of
the amoeboid type cells (amoebs, somatic cells, fibroblasts, leykocites,

cancer cells etc.) have been developed only lately. Two such classes of the models are known now.

The first class of models takes into account an influence of hydrodynamical streamings of protoplasm brought about by nonstationary gradients of the intracellular pressure. This pressure is generated by actin-myosin interaction. Reorganization of the contractile apparatus does not occur in this case. This process is similar to actin-myosin interaction in muscullar cells [1–5].

The second class of models describes dynamics of delation, contraction and solution in protoplasm as two-phase medium. In this case the reorganization of the contractile apparatus is noticed. A cell can change its shape, form the pseudopods and migrate over the surface [6–9].

In this paper we present a mathematical model of the first class for rhythmic motility in the strands of the *Physarum* plasmodium. This model is constructed on the notion that the cytoskeleton is the essential element of autooscillatory system of cells. In [3,5,10–13] we have discussed some autonomous and nonautonomous versions of such models. Here we present a mathematical description of *Physarum* reaction to external light stimuli.

At first we quote a brief information on autowave phenomena in *Physarum Polycephalum* plasmodium.

2. MECHANO-CHEMICAL PROPERTIES OF *PHYSARUM*

The plasmodium *Physarum Polycephalum* is a sheet-like multinuclear mass of protoplasm characterized by its large size up to a square meter. The plasmodium is not subdivided into cells. It is differentiated into a leading frontal zone and a network of interconnected protoplasmic strands (see fig. 1).

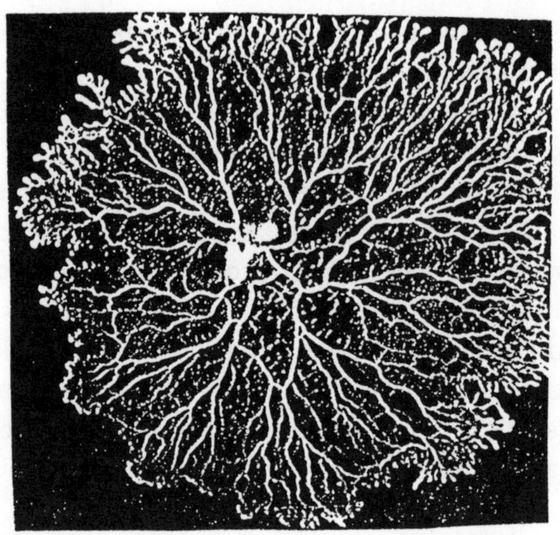

Fig. 1. The migrating plasmodium *Physarum Polycephalum*.

In a large plasmodium the lengths of strands can reach a few tens of centimetres with the diameters of about two millimetres. The strands are differentiated into the outer tube-like gel ectoplasm and the more fluid inner sol endoplasm.

There are the following kinds of autowave phenomena in *Physarum* [3–5,14–18]:

1. A wave-like propagation of thickness changes in the frontal zone with velocities of about 1 cm/sec. Sometimes circulations of such waves can be observed. Their periods are in the range of minutes.

2. Quasi-stochastic oscillations of sheet thickness.

3. Nearly synphasic radial pulsations of the strands.

4. Radial pulsation of long strands in the form of running and standing waves.

5. Synphasic longitudinal homogeneous contractions of any segment in a strand excited from the plasmodium.

6. Variations of the autowave amplitude and period under variations of external light, thermal, chemical and mechanical stimuli.

3. DISTRIBUTED MATHEMATICAL MODEL

Events occuring in the plasmodium strands can be described as follows. The protoplasmic strand is a long cylindrical tube with visco–elastic walls which are filled with an incompressible fluid. Let acto-myosin fibrils be homogeneously distributed in the wall. Owing to an increase of the local concentration of calcium ions, the myosin oligomers can interact with active centres of neighbouring actin filaments. This leads to a radius decrease of the given strand section. The endoplasm is evacuated from the contracting region, streams into the neighbouring regions and stretches these parts of the strand. This stretching can promote an increase in local calcium concentration and thus conditions for a new active contraction are created. It is assumed that there is a sufficient local supply of ATP molecules.

Corresponding differential equations look as [3–5,10,12]:

$$u_t = (EhR\,u_{xx} + \eta hR\,u_{txx} + R^3 p_{xx})\,/\,16\mu\,, \tag{1}$$

$$p_t = k_1\,F(c)\,(p_m - p) - k_2\,p\,, \tag{2}$$

$$c_t = k_3\,u - k_4\,c\,, \quad t > 0,\ x \in [0,\ell]\,. \tag{3}$$

These equations are complemented by Neumann's boundary conditions for the functions u and p :

$$\left.\frac{\partial u}{\partial n}\right|_{x\,=\,0,\,\ell} = \left.\frac{\partial p}{\partial n}\right|_{x\,=\,0,\,\ell} = 0,\ t > 0, \tag{4}$$

and by initial conditions:

$$u\,|_{t=0} = u_0(x),\ p\,|_{t=0} = p_0(x),\ c\,|_{t=0} = c_0(x),\ x \in [0,\ell]\,. \tag{5}$$

Here $u(t,x)$ – small deviation of the strand radius from the rest level R in the absence of a passive stress, $p(t,x)$ – active pressure, $c(t,x)$ – calcium ion concentration, p_m – maximal value of p, E – elastic modul of the strand wall, η –viscous modulus of the strand wall, h – wall thickness, kinetic coefficients $k_1 - k_4$ specify velocities of the mechano-chemical interactions, $F(c)$ – dimensionless activation function.

Mathematical correctness of this model has been verified. In particular the existence and uniqueness of classical [5] and generalized [13] solutions and their continuous dependence on initial data have been proved. Also the convergence of the finite-dimensional projection-difference approximations has been established [13]. Besides the uniqueness of the stationary and spatially uniform solution of the system (1) – (4) has been proved. A stability of the stationary solution has been investigated by applying the eigensolutions of the linearized system of equations as small perturbations which have the wave form. We quote some results of the computer simulation [5,12].

a. The periods of autowaves are almost independent on the waveband from 0.5 cm to 10 cm.

b. The periods determined by a computer are larger than the ones determined by a linear analysis.

c. The parameters of the solution: T, amplitudes of the u, p, c are not far from experimental data.

d. If the first four modes are unstable then the deformations occurs only on the ends of the strand. It is also observed in the experiment.

In this paper we present below a scheme which simulates the plasmodium responses to external light stimuli and extends the model (1) – (5). Before we shall consider experimental data.

4. PHOTOREACTIONS OF *PHYSARUM* PLASMODIUM

We shall use the experimental results which are described in [14]. *Physarum* displays the most powerful reactions to the light in the wavelength range of 450 - 500 nm.

The measurement of the protoplasm shuttle streaming period have been performed by the laser Doppler spectroscopy (fig. 2) [15, 18].

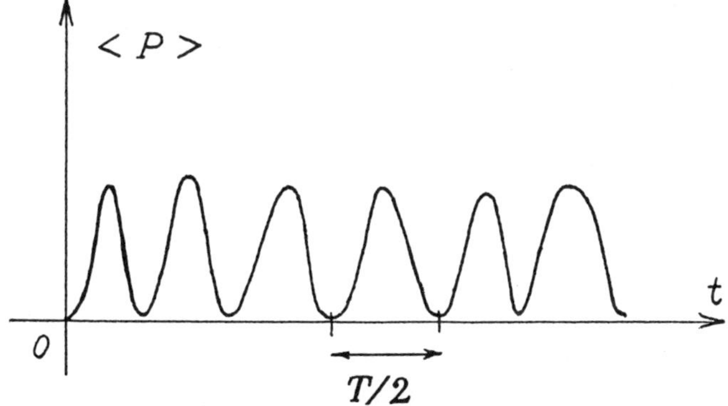

Fig. 2. The alteration of the electric signal $< P >$ value, which is proportional to the average velocity of protoplasm streaming in a given strand cross-section. The distance between two minima corresponds to the semiperiod of the push-pull movement.

The photoresponse kinetics by the continuous monochromatic light ($\lambda = 466$ nm) is shown on fig. 3.

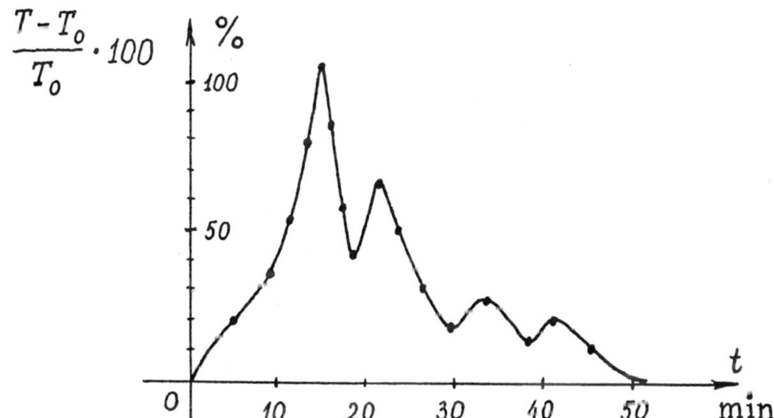

Fig. 3. The variation of the protoplasm flow period by monochromatic light ($\lambda = 466$ nm). $t = 0$ corresponds to the beginning of the light irradiation; T_0 – velocity oscillation period in the dark.

The following experimental results have been obtained [14]:

1. existence of three oscillation period maxima in the photoreaction evolution phase;

2. their positions are independent on the wavelength in the range of $450 - 500$ nm and on the power density variation from $2\,\mathrm{W/m^2}$ to $10\,\mathrm{W/m^2}$ of irradiating light.

5. NONAUTONOMOUS MATHEMATICAL MODEL

Physarum is turned to a nonautonomous autowave system by external stationary and nonstationary mechanical, thermal or light stimuli. The corresponding mathematical models are changed to nonautonomous ones.

We try to find such functions $k_i(t)$ which make the period T dependence on t the closest to the experimental data. A preliminary analysis has shown that T_0 depends most strongly on k_1 (see fig. 4).

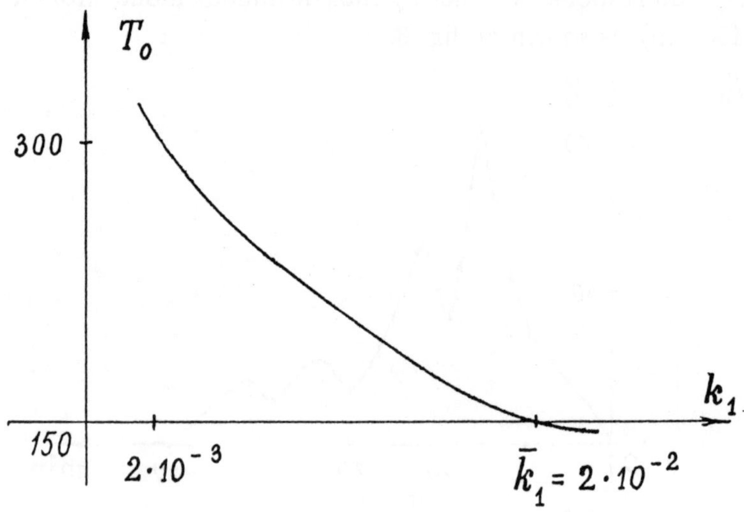

Fig. 4. Dependence $T_0(k_1)$. Here $\overline{k}_1$ is the value of koefficient k_1 in the absence of the light [5].

The approximation of the experimental photoreaction curve (fig. 3) and the dependence $T(k_1)$ (fig. 4) makes possible   to construct time function $k_1(t)$ which is shown on fig. 5.

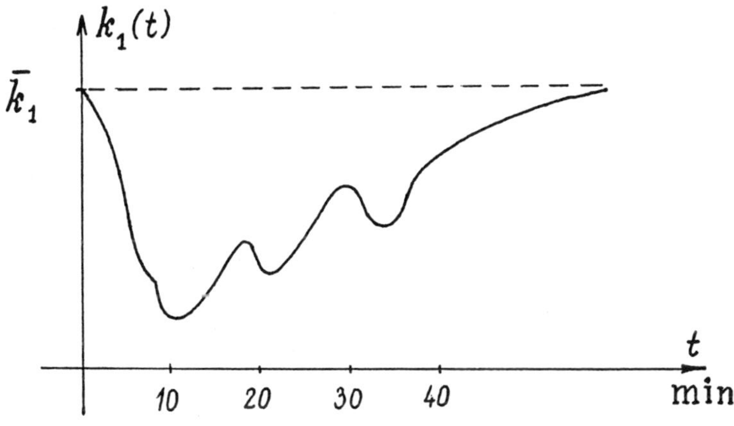

Fig. 5. Time function $k_1(t)$.

The computer experiment has displayed that a substitution of $k_1(t)$ into initial model leads to the following result: the period $T(t)$ variation is qualitatively the same as in the experiment (compare fig. 3 with fig. 6).

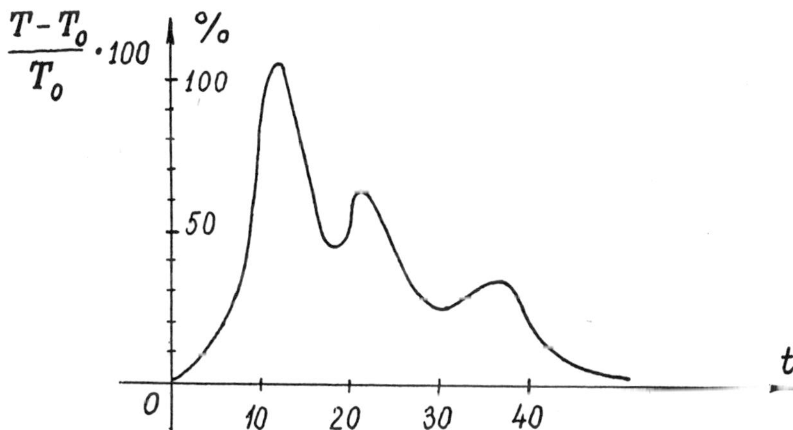

Fig. 6. $T(t)$ alteration. $k_1 = k_1(t)$.

6. MATHEMATICAL MODEL OF PHOTOREACTION

The next step in photoreaction simulation is the description of light acting mechanism on the initial model parameters. We propose the following scheme of the light stimulus effect:

$$\dot{y}_1 = R_0 \exp(-t/\tau) - \alpha_1 y_1 y_2 , \tag{6}$$

$$\dot{y}_2 = \beta_2 y_1 y_2 - \alpha_2 y_2 , \tag{7}$$

$$\dot{k}_1 = \gamma_1 y_2 + \frac{\gamma_2}{\gamma_3 + \gamma_4 y_2} - \delta k_1 , \tag{8}$$

$$y_1(0) = y_{10},\ y_2(0) = y_{20},\ k_1(0) = \overline{k}_1 . \tag{9}$$

Here R_0 depends on irradiation power density and on photoreceptors activity. Parameter τ is the photoreceptor relaxation characteristic time. y_1, y_2 are the enzymes concentrations. Thus y_1, y_2 play the roles of activator and inhibitor of these chemical processes. Flavins and carotenoids claim for the photoreceptor role equally. This nonlinear system is added to initial model.

The result of computer simulation for the linear analog of the scheme (6) – (9):

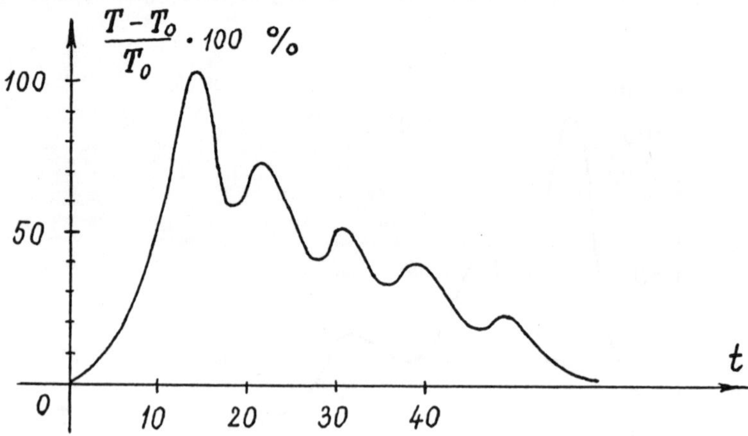

Fig. 7. $T(t)$ alteration. $k_1 = \tilde{k}_1(t) + \overline{k}_1$ and $\tilde{k}_1(t)$ is defined from the linear analog of (6) – (9).

7. CONCLUDING REMARKS

As you can see above the results of the photoresponse simulation suggest   certain optimism. The scheme presented is undoubtedly not final; it is perhaps an elementary step towards more detailed explanation of the cell photoreactions mechanisms. This model will be extended or modified with the help of a computer simulation and as new experimental data accumulate.

We suppose that our model will be   successful   when associated with the model by M. Dembo and W. Alt [6 - 9]. This association permits to describe the cell migration etc.

ACKNOWLEDGEMENTS

We would like to thank Dr. A.V. Priezzhev, Dr. M.Yu. Poroshina and Dr. V.A. Teplov for helpful remarks and special thanks to O.A. Latushkin for computer programs.

REFERENCES

1. G.M. Odell, "A mathematically modelled cytogel cortex exhibits periodic Ca^{2+}-modulated contraction cycles seen in *Physarum* shuttle streaming", J. Embryol. Exp. Morphol. **83**, 261–287 (1984).

2. J.F. Oster and G.M. Odell, "Mechanics of cytogels I: Oscilations in *Physarum*", Cell Motility **4**, 469–503 (1984).

3. Yu.M. Romanovsky and V.A. Teplov, "Autowave mechanochemical model for *Physarum* shuttle streaming", Thermodynamics and Pattern Formation in Biology, I. Lamprecht and A.I. Zotin (eds.), De Gruyter, Berlin, 395–414 (1988).

4. S.I. Beylina, N.B. Matveeva, A.V. Priezzhev, Yu.M. Romanovsky, A.P. Sukhorukov and V.A. Teplov, "Plasmodium of the myxomycete *Physarum Polycephalum* as an autowave self-organizing system", Self-Organization. Autowaves and structures far from equilibrium, V.I. Krinsky (eds.), Springer-Verlag, Berlin, 218–221 (1984).

5. V.A. Teplov, Yu.M. Romanovsky and O.A. Latushkin, "A continuum model of contraction waves and protoplasm streaming in

162

strands of *Physarum* plasmodium, Biosystems, $\underline{24}$, 269–289 (1991).

6. W. Alt, "Contraction and oscillations in a simple model for cell plasma motion", Temporal Order, L. Rensing and N. Jaeger (eds.) Synergetics $\underline{29}$, Springer, Berlin, 163–174 (1985).

7. W. Alt, "Mathematical models in actin-myosin interaction", Prog. Zool. $\underline{34}$, 219–230 (1987).

8. M. Dembo, "Mechanics and control of the cytoskeleton in *Amoeba Proteus*", Boiph. J. $\underline{58}$, 1053–1080 (1989).

9. M. Dembo, F. Harlow and W. Alt, "The biophysics of cell surface motility", Cell Surface Dynamics. Concepts and models, A. Parelson, Ch. DeLisi and F. Wiegel (eds.), Marcel Dekker, New York, 495–542 (1984).

10. V.A. Vasilyev, Yu.M. Romanovsky, D.S. Chernavsky and V.G. Yakhno, "Autowave processes in kinetic systems", Spatial and Temporal Self-Organization in Physics, Chemistry, Biology and Medicine, D. Reidel (ed.), Dordrecht (1987).

11. O.A. Latushkin, N.B. Netrebko, Yu.M. Romanovsky and V.A. Teplov, "Mathematical two-dimensional model of autowave processes in living cell", Collective dynamics of excitations and structure formation in biological tissues, Inst. Appl. Phys. USSR Acad. Sci. Press, Gorky, 109–119 (1988) (in Russian).

12. Yu.M. Romanovsky, "Mathematical models of intracellular motility", submitted to Math. Model. (in Russian).

13. D.A. Pavlov and M.M. Potapov, "The distributed model of intracellular motility and its projection-difference approximation", submitted to Vestnik Mosk. Univ., Ser.15, Vychisl. Matem. i Kibern. (in Russian).

14. M.Yu. Poroshina, A.V. Priezzhev and Yu.M. Romanovsky, "Photoreception and autooscilating motility of the living cell", Biofizika, $\underline{34}$, 980-984 (1989) (in Russian).

15. V.G. Kolin'ko, T.A. Arkhangelskaya and Yu.M. Romanovsky, "*Plasmodium mixomicete Physarum* protoplasm motion under changing temperature conditions, Studia Biophysica $\underline{106}$, No 3, 215–222 (1985) (in Russian).

16. V.A. Teplov, "Autooscillations in *Physarum* plasmodium. Cor-

relation between force generation and viscoelasticity during rhythmical contractions of protoplasmic strand". Protoplasma 1, 81–88 (1988).

17. Z. Baranovsky and V.A. Teplov, "Endoplasmic streaming mediates integrity of autooscillations in *Physarum Polycephalum* plasmodium, Cell Biology International Reports 16, No 11, 1091–1096 (1992).

18. A.V. Priezzhev, Yu.M. Romanovsky and E.B. Chernyaeva, "Laser Doppler spectroscopy of living cells and models of intracellular motility", Laser Scattering Spectroscopy of Biological Objects, J.Štěpánek, P. Anzenbacher, B. Ledláček (eds.), Elsevier, Amsterdam, 503–530 (1987).

FROM THE HAMILTONIAN MECHANICS TO CONTINUOUS MEDIA. UNIFIED DESCRIPTION OF KINETIC, HYDRODYNAMIC AND DIFFUSION PROCESSES IN ACTIVE MEDIA

Yu.L. Klimontovich

Department of Physics, Moscow University, 117234 Moscow, Russia

ABSTRACT

This paper presents main ideas and results of the modern statistical theory of macroscopic open systems.

We begin from the demonstration of the necessity and the possibility of the unified description of kinetic, hydrodynamic and diffusion processes in nonlinear macroscopic open systems on the base of generalized kinetic equations.

A derivation of the generalized kinetic equations for a continuous medium is presented . A "point" of a continuous medium is determined by definition of physically infinitesimal scales. The definition of the Gibbs ensemble is given on the same basis. The Boltzmann gas and a fully ionized plasma are used as the test systems.

The dynamic instability of the motion of particles plays a constructive role in the transition from the reversible Hamilton equations to the generalized kinetic equations.

The kinetic equation for the Boltzmann gas consists of the two dissipative terms: 1) the "collision integral" is defined by processes in a velocity space; 2) the additional dissipative term of the diffusion type in the coordinate space. Owing to the later the unified description of the kinetic, hydrodynamic and diffusion processes for all values of the Knudsen number becomes possible.

The H-theorem for the generalized kinetic equation is proved. An entropy production is defined by the sum of two independent positive terms corresponding to redistribution of particles in velocity and coordinate space respectively.

An entropy flux also consists of two parts. The one is

proportional to the entropy, and the other one is proportional to the gradient of entropy. The existence of second term allows us to give the general definition of the heat flux for any values of the Knudsen number. This general definition for small Knudsen number and a constant pressure leads to the Fourier law.

The equations of gas dynamic for any class of distribution functions follow from the generalized kinetic equation without perturbation theory with respect to the Knudsen number. These equations differ from the traditional ones by taking the self-diffusion processes into account.

The generalized kinetic equation for description of the Brownian motion and of autowave processes in active media are considered. The connection with reaction diffusion equations – the Fisher – Kolmogorov – Petrovski – Piscunov and Ginzburg – Landau equations is established. These equations are valid within a single moment approximation and therefore they are not suitable for bifurcation regions.

We discuss the connection between the diffusion of particles in a restricted system with the natural flicker (1/f) noise in passive and active systems.

In the conclusion the simplest dynamic model for associative memory and pattern recognition is considered. This model can serve as the base for the corresponding kinetic theory.

1. The Transition from Reversible Equations of Mechanics to Irreversible Equations of the Statistical Theory.

1 1 Physical Definition of Continuous Medium.

In order to describe the transition to irreversible equations of the statistical theory it is necessary to take into account the structure of the "continuous medium". In other words, it is necessary to give the concrete definition of the infinitesimal scales in order to give the physical definition of the notion "point" [1-6].

Obviously, it is impossible to give unified definition of the physically infinitesimal scales for all systems. We shall introduce these scales for two simplest cases: the Boltzmann gas, and a fully ionized plasma. Such definitions are different for the kinetic and hydrodynamic region of scales.

The kinetic region. A rarefied (Boltzmann) gas and a rarefied fully ionized plasma are characterized, respectively, by the dimensionless small density and plasma parameters

$$\varepsilon = nr_o^3; \qquad \mu = 1/nr_D^3, \qquad\qquad (1.1)$$

where r_o is the diameter of an atom, r_D is the Debye radius and n is the mean density of the number of particles. These parameters determine the connection between the infinitesimal scales τ_{ph}, l_{ph} and the corresponding "collision" parameters τ, l of the Boltzmann gas and the Debye plasma.

We denote by $N_{ph} = nV_{ph}$ the number of particles in the physically infinitesimal volume and define the physically infinitesimal time as

$$\tau/N_{ph} = \tau_{ph}, \qquad N_{ph} = nV_{ph} \sim nl_{ph}^3, \qquad \tau_{ph} = l_{ph}/v_T. \qquad (1.2)$$

Thus defined τ_{ph} is the time within which any particle

out of the number N_{ph} in the volume V_{ph} undergoes a collision.

The physically infinitesimal scales for the Boltzmann gas follow from the expressions (1.2) ([1-6]).

$$\tau_{ph} \sim (\varepsilon\tau)^{1/2} \ll \tau, \qquad l_{ph} \sim (\varepsilon l)^{1/2} \ll l, \qquad N_{ph} \sim (\varepsilon\tau)^{-1/2} \gg 1. \qquad (1.3)$$

The corresponding kinetic characteristics for a rarefied plasma, i.e., for the plasma parameter $\mu \ll 1$, are (a stands for electron or ion)

$$\tau_{ph}^{(a)} \sim \mu\tau \quad , \qquad l_{ph}^{(a)} \sim r_D \sim \mu l, \qquad N_{ph}^{(a)} \sim 1/\mu \gg 1. \qquad (1.4)$$

We see that for the Boltzmann gas and for a rarefied plasma the general conditions imposed on the physically infinitesimal scales are satisfied.

The gas dynamic region. In this case the relaxation time is defined by the characteristic scale of the system L: $\tau_D = L^2/D$, where D may denote the viscosity ν, the thermal conductivity χ, and the self-diffusion coefficient D. Here $D = \nu = \chi$. The physically infinitesimal characteristics are defined now by relations

$$\tau_{ph}^{G} \sim \frac{\tau_{D}}{N_{ph}^{G}}, \qquad l_{ph}^{G} \sim \frac{L}{N^{1/5}}, \qquad N_{ph}^{G} \sim N^{2/5}. \qquad (1.5)$$

where $N = nL^3$

The definition of physically infinitesimal scales allows us to use the notion of the "continuous medium", respectively, for the kinetic and hydrodynamic description. In order to illustrate the importance of this notion, we consider the following example.

The maximum of the Reynolds number in the Kolmogorov theory.

In the Kolmogorov theory of the developed turbulence the number of collective (turbulent) degrees of freedom $N_{turb} = (L/L_o)^3$ is connected with the Reynolds number by the relation [7,8,4]

$$N_{turb} \sim Re^{9/4}. \qquad (1.6)$$

Here L, and L_o are the largest and the smallest scales of the developed turbulent motion, respectively.

It follows from the definition of N_{turb} follows that the maximum value of the Reynolds number in the Kolmogorov theory is restricted by the condition $L > l^G$ and, as a results

$$(Re)_{max} \sim N^{4/15} ; \quad N = nL^3 \qquad (1.7)$$

We see that, as a consequence of (1.7), the limit $Re \Rightarrow \infty$, which is used frequently in the theory of turbulent motion, for the "physical continuous medium" cannot be realized.

1.2. The Gibbs Ensemble for Nonequilibrium Processes.

The statistical ensemble was introduced for describing the equilibrium state. In this case the number of controllable degrees of freedom is very small, and hence the indeterminacy in the microscopic states of the system is extremely high. This way of defining of a statistical ensemble in the case of arbitrary nonequilibrium states is not suitable. What then can be done?

Let us assume that we have chosen a concrete definition of the physically infinitesimal scales. We can then assume that the indeterminacy of defining the microscopic states of a system in the Gibbs ensemble is governed by the indeterminacy in the states of particles confined within the volume V_{ph}. This allows us to carry out the operation of averaging, or smoothing of dynamic distributions over a physically infinitesimal volume.

1.3. The Unified Definition of "Continuous Medium". Averaging over a Physically Infinitesimal Volume [27,28].

In order to develop the unified description of kinetic and hydrodynamic processes, it is necessary to use the equation $(V_{ph}^G)_{min} = V_{ph}$ to express the minimal physically infinitesimal volume $(V_{ph}^G)_{min} \equiv L_{min}^3$ via a small density parameter ε:

$$l_{min} \sim (N_{ph})^{1/2} l_{ph} \sim 1/(N_{ph})^{1/2}, \qquad N_{min} = nL_{min}^3 \sim \varepsilon^{-5/4}. \qquad (1.8)$$

Thus, for example, at atmospheric pressure the density parameter $\varepsilon \approx 10^{-4}$ and therefore the number of particles in the "point" N_{min} is order of 10^5.

Let us consider a local random function $N(r,p,t)$ which denotes the microscopic phase density in the six-dimensional space of coordinates and momenta

$$N(r,p,t) = \sum_{1 \leq i \leq N} \delta(r-r_i(t))\delta(p-p_i(t)) \qquad (1.9)$$

and let us introduce a new function smoothed over the distribution

$$F(\rho) = (2\pi L_{min}^2)^{-3/2} \exp\left(- \frac{\rho^2}{(L_{min})2} \right) \qquad (1.10)$$

$$\tilde{N}(r,p,t) = \int N(r-\rho,p,t)F(\rho)d\rho.$$

At the beginning let us consider the following dynamical reversible equations for the microscopical function (1.9) [1-6,9]

$$\frac{\partial N}{\partial t} + v\frac{\partial N}{\partial r} + F^m(r,t)\frac{\partial N}{\partial p} = 0,$$

$$F^m(r,t) = F_o - \frac{\partial}{\partial r} \int \Phi(|r-r'|) N(r',p',t)dr'dp'. \tag{1.11}$$

Here F^m is the microscopic force.

In order to obtain the generalized kinetic equation on the basis of this dynamic reversible equation, it is necessary, in the first step towards irreversible equation to take into account, the dynamic instability of the motion of atoms.

1.4. The Constructive Role of the Dynamic Instability of the Motion of Atoms.

The kinetic theory of gases is traditionally constructed without using the principal ideas of the dynamic theory - the concepts of dynamic instability, dynamic chaos, K-entropy, mixing (the Bogolubov-Born-Green-Kirkwood-Yvon theory, for example) . Let us show that these enable us to elucidate more fully the causes of irreversibility and to obtain the generalized kinetic equation.

The first steps in this direction were made by N.S.Krylov in 1950 in Ref.[10]dealing with the substantiation of statistical physics. The importance of the dynamic instability of motion in justifying the irreversible equations of the statistical physics is discussed elsewhere by Prigogine [11,12], and in Refs. [13,14].

Let us make some elementary quantitative estimates which connect the time interval τ_{ph}, introduced here to characterize physically infinitesimal time scale, with the minimum time of development of dynamic instability .

The motion of spheres representing atoms in a Boltzmann gas, like the motion of balls in the Sinai billiards, is dynamical unstable. For a single atom the time of development of instability, or, in other words, the characteristic time of exponential divergence of initially close trajectories of two atoms, is of the order of the free path time[13,14]:

$$\tau_{inst} \sim \tau. \tag{1.12}$$

This estimate relates to the path of a single chosen atom. We take into account that in the process of smoothing over the physically infinitesimal volume V_{ph} all N_{ph} particles contained therein are indistinguishable. We can, therefore, introduce along with τ_{inst}, the characteristic time of development of instability for a particle within V. Let us denote the new quantity by $(\tau_{inst})_{min}$. This time is N_{ph} times shorter than τ_{inst}, and hence, given the definition (1.12), we come to the following estimate for the minimum characteristic time of development of dynamic instability related to the motion of atoms in the Boltzmann gas:

$$(\tau_{inst})_{min} \sim \tau_{ph}. \tag{1.13}$$

Thus, the characteristic time for the development of instability of motion per particle within N_{ph} is of the order of the physically infinitesimal time scale τ_{ph}. This is another argument in favor of our method of defining τ_{ph}.

The dynamic instability of motion of atoms in the Boltzmann gas leads to mixing and thus facilitates the transition from the reversible Hamiltonian equations to the much simpler irreversible Boltzmann kinetic equation. Thus, we see the constructive role of the dynamic instability of motion in the formulation of the statistical theory of nonequilibrium processes.

Macroscopic characteristics as, for instance the distribution function, can be termed the *functions of order*, insofar as they single out and describe a more ordered motion against the background of molecular chaotic motion. In other words, they reveal the statistical laws.

This illustration, drawn for the Boltzmann gas, in no way exhausts the constructive role of dynamic instability.

In fact, in statistical theory the dynamic instability can be associated not only with the atoms, but also with the macroscopic characteristics of the system. The latter kind of instability was first discovered with the model equations of thermal convection in a fluid, with the Lorenz equations [13].

An example of a physically feasible system, in which a large amount of feedback gives rise to dynamic instability of the macroscopic characteristics, are the lasers and some electrical devices [15-18, 5, 6].

In connection with the view presented above and concerning the constructive role of dynamic instability of motion of atoms in a Boltzmann gas, some questions arise inevitably. Can the dynamic instability of motion of the macroscopic characteristics also play a constructive role? Will this kind of instability lead to more sophisticated dissipative structures, or to the "dynamic chaos"? In what relation does a physical chaos stand to the dynamic chaos?

To answer these questions, we need criteria for the relative degree of order or organization (or, alternatively, chaoticity) for nonequilibrium states of open systems. We shall use the Boltzmann-Gibbs-Shannon entropy, renormalized to a given mean effective energy - the effective Hamiltonian function of an open system.

The quantitative assessment of the relative order may be based on the S-Theorem. The S-theorem will be used to check on the proper choice of controlling parameters, and then to assess the relative degree of order. It is possible also to organize the optimization of the search for the most ordered states in the space of the controlling parameters of an open system.

Basing on our arguments on the criteria of the relative degree of order, we shall use some examples to demonstrate that the processes of self-organization are also possible in the presence of the dynamic instability of motion of the macroscopic characteristics of open systems. However, we shall first continue to discuss the problem of the generalized kinetic equation.

2. The Unified Description of Kinetic and Hydrodynamic Motion. The Generalized Kinetic Equation [5, 27, 28].

In order to take into account the existence of the dynamic instability of the motion of atoms, we can introduce in the reversible dynamic equation(1.11) the term [27, 28]

$$- \frac{1}{\tau_{ph}} (N(r, p, t) - \tilde{N}(r, p, t)), \tag{2.1}$$

172

which describes the relaxation from the dynamic microscopic phase density $N(r,p,t)$ to the smoothed distribution $\tilde{N}(r,p,t)$. After the averaging over the Gibbs ensemble we can obtain the equation for the distribution function $f(r,p,t) = \langle N(r,p,t)\rangle/n$ with the additional relaxation term

$$\frac{\partial f}{\partial t} + v\frac{\partial f}{\partial r} + F(r,t)\frac{\partial f}{\partial p} = -\frac{1}{n}\frac{\partial\langle\delta F\delta N\rangle}{\partial p} - \frac{1}{\tau_{ph}}(f(r,p,t) - \tilde{f}(r,p,t)$$

(2.2)

$$F(r,t) = F_0 - n\int\frac{\partial\Phi(|r-r'|)}{\partial r}f(r',p',t)dr'dp'.$$

In the right hand side of this equation there are two dissipative terms now. The first one describes the dissipation by processes in the velocity space - by "collisions" [1-6,8]. This term can be represented as the Boltzmann collision integral for a rarefied gas or by the Landau or the Balescu–Lenard "collision integral" for a fully ionized plasmas.

In order to obtain the generalized kinetic equation for the unified description of kinetic and hydrodynamic processes it is naturally to use the more simple form for this "collision integral. Indeed, we see that number of particle $N_{min} \sim \varepsilon^{-5/4}$ and for a rarefied gas is much larger than unity $(N_{min} \approx 10^5$ at atmospheric pressure).Therefore, it is possible to use not the Boltzmann form but the nonlinear Fokker–Planck form for the "collision integral"

$$-\frac{1}{n}\frac{\partial\langle\delta F\delta N\rangle}{\partial p} \equiv I_{(v)}(r,p,t) =$$

(2.3)

$$\frac{\partial}{\partial v}[D_{(v)}(r,t)\frac{\partial f}{\partial v}] + \frac{\partial}{\partial v}[\frac{1}{\tau}(v - u(r,t))f].$$

The diffusion coefficient in the velocity space is defined by the local temperature $T(r,t)$

$$D_{(v)}(r,t) = \frac{1}{\tau}\kappa T(r,t) = \frac{m}{3}\int(v - U(r,t))^2 f(r,p,t)dv, \qquad (2.4)$$

where $u(r,t)$ is a local hydrodynamic velocity, τ is a "collision" time.

In order to obtain the final form for the second "collision integral", we can use the expansion in the "physical Knudsen number"

$$(Kn)_{ph} = \frac{l_{ph}}{L}. \tag{2.5}$$

Here $l_{ph} = L_{min}$. If the mean force $F(r,t)$ is not zero, then it is necessary to change the definition (1.10) for the smoothing function: the mean value $\langle \rho \rangle = (b/m)F\tau_{ph}$ where b is the mobility of atoms.

By the expansion of the additional dissipative term in the equation (2.2) in the "physical Knudsen parameter" (2.5), we have the following expression for the new "collision integral"

$$-\frac{1}{\tau_{ph}}(f - \tilde{f}) \equiv I_r(r,p,t) = \frac{\partial}{\partial r}\left(\frac{L^2_{min}}{\tau_{ph}}\frac{\partial f}{\partial r}\right) - \frac{\partial}{\partial r}\left(\frac{b}{m}Ff\right). \tag{2.6}$$

In order to obtain the final form for the new "collision integral", we must take into account the definitions (1.3), (1.8) for the scales τ, L. Finally we have:

$$\frac{L^2_{min}}{\tau_{ph}} \approx \frac{l^2}{\tau} = D = b\frac{\kappa T}{m}. \tag{2.7}$$

Here D is the space self-diffusion coefficient which is connected with the mobility coefficient by the Einstein relation. For non-equilibrium states the Einstein relation connects local characteristics, therefore the final form of the new "collision integral" is

$$I_r(r,p,t) = \frac{\partial}{\partial r}[D(r,t)\frac{\partial f}{\partial r} - \frac{b}{m}Ff], \quad D(r,t) = b\frac{\kappa T}{m}. \tag{2.8}$$

It is useful to remind that D is one of the three kinetic coefficients: kinematic viscosity ν, temperature conductivity χ, or self-diffusion D. For our model these three coefficients are equal, D $= \nu = \chi$. It is possible to take into account the distinction between

D, ν, χ in the equation of hydrodynamics.

The inclusion of self-diffusion into the equations of hydrodynamics has been suggested more than once (see [5,6]). It was argued that, in the first, self-diffusion distorts the conventional structure of the equations of hydrodynamics, and, in the second, that from the kinetic Boltzmann equation it follows that self-diffusion is absent in the equations of hydrodynamics, since the transfer of matter is completely determined by the convective flow ρu. (Landau and Lifshitz ([8], p. 274 of the Russian edition) observed that the inclusion of self-diffusion into hydrodynamic equations may result in the violation of the condition that the entropy production in a closed system should be positive, thus violating the second law of thermodynamics.

Of course, the last two arguments cannot be dismissed, and in the next sections we shall discuss them.

At this point we just note the following.

In the equilibrium state the kinetic equation (2.2) is satisfied by the Maxwell-Boltzmann distribution. The left-hand side, determined by the nondissipative terms, goes to zero independently of the dissipative terms on the right-hand side, each of which also goes to zero.

The first of the latter, which is either the Boltzmann collision integral or more simple "collision integral" (2.3) through canceling out the collisions of two types, forms the Maxwell distribution.

The second dissipative term on the right-hand side also describes the balance of two dissipative flows. One of these is caused by the external force and is proportional to the mobility, while the second is the flow of matter due to self-diffusion. In this way the existence of the equilibrium Boltzmann distribution can be regarded as a manifestation of self-diffusion.

3. The Equation of Entropy Balance. The Heat Flow [27,28].

The local Boltzmann entropy is defined by the expression

$$S(r,t) \equiv \rho(r,t)s(r,t) = - \kappa n \int \ln(nf(r,p,t))f(r,p,t)dp. \tag{3.1}$$

At the condition F = const it is possible to present the equation of entropy balance for the generalized kinetic equation (2.2) in the following elegant form [27,28]

$$\frac{\partial \rho s}{\partial t} + \frac{\partial}{\partial r}[(\rho u - D\frac{\partial \rho}{\partial r} + \frac{b}{m} F)s] = \frac{\partial}{\partial r}(D\rho \frac{\partial s}{\partial r}) + \sigma(r,t). \tag{3.2}$$

The full entropy production is defined by the expression

$$\sigma(t) = \kappa n \int D_v f(\frac{\partial}{\partial v}\ln\frac{f}{f_{lo}})^2 drdp + \kappa n \int Df(\frac{\partial}{\partial r}\ln\frac{f}{f_o})^2 drdp \geq 0. \tag{3.3}$$

Here f_o is the Maxwell distribution, and the distribution f_{loc} is the local one.

We see that the entropy production is a sum of two positive terms defined, respectively, by the changing of the distribution function in the velocity and in the coordinate spaces.

The entropy flow j_s consists of two parts:

$$j_s(r,t) = j_{con} + j_{dif}. \tag{3.4}$$

The convective part j_{con} is proportional to the full flow of matter and the entropy $s(r,t)$. The diffusion part j_{dif} is proportional to the gradient of entropy $s(r,t)$. It is natural to define the heat flow by the expression [27,28]

$$q(r,t) = T(r,t)j_{dif} = - T(r,t)D\rho \frac{\partial s}{\partial r}. \tag{3.5}$$

For the region of small values of the Knudsen number, if the local equilibrium exists, the heat flow is defined by the gradients of density and temperature:

$$q(r,t) = (\frac{\kappa}{m} D \frac{\partial \rho}{\partial r} - c_v \rho \chi \frac{1}{T} \frac{\partial T}{\partial r})T(r,t), \qquad c_v = \frac{3}{2} \frac{\kappa}{m}. \tag{3.6}$$

Only for slow processes, when the pressure $p = $ const and the gradient of density is proportional to the temperature gradient, the heat flow is defined by the Fourier law

$$q(r,t) = -\lambda \frac{\partial T}{\partial r}, \qquad \lambda = c_p \rho \chi, \qquad c_p = \frac{5}{2}\frac{\kappa}{m}. \tag{3.7}$$

where c_p is a heat capacity at constant pressure.

4. Equations of Hydrodynamics with Self-Diffusion [5,27,28].

The representation of the "collision integral" $I_v(r,p,t)$ in the form (2.3) is based on the fact that the number of the particles in a "point" $N_{min} \sim \varepsilon^{5/4} \gg 1$. By same reason it is possible to restrict the class of distribution functions $f(r,p,t)$ by condition

$$f(r,p,t) = f(r,m|v - u(r,t)|,t). \tag{4.1}$$

Then the closed system of equations for the gas-dynamical functions follows from the generalized kinetic equation (2.2), with "collision integrals" (2.3), and (2.7), without the perturbation expansion for the Knudsen number.

In order to obtain the gas-dynamical equations for the functions $\rho(r,t)$, $u(r,t)$, and $T(r,t)$, we substitute distribution (4.1) into kinetic equation (2.2) and go over to equations for the relevant moments of the distribution (4.1) As a result, we obtain the set of equations of hydrodynamics with due

$$\frac{\partial \rho}{\partial t} + \frac{\partial}{\partial r} j(r,t) = 0, \qquad J = \rho u - D \frac{\partial \rho}{\partial r} + \frac{b}{m} F\rho, \tag{4.2}$$

$$\frac{\partial \rho u_i}{\partial t} + \frac{\partial}{\partial r}(j_j u_i) = -\frac{\partial p}{\partial r} + \rho v \frac{\partial^2 u_i}{2} + \frac{\rho}{m} F, \qquad p = \frac{\rho}{m}\kappa T, \tag{4.3}$$

$$\frac{\partial}{\partial t}\left(\frac{\rho u^2}{2} + \varepsilon(r,t)\right) + \frac{\partial}{\partial r_i}\left[j_i\left(\frac{\rho u^2}{2} + \varepsilon\right) + u_i p - \rho v u_j \frac{\partial u_i}{\partial r_j} - c_v \rho \chi \frac{\partial T}{\partial r_i}\right] = \frac{\rho}{m} Fu.$$

$$\varepsilon(r,t) = \frac{\rho}{m}\frac{3}{2}\kappa T. \tag{4.4}$$

From the equation of continuity it follows that the transfer of matter is now determined by three flows: convective transfer, self-diffusion and the flow defined by the mobility of atoms.

If we take the external force into account, the equation takes the form

$$\frac{\partial \rho}{\partial t} + \frac{\partial}{\partial r}(\rho u - D\frac{\partial \rho}{\partial r} + \frac{b}{m}F\rho) = 0. \tag{4.5}$$

Equations (4.2)-(4.5) take into account the fact that the kinetic coefficients D, v, χ may be not all the same. Observe also that in this approximation the tensor of viscous stress has a different form from the conventional representation. Namely. here

$$\pi_{ij} = -\eta\, \partial u_i/\partial r_j, \quad \eta = \rho v.$$

In the approximation of incompressible fluid, when ρ = const, F = const, and the temperature (the variance of the velocity)is zero, the distribution function $f(r,p,t)$ takes the form

$$f(r,v,t) = \delta[v - u(r,t)], \tag{4.6}$$

and the equations of hydrodynamics coincide with the Navier-Stokes equations

$$\frac{\partial u}{\partial t} + (u\ \text{grad})\ u = -\frac{1}{\rho}\ \text{grad}\ p + v\Delta u + \frac{1}{m}F,$$

$$\tag{4.7}$$

$$\text{div}\ u = 0, \quad \Delta p = -\rho\frac{\partial u_i}{\partial R_j}\frac{\partial u_j}{\partial R_i}.$$

The equation for the pressure follows from the equation of continuity.

We can remark here that the Navier-Stokes equations, although extremely efficient, are intrinsically contradictory, because in this approximation the entropy in a closed system does not change, and at the same time the production of entropy (3.3) is not zero.

For the class of distribution functions (4.1) in order to calculate the entropy production it is necessary to know the solution of kinetic equation. Only for the local Maxwell distribution the entropy production is completely is defined by functions $\rho(r,t)$,

$u(r,t)$, and $T(r,t)$:

$$\sigma(r,t) = \frac{\kappa}{m} \left[D\rho \left(\frac{\text{grad}\rho}{\rho}\right)^2 + \nu\rho \frac{m}{\kappa T} \left(\frac{\partial u_i}{\partial r_j}\right)^2 + \frac{3}{2} \chi\rho \left(\frac{\text{grad}T}{T}\right)^2 \right] \geq 0. \qquad (4.8)$$

Thus, the production of entropy is non-negative also when self-diffusion is taken into account, in full compliance with the second law of thermodynamics.

Observe finally that in the incompressible fluid approximation, from (4.1),(4.8) it follows that the entropy production· is non-zero while the entropy itself remains constant. Therefore, for describing thermal processes we must go beyond the incompressible fluid approximation, for instance, to the Boussinesq approximation: to solve the Navier-Stokes equations together with the heat transfer equation

$$\frac{\partial T}{\partial t} + (v\text{grad})T = \chi \frac{\partial^2 T}{\partial^2 R} + \frac{2}{3} \frac{m}{\kappa} \nu \left(\frac{\partial u_i}{\partial R_j}\right)^2. \qquad (4.9)$$

We should like to underline that the Boussinesq approximation is valid only for the local Maxwell distribution. In more general case it is naturally to use the generalizedzed kinetic equation.

5. Effect of Self-Diffusion on the Spectra of Hydrodynamic Fluctuations [5,6].

To calculate the equilibrium hydrodynamic fluctuations we use the linearized equations of hydrodynamics (4.2)-(4.4). In the linear approximation the velocity field for the Fourier components is represented as a sum of the transverse (with respect to the wave vector k) and the longitudinal parts. The inclusion of self-diffusion has no effect on the fluctuations of the transverse field of velocity, so the width of the spectrum of fluctuations is again determined by νk^2, where ν is the kinematic-tic viscosity.

The set of equations for the Fourier components of the density, the longitudinal velocity, the temperature and the pressure accounts for self-diffusion has the form:

$$(-i\omega + Dk^2)\delta\rho + \rho i(k\delta u) = 0 \; ;$$

$$(-i\omega + \nu k^2)i(k\delta u)\rho = k^2\delta p \; ;$$

$$-i\omega\delta p + pDk^2 \; \frac{\delta\rho}{\rho} = \frac{p}{T} \chi k^2\delta T + \frac{5}{3} pi(k\delta u) \; ;$$

$$p = \frac{\rho}{m} \kappa T, \quad \delta p = \frac{\kappa}{m} \rho\delta T + \frac{\kappa T}{m} \delta\rho.$$

(5.1)

In calculations of fluctuations, two regions are usually distinguished: 1) low frequencies, when (for $D = \nu = \chi$) ω, $Dk^2 <<$ v_{sound}, and 2) high frequencies, when $Dk^2 << \omega \sim kv_{sound}$.

Calculations of equilibrium hydrodynamic fluctuations on the basis of equations of hydrodynamics can be found elsewhere [19],[4]. Here we shall only indicate the basic changes in the spectra which occur when self-diffusion is taken into account.

For low frequencies, the spectra of fluctuations of density, temperature and entropy are now given by [5,6]

$$(\delta\rho\delta\rho)_{\omega,k} = \frac{2Dk^2}{\omega^2 + (Dk^2)^2} \frac{\kappa\rho}{c_p} \; ; \qquad (\delta T\delta T)_{\omega,k} = \frac{2\chi k^2}{\omega^2 + (\chi k^2)^2} \frac{\kappa T^2}{\rho c_p} \; ;$$

(5.2)

$$(\delta S\delta S)_{\omega,k} = \left[\left(1 - \frac{c_v}{c_p}\right) \frac{2Dk^2}{\omega^2 + (Dk^2)^2} + \frac{c_v}{c_p} \frac{2\chi k^2}{\omega^2 + (\chi k^2)^2}\right] c_p \frac{\kappa}{n} \; .$$

(5.3)

When self-diffusion is neglected, the widths of all these spectra are the same and they are determined by the heat conductivity χ. The results are in a rigid correlation between the fluctuations of density and temperature at all temperatures $[\delta\rho(\omega,k)/\rho - \delta T(\omega,k)/T]$. When self-diffusion is included, this condition is removed: the width of the spectrum of density fluctuations is determined by the coefficient of self-diffusion, and that of the temperature fluctuations depends on the coefficient of temperature conduction χ.

180

Total correlation is only observed for characteristics which are integral with respect to ω. This correlation is dictated by the condition of local thermodynamic-dynamic equilibrium.

The spectrum of fluctuations of the entropy is represented as a sum of two spectral terms, the relative contributions of which depend on the ratio between the heat capacities c_v and c_p.

Self-diffusion also affects the spectra of fluctuations at high frequencies. The line width is now determined by the combination-nation of the three dissipative characteristics D, ν, χ:

$$\gamma = \frac{1}{2}\,[\nu + \frac{c_v}{c_p}\,D + [1 - \frac{c_v}{c_p}]\chi]\,k^2.\tag{5.4}$$

Thus, the old problem - whether to include the contribution of self-diffusion into the equations of hydrodynamics - can also be resolved by analyzing experimental data on the spectra of molecular scattering in liquids.

The kinetic approach to the description of hydrodynamic motion described here can be extended to the case of turbulent motion.

6. The Kinetic Approach in the Theory of Self-Organization - Synergetics [26-28,47].

Basic Mathematical Models.

Van der Pol generators, along with more complicated oscillators, can serve as elements of active (excitable) media. The first mathematical models of active media were proposed four decades ago in the well-known works of Wiener and Rosenbluth; Gelfand and Tsetlin (see in [13]). Landau's work (1944) on the origins of turbulence also made a fundamental contribution to the theory of excitable media.

Currently the theory of active media relies mainly on the equations of reaction-diffusion type [20 -28,47]:

$$\frac{\partial \mathcal{X}(R,t)}{\partial t} = F[\mathcal{X}(R,t)] + \frac{\partial}{\partial R_i}\left(\mathbb{D}_{ij}(\mathcal{X})\,\frac{\partial \mathcal{X}}{\partial R_j}\right).\tag{6.1}$$

Here $\mathbb{X}(R,t)$ is a set of macroscopic functions characterizing the system – for instance, the concentrations of chemical reactants; $F(\mathbb{X})$ are nonlinear functions determined by the structure of the system and the nature of the processes; $\mathbb{D}_{ij}$ are the coefficients of spatial diffusion of elements of the system. Here and bellow $r \equiv R$.

Specific examples of equations like (6.1) have been proposed and analyzed by Fisher (see [24]), Kolmomogorov, Petrovsky, Piscunov (1937), Zel'dovich ,Turing. Various modifications of the Ginzburg-Landau equation, which is widely employed in the theory of equilibrium and nonequilibrium phase transitions, also belong to the same class.

Equations of reaction-diffusion type describe a broad class of physical, chemical and biological phenomena. We begin with an example of the spatial diffusion of the independent Van der Pol oscillators (generators).

If in the description of the generators we restrict ourselves to information concerning the energy of oscillations $E(R,t)$, then appropriate Kolmogorov-Petrovsky-Piskunov (KPP) equation for this function with the constant coefficient of space diffusion D has the form

$$\frac{\partial E(R,t)}{\partial t} = (a - bE(R,t))E(R,t) + \mathbb{D}\,\frac{\partial^2 E}{\partial R^2}\,, \qquad a = a_f - \gamma. \qquad (6.2)$$

Here γ, and b are linear and nonlinear friction coefficients, respectively and a_f is the feedback parameter.

This equation itself, however, is not sufficient to give a complete statistical description of the distributed system of generators, since, apart from the spatial diffusion, there is another factor that disturbs the dynamic regime of generation. This is due to the effects of noise on internal degrees of freedom of generators with the intensity η_E. Tho combined influence of both factors is taken into account in the equation for the distribution function $f(E,R,t)$ $(\int dE \frac{dR}{V} f = 1)$ [5,6,26]

$$\frac{\partial f}{\partial t} = \frac{\partial}{\partial E}(D_{(E)} E \frac{\partial f}{\partial E}) + \frac{\partial}{\partial E}[(-a + bE)Ef] + \mathbb{D}\frac{\partial^2 f}{\partial R^2}. \tag{6.3}$$

In the special case only, when the noise acting on the internal degrees of freedom of a generator is negligibly small ($D \equiv 0$), then the equation (6.3) has a particular solution

$$f(E, R, t) = \delta(E - E(R, t)); \quad <E> = \int_0^\infty EfdE = E(R, t), \tag{6.4}$$

which corresponds to the one-moment approximation. The function $E(R, t)$ satisfies the FKPP equation (6.2).

7. Kinetic and Hydrodynamic Description of the Heat Transfer in Active Medium [27,28].

Let us go back to the generalized kinetic equation (2.2) with the collision integrals (2.3), (2.7). Let the mean $u(r, t) = 0$ and therefore the distribution function $f = f(r, |v|, t)$ (see (4.1)). From the kinetic equation the heat transfer equation follows

$$\frac{\partial <E>}{\partial t} = \chi \frac{\partial^2 <E>}{\partial R^2}, \qquad <E> = \frac{3}{2} \kappa T(r, t). \tag{7.1}$$

From the kinetic equation the corresponding equation for the dispersion of temperature may be obtained

$$\frac{\partial}{\partial t}<(\delta E)^2> = \frac{4}{\tau}\left(\frac{2}{3} <E>^2 - <(\delta E)^2>\right) + \chi\frac{\partial^2 <(\delta E)^2>}{\partial R^2} + 2\chi\left(\frac{\partial <E>}{\partial R}\right)^2. \tag{7.2}$$

We see that the source of temperature fluctuations is defined by the gradient of mean temperature $T(r, t)$. Therefore, for the complete description of the heat transfer, it is necessary to use the kinetic equation for the distribution $f(R, |v|, t)$.

If a nonlinear source of heat exists, then the constant $1/\tau$ is replaced by nonlinear dissipative coefficient $\gamma(E)$. We assume (for the illustration only) that

$$\gamma(E) = -a_s + \frac{1}{\tau} + bE, \quad \text{and} \quad D_v(E) = (\frac{1}{\tau} + bE)\kappa T_o. \tag{7.3}$$

Here the coefficient a_s characterizes the source of heat, and T_o is the thermostat's temperature. In the equilibrium state it is Maxwell distribution is with the temperature T_o.

Now we can represent the generalized kinetic equation (2.2) in the form

$$\frac{\partial f(R,|v|,t)}{\partial t} + v\frac{\partial f}{\partial R} = \frac{\partial}{\partial v}[D_{(v)}(E)\frac{\partial f}{\partial v}] + \frac{\partial}{\partial v}[\gamma(E)vf] + \chi\frac{\partial^2 f}{\partial R^2}. \tag{7.4}$$

In one-moment approximation we have the equation of the nonlinear heat transfer

$$\frac{\partial \langle E\rangle}{\partial t} = 2[\frac{3}{2}mD\langle E\rangle - \gamma(\langle E\rangle)\langle E\rangle] + \chi\frac{\partial^2 \langle E\rangle}{\partial R^2}. \tag{7.5}$$

Thus, in this approximation we obtain the reaction diffusion equation type (6.1). Here, however, not only the space diffusion D, but also the internal diffusion $D_{(v)}$ are taken into account. It is possible, therefore, to obtain the solution for all values of the bifurcation parameter a_s.

In the two-moment approximation the corresponding reaction diffusion equation follows from the kinetic equation (7.4).

In order to illustrate the effectiveness of the kinetic approach for description of processes in active medium it is useful to consider a stationary solution. If the functions $\gamma(E)$ and $D(E)$ are defined by the expression (7.3) then we can represent the stationary solution in the form

$$f(v) = exp[-\frac{1}{\kappa T}(\frac{m v^2}{2} - \frac{a_s}{b}\ln(1 + \frac{b}{\gamma}\frac{mv^2}{2}))]. \tag{7.6}$$

It describes the velocity distribution for all values of the control parameter a_s. At $a_s = 0$ it coincides with the Maxwell distribution.

184

8. Kinetic Equation for Active Medium of Bistable Elements.

We have considered some examples of the kinetic equations for active media taking into account both the spatial diffusion and the diffusion with respect to the internal variables of the elements of the medium. Let us suppose that the nonlinear force is defined de the expression

$$F(x) = -m\omega_o^2(1 - a_f + bx^2)x \tag{8.1}$$

Here a_f, and ω_o are the "feedback" parameters and eigenfrequency of the linear oscillator respectively.

Let us suppose also that $f(x,v,R,t)$ is the distribution function of the internal variables x, v, and a space position R of the bistable element. Then we can write the generalized Fokker–Planck equation in the following form:

$$\frac{\partial f}{\partial t} + v\frac{\partial f}{\partial x} + \frac{F(x)}{m}\frac{\partial f}{\partial v} = \frac{\partial}{\partial v}\left(D_{(v)}\frac{\partial f}{\partial v}\right) + \frac{\partial}{\partial v}(\gamma vf) + \tag{8.2}$$

$$\frac{\partial}{\partial x}[D_{(x)}\frac{\partial f}{\partial x} - \frac{F(x)}{m\gamma}f] + D\frac{\partial^2 f}{\partial R^2}.$$

We see that there are two internal diffusion coefficients:

$$D_{(v)} = \gamma\frac{\kappa T}{m}, \qquad D_{(x)} = D_{(o)}(1 + b x^2) \qquad (D_{(o)} = \frac{\kappa T}{m\gamma}) \tag{8.3}$$

and the diffusion coefficient D of elements in space R .For the equilibrium state the Maxwell–Boltzmann distribution is the solution of the kinetic equation (8.2).

If we restrict the class of the distribution functions (like in (4.1)) by condition

$$f(x,v,t) = f(x,|v|,t), \tag{8.4}$$

then one can obtain, by integration with respect to v, the corresponding generalized Einstein–Smoluchowsky equation for the function $f(x,,t)$

$$\frac{\partial f}{\partial t} = \frac{\partial}{\partial x}[D_{(o)}(1+bx^2)\frac{\partial f}{\partial x}] - \frac{\partial}{\partial x}[m\omega_o^2(1-a_f+bx^2)xf] + D\frac{\partial^2 f}{\partial R^2}. \tag{8.5}$$

It is useful to remind that in the theory of stochastic processes the transition from the Fokker-Planck (the Kramers) equation to the Einstein-Smoluchowsky is done by applying perturbation method with respect to the small parameter $F'/m\gamma^2$ ($\omega_0^2/\gamma^2 \ll 1$) [29-31,6]). This method corresponds to the Hilbert, Chapman-Enskog, and Grad perturbation theories for the Boltzmann equation on small Knudsen number.

We see that also in the stochastic theory it is possible to avoid using of the perturbation theory on the corresponding small "Knudsen parameter".

It is interesting to consider some limiting cases.

1). The distribution function for a stationary homogeneous state

$$f_{st}(x) = C\exp[\frac{U_{eff}(x)}{\kappa T}], \qquad U_{eff} = \frac{m\omega_0^2}{2}[x^2 - \frac{a_f}{b}\ln(1+bx^2)]. \qquad (8.6)$$

At $a_f = 0$ it coincides with the Boltzmann distribution for $U = m\omega_0^2/2$.

2). The space distribution function $f(R,t)$. From (8.5) the self-diffusion equation follows:

$$\frac{\partial f}{\partial t} = D\frac{\partial^2 f}{\partial R^2}, \qquad \int f(R,t)\frac{dR}{V}=1. \quad (8.7)$$

3). The one-moment approximation for incompressible medium

If $D \equiv 0$, then the distribution function have the form

$$f(x,R,t) = \delta(x - x(R,t)), \qquad x(R,t) = \langle x\rangle. \qquad (8.8)$$

For the "field function" $x(r,t)$ from the generalized Einstein-Smoluchowsky equation one gets the FKPP type equation:

$$\frac{\partial x(R,t)}{\partial t} = \underline{G}[a_f - 1 + bx^2(R,t)]x(R,t) + D\frac{\partial^2 x(r,t)}{\partial R^2}. \qquad \underline{G} = \frac{\omega_0^2}{\gamma} \qquad (8.9)$$

In this approximation the internal diffusion coefficient $D_{(x)} = 0$. The diffusion $D_{(x)}$ may play, however, an important role.

Let us suppose that the space diffusion is one-dimensional, in the direction ζ, then there exists the following well-known

186

stationary solution of the equation (8.9)

$$x(\xi) = (\frac{a_f - 1}{b})^{1/2} \text{ th } [(\frac{a_f - 1}{2D})^{1/2}\xi].$$

(8.10)

The relative width of the front is determined by the expression: $(2D/(a_f - 1))^{1/2}$. We see that this result is not valid in the region near the bifurcation point $a_f = 1$.

In order to obtain a more general solution for all values of the bifurcation parameter, it is necessary to use the kinetic equation [26].

9. Kinetic fluctuations in active media.

9.1. The Langevin source in the kinetic equation.

The "collision integrals" are defined by the small-scale (kinetic) fluctuations. We shall now consider the large-scale (kinetic) fluctuations, which dynamics is determined by the kinetic equations themselves.

There are two well-known methods of calculating the kinetic fluctuations [1-6, 19, 29, 32-35]. One of them is based on the approximate solution of the set of equations for the moments of random (pulsating) distribution functions. The second one is based on solving the corresponding Langevin equations for the random distribution function $\tilde{f}(x, R, t)$ and this is what we are going to use here.

First we introduce a Langevin source $y(r, x, t)$ into the kinetic equation (8.5) for the active medium of bistable elements. In the Gaussian approximation the two moments of a Langevin source are given by the following expressions [5, 26]:

$$\langle y(x, R, t)\rangle = 0, \quad \langle y(x, R, t)y(x', R', t')\rangle =$$

(9.1)

$$2[\frac{\partial^2}{\partial x \partial x'} D_{(x)} + D \frac{\partial^2}{\partial R \partial R'}] \frac{1}{n} \delta(x-x')\delta(R-R')f(x, R. t)\delta(t-t').$$

The distribution function $f=\langle f\rangle$ is defined by the equation (8.5).

These general expressions gives possibility to find the Langevin sources in the diffusion equation for the space distribution $\tilde{f}(R,t)=\int \tilde{f}(x,R,t)dx$ and in the FKPP equation (8.9).

9.2. Spatial Diffusion. "Tails" in the Time Correlations.

From the kinetic equation (8.5) with the Langevin source (9.1) it follows the equation for fluctuations of space distribution

$$\tilde{n}(R,t) = n\tilde{f}(R,t) \quad (n = N/V, \quad n(R,t) = \langle\tilde{n}(R,t)\rangle)$$

$$\frac{\partial\delta n}{\partial t} = D \frac{\partial^2 \delta n}{\partial R^2} + y(R,t), \qquad \langle y(R.t)\rangle = 0,$$

$$\langle y(R,t)y(R',t')\rangle = 2D \frac{\partial^2}{\partial R\partial R'} \delta(R-R') \, n(R,t)\delta(t-t') \tag{9.2}$$

In this way, we have come to the well-known equation of spatial diffusion with a random source. The distribution $n(R,t)$ is a solution of the diffusion equation (8.7).

In the equilibrium state the spectral density of fluctuation is defined by the well-known expression

$$(\delta n\delta n)_{\omega,k} = \frac{2Dk^2}{\omega^2 + (Dk^2)^2} \, n. \tag{9.3}$$

Using the Fourier transformation over ω and k we find the expression for the space-time correlation of the fluctuations δn:

$$\langle\delta n\delta n\rangle_{r,\tau} = \frac{n}{(4\pi D|\tau|)^{3/2}} \exp\left(-\frac{r^2}{4D|\tau|}\right), \qquad r = R-R', \quad \tau = t-t'. \tag{9.4}$$

We see that at $r=0$ the time correlations decay according to a power law: $\propto 1/|\tau|^{3/2}$.

It is necessary to underline, that the results presented here were obtained without taking into account the boundary conditions. In other words, they are valid only for an infinite medium. If the size of system L is finite, then these results are valid only at conditions

$$\omega \gg \tau_D^{-1}, \qquad \tau \ll \tau_D = L^2/D. \tag{9.5}$$

The role of the boundaries is considered in the section 10.

9.3. The Langevin Source in Reaction Diffusion (FKPP) Equation.

For the incompressible medium the integral $\int f(x, R, t) = \text{const}$, therefore the correlator of the source $y(R, t)$ in the diffusion equation -the correlator (9.2), now is zero, but the moments of the Langevin source $y_{(x)}(R, t) = \int xy(x, R, t)dx$ are not zero and they are defined by the expressions

$$\langle y_{(x)}(R, t)\rangle = 0, \qquad \langle y_{(x)}(R, t)y_{(x)}(R', t')\rangle =$$

$$(9.6)$$

$$2[D_{(0)}(1 + b\langle x^2\rangle_{R, t}) + D\frac{\partial^2}{\partial R\partial R'}\langle x^2\rangle_{R, t}] \frac{1}{n} \delta(R-R')\delta(t-t').$$

In order to obtain a closed expression for this correlator it is necessary to use the solution of the generalized Einstein-Smoluchowsky equation (8.5). The situation is much more simple when it is possible to use the stationary solution (8.6). In this case $\langle x^2\rangle$ in (9.6) is defined for all values of the bifurcation parameter a_f by the following expression:

$$\langle x^2\rangle = C \int x^2\exp[- \frac{m\,\omega_o^2}{2\kappa T}(x^2 - \frac{a_f}{b}\ln(1 + bx^2)]dx. \qquad (9.7)$$

Thus, in the stationary state (9.9) the intensity of noise is defined by the mean-square value $\langle x^2\rangle$, therefore to calculate the fluctuation $\delta x(R, t)$ it is possible to use the self-consistent approximation on $\langle x^2\rangle$ [3, 4, 26]

$$\frac{\partial\delta x(R, t)}{\partial t} + \underline{G}_{(x)}\delta x(R, t) - D\frac{\partial\delta x(R, t)}{\partial R^2} = y(R, t). \qquad (9.8)$$

With the help of this equation and the definition (9.6) for the intensity of noise we obtain the following expression for the spectral density of fluctuations

$$(\delta x \delta x)_{\omega, k} = \frac{2(\gamma_{(x)} + Dk^2)}{\omega^2 + (\gamma_{(x)} + Dk^2)^2} \frac{\langle x^2 \rangle}{n}. \tag{9.9}$$

We introduced here the special definition for the half width of a spectral line determined by the internal diffusion $D_{(o)}$ in any bistable element:

$$\gamma_{(x)} = \frac{D_{(o)}}{\langle x^2 \rangle} (1 + b\langle x^2 \rangle). \tag{9.10}$$

This definition is valid for all values of the bifurcation parameter a_f.

It is necessary to underline that the results presented in this section and the corresponding results of the previous section, are valid only for an infinite medium - the condition (9.5).

10. Natural Flicker Noise ("1/f Noise").

10.1. Natural Flicker Noise for Diffusion Processes [4,5,36-38].

For a fluctuative diffusion process we can distinguish two domains: $\tau_{cor} \ll \tau_D$; $\tau_{cor} \gg \tau_D$. For the first region the size of system has little effect on the fluctuations. The spectrum in this case practically coincides with the spectrum(9.3).

Natural flicker noise exists in the low-frequency region

$$\frac{1}{\tau_{obs}} \ll \omega \ll \omega_{max} = \frac{1}{\tau_D} = \frac{D}{L^2}. \tag{10.1}$$

Thus, the upper limit of the region is determined by the diffusion time. The minimum frequency is determined by the time of observation τ_{obs} which is limited only by the lifetime of the device.

From (10.1) it follows that the actual volume of diffusion $V_\omega \equiv L_\omega^3$ depends upon the frequency and is much larger than the volume V of the sample,

$$V_\omega \equiv L_\omega^3 = (D/\omega)^{3/2} \gg V. \tag{10.2}$$

190

Under this condition the size of a system is not important, and in the limit a sample can be treated as a point (dimension zero).

In order to find the spectral density of the Langevin source, it is necessary to take into account that the diffusion volume V_ω for the region, where flicker noise is V_ω/V times larger than volume of the sample V. Accordingly, the mean concentration n is replaced by the effective concentration [36]

$$n \to n_{eff} = A V_\omega \langle \delta n_V \delta n_V \rangle. \tag{10.3}$$

Here we have used δn_V to denote the one-point correlator of fluctuations, smoothed over the volume of the considered sample V. The constant A will be defined below. As a result, the expression for the spectral density of the Langevin source now takes the form

$$(yy)_{\omega,k} \equiv 2Dk^2 n_{eff} \exp\left(-\frac{L_\omega^2 k^2}{2}\right). \tag{10.4}$$

The corresponding expression for the spectral density of δn:

$$(\delta n \delta n)_{\omega,k} = \frac{2Dk^2}{\omega^2 + (Dk^2)^2} A V_\omega \langle \delta n_V \delta n_V \rangle \exp\left(-\frac{L_\omega^2 k^2}{2}\right). \tag{10.5}$$

From these expressions, the variance of the distribution of wave vector is $\langle(\delta k)^2\rangle = L_\omega^{-2} = \omega/2D$ and tends to zero as $\omega \to 0$. In this way, we come to the coherent distribution in the space of wave numbers.

After integrating over k, we find the expression for the temporal spectral density of natural flicker noise

$$(\delta n \delta n)_\omega = \frac{\pi}{\ln(\tau_{obs}/\tau_D)} \frac{\langle \delta n_V \delta n_V \rangle}{\omega}, \qquad \tau_{obs}^{-1} \ll \omega \ll \tau_D^{-1}. \tag{10.6}$$

The constant A in (10.5) is defined here from the condition

$$\int_{1/\tau_{obs}}^{1/\tau_D} (\delta n \delta n)_\omega \frac{d\omega}{\pi} = \langle \delta n_V \delta n_V \rangle. \tag{10.7}$$

Thus we assume that the main contribution to the integral with respect to ω comes from the region of flicker noise. This expression can be written in the form

$$\frac{(\delta n \delta n)_\omega}{n^2} = \frac{2\pi a}{N\omega}, \qquad a = \frac{1}{2\,\ln(\tau_{obs}/\tau_D)} \cdot \frac{\langle \delta n_V \delta n_V \rangle}{n/V}. \qquad (10.8)$$

Here we have introduced the notation a for Hoog's constant [39]:

The time correlator for the region of natural flicker noise is given by the expression [36]

$$\langle \delta n \delta n \rangle_\tau = [C - \ln(\tau/\tau_D)/\ln(\tau_{obs}/\tau_D)]\langle \delta n_V \delta n_V \rangle \ ;$$

$$\tau_D \ll \tau \ll \tau_{obs}, \qquad C = 1 - \gamma/\ln(\tau_{obs}/\tau_D), \qquad \gamma = 0,577. \qquad (10.9)$$

We see that the dependence on τ in the region of flicker noise is very weak (logarithmic with a large argument), and therefore we may speak of the residual time correlations.

The Langevin equation for the Fourier component $\delta n(\omega, k)$ can be presented in the form

$$(-\,i\omega + Dk^2)\delta n(\omega, k) = y(\omega, k). \qquad (10.10)$$

The spectral density of the source is given by (10.4).

Since the distribution over the wave numbers in the region of flicker noise is very narrow, in (10.10) we may replace

$$k^2 \to \overline{k^2}^{\,k} = L_\omega^{-2} = \omega\,/\,D \qquad (10.11)$$

and use a simpler equation for the fluctuations $\delta n(\omega)$:

$$(-\,i\omega + \gamma_\omega)\,\delta n(\omega) = y(\omega); \qquad \gamma_\omega - |\omega|. \qquad (10.12)$$

The expression forthe spectral density of the Langevin source:

$$(yy)_\omega = 2\gamma_\omega \frac{\pi}{\ln(\tau_{obs}/\tau_D)}\langle \delta n_V \delta n_V \rangle; \qquad \gamma_\omega = |\omega|. \qquad (10.13)$$

Thus, for the region of natural flicker noise the dissipative coefficient $\gamma_\omega = |\omega|$ and tends to zero as $\omega \to 0$. This expression relates the spectral density to the dissipative coefficient and the one-time correlator, and is therefore an FDR for the region of natural flicker noise.

In this way the principal results of the author's theory of flicker noise [36-38,4,5] have been formulated. The natural flicker noise arises, whenever the final stage of relaxation towards the equilibrium state is associated with spatial diffusion.

The dependence on the actual structure of the system enters only via two parameters: the time of diffusion τ_D and the one time correlator $\langle \delta n_v \delta n_v \rangle$. Recall that we are considering the diffusion of a physical entity of any kind.

This theory has been employed elsewhere to explain the existence of natural flicker noise in music, as observed by Voos and Clark [40]. We considered the possible connection between natural flicker noise and superconductivity [38,5]. So far, these two fundamental phenomena have been treated independently. We demonstrated, however, that the existence of low-frequency natural flicker noise in a system of charged Bose particles results in vanishing electrical resistance and thus facilitates the existence of a permanent superconductivity current and the Meissner-Oxenfeld effect.

The transition from the normal to the superconducting state is a well-known example of a phase transition of the second kind, which results in the appearance of a macroscopic quantity of charged Bose particles (Cooper pairs).

In order to understand the important role of natural flicker noise in a system of Cooper pairs for the existence of permanent superconducting electrical current, it is necessary to take into account that the appearance of natural flicker noise, according to this theory, implies the creation of a coherent state in the space of wave numbers at $\omega \Rightarrow 0$. In this way, we are talking about the link between two coherent processes.

We can remark that the natural flicker noise exists not only in a system of Cooper pairs. In a system of neutral Bose particles (He^4) the existence of natural flicker noise makes superfluidity in narrow gaps possible. The coefficient of diffusion is then of the order of Planck's constant h.

Naturally, there are also "technical " noise with 1/f spectrum, which are associated, for example, with mobile defects.

It is very interesting also to remark that after integrating over ω (not over k!), from (10.5) we find the spatial natural flicker noise "1/k". Thus, it is exists not only the temporal, but also and the spatial natural flicker noise - the noise"1/|k|".

10.2. Natural Flicker Noise for Reaction-Diffusion Processes.

We see that for statistical description of processes in active media it is more naturally to use the corresponding generalized kinetic equations, instead of the reaction-diffusion equation (6.1). We see also that such approach gives possibility for description of kinetic fluctuations.

One of the methods of description of kinetic fluctuations is based on the solution the corresponding Langevin equations. For example, the moments of the Langevin source in the Einstein-Smoluchowsky equation for active media of bistable elements are defined by the expressions (9.1), (9.2).

We used these general expressions in order to define the Langevin source in the generalized reaction -diffusion equation for the "field" variable $x(R,t)$. As a result, we obtained the expression (9.9) for the spectral density of fluctuations. This result is valid for all values of the bifurcation parameter a_f.

As for the diffusion processes (see (9.3)) this result was obtained without taking the boundary conditions into account, and it is valid only for an infinite medium - for the region of frequencies $\omega \gg \tau_D^{-1} - D/L^2$.

For diffusion systems the natural flicker noise exists in the low-frequency region (10.1). For the reaction-diffusion systems there exists not only dissipation determined by diffusion, but also an additional internal dissipation. If the corresponding coefficient $\gamma_{(x)} \ll \tau_D$, then it exists the low-frequency region of the flicker noise

$$\gamma_{(x)} = \omega_{min} \ll \omega \ll \omega_{max} = 1/\tau_D = D/L^2 \qquad (10.14)$$

with ω_{min} is defined by the friction coefficient $\gamma_{(x)}$.

In order to find the corresponding spectral function, it is necessary to change the second term in (9.6) in correspondence with the definition (10.4). After integration over k we again obtain the expression for the temporal spectral density in the low-frequency region (10.14)

$$(\delta x \delta x)_\omega = \frac{2(\gamma_{(x)} + |\omega|)}{\omega^2 + (\gamma_{(x)} + |\omega|)^2} A <\delta x_v \delta x_v>. \qquad (10.15)$$

The constant A is defined by the expression:

$$A = \frac{4}{\pi} \frac{1}{\gamma_{(x)}} \int^{1/\tau_D} \frac{\gamma_{(x)} + |\omega|}{\omega^2 + (\gamma_{(x)} + |\omega|)^2} d\omega. \qquad (10.16)$$

The dissipative coefficient $\gamma_{(x)} \equiv \omega_{min}$ is defined by the expression (9.12) for all values of the bifurcation parameter a_f.

11. Conclusion. Associative Memory and Pattern Recognition.

Since the well-known work by Hopfield (1982) the problems of associative memory and pattern recognition occupy an important place in statistical physics, in particular, in the theory of spin glasses and in optical systems.

The models considered in the theory of associative memory and pattern recognition are mostly based on discrete neural networks, each neuron possesses a number of discrete states. Haken (1988) suggested to treat the recognition of patterns as a process similar to the formation of dissipative structures in synergetic systems. In the proposed models the dynamics of the system is determined by a potential function which depends on some parameters. This dependence may be used for storing the information concerning a pattern to be recognized. The test patterns are introduced via the initial components of the vector of a dynamic system. Systems of this kind exhibit the phenomenon of associative memory since they are capable of restoring the complete pattern in the process of evolution towards a stationary state, starting with the incomplete information contained in the initial conditions [41,42].

Let us demonstrate that the simplest medium capable of associative memory can be constructed from active elements, which are Van der Pol generators interacting via common feedback. Such a system is practically feasible.

Consider the following set of equations [26]:

$$\frac{dX_i}{dt} + \frac{1}{2} \sum_{j=1}^{N} (-a_j + b_j E_j) X_j = V_i, \quad i = 1, 2, \ldots, N),$$

(11.1)

$$\frac{dV_i}{dt} + \frac{1}{2} \sum_{j=1}^{N} (-a_j + b_j E_j) V_j + \omega_i^2 X_i = 0, \quad E_i = \frac{1}{2}\left(V_i^2 + \omega_i X_i^2\right)$$

The generators are linked together via the common feedback. In the general case all the coefficients a_i, b_i are different. Observe that the excitation of all generators in the system only requires that X_{oi}, V_{oi} should be nonzero for at least one generator.

Introducing the "dissipative potential" of a system of generators

$$U = \sum_i U_i, \quad U_i = \frac{b_i}{2}\left(E_i - \frac{a_i}{b_i}\right)^2,$$

(11.2)

we may write these equations in the form

$$\frac{dX_i}{dt} + \frac{1}{2} \sum_j \frac{\partial U}{\partial X_j} \frac{1}{\omega_j^2} = V_i; \quad \frac{dV_i}{dt} + \frac{1}{2} \sum_j \frac{\partial U}{\partial V_j} + \omega_i^2 X_i = 0.$$

(11.3)

In the stationary state the dissipative potential (11.2) goes to zero.

The stationary solution with fixed initial phases ϕ_{oi} depends on N parameters a_i/b_i. This dependence can be used for storing information about a pattern of N points. For restoring the pattern it is sufficient to introduce into the initial parameters the information about at least one point of the pattern. Then in the process of evolution towards the stationary state the system will restore complete information about the pattern, installed into the stationary state via the values of the parameters a_i/b_i.

Obviously, this model is just a very simple illustration.

Possible generalizations may proceed in various directions. One of then is, for instance, the statistical generalization, when a state of the system is characterized by the "N-particles" distribution function f_N. We enter then the domain of the statistical theory of "nonideal" active systems with interaction of different "particles", which can simulate the processes of associative memory and pattern recognition due to fluctuative processes involving both the internal variables of the system's macroscopic elements and the variables which describe the motion of each element as a whole [5,6].

References

1. Klimontovich Yu.L.,TMF, 8, 109 (1971).

2. Klimontovoch Yu.L., Kinetic Theory of Non Ideal Gases and Non Ideal Plasmas. Nauka, Moscow 1975; Pergamon Press Oxford 1982.

3. Klimontovich Yu.L., The Kinetic Theory of Eltctromagnetic Processes. Nauka, Moscow 1980; Springer, Berlin Heidelberg 1983.

4. Klimontovich Yu.L., Statistical Physics. Nauka Moscow 1982; Harwood Academic Publishers, New York 1986.

5. Klimontovich Yu.L., Turbulent Motion and the Structure of Chaos. Nauka, Moscow 1990; Kluwer Acad. Pub., Dordrecht 1991.

6. Klimontovich Yu.L., Statistical Theory of Open Systems. Kluwer Academic Publishers, Dodrecht (in press).

7. Monin A.S., Yaglom A.M., Statistical Fluid Mechanics. Nauka, Moscow 1965; MIT 1971.

8. Landau L.D., E.M.Lifshits. Fluid Mechanics. Nauka, Moscow 1986; Pergamon Press, Oxford, 1959.

9. Klimontovich Yu.L., Statistical Theory of Non Equilibrium Processes in a Plasma. Nauka, Moscow 1964; Pergamon press, Oxford 1967.

10. Ktylov N.S., Works for the Foundation of Statistical Physics, Moscow, Nauka;

11. Prigogine I., From Being to Becoming. Freemam, San Francisco 1980.

12. Prigogine I., Stengers I., Order out of Chaos, Heinemann, London, 1984.

13. Romanovski Yu.M., Stepanova N.V., Chernavsky D.S., Mathemastical

Biology. Nauka, Moscow 1984.

14. Klimontovich Yu.L., Physica, 142A, 390 (1987).

15. Lorenz E., J.Atm. Sci. 20, 167 (1963).

16. Haken H., Synergetics. Springer, Berlin, Heidelberg 1978.

17. Anishchenko V.S., Complex Oscillations in Complex Systems. Nauka, Moscow 1990.

18. Neimark Yu.I., Landa P.S. Stochastic and Chaotic Oscillations. Nauka, Moscow 1987; Kluwer Acad.Publ., Dordrecht 1992.

19. Lifsits E.M., Pitaevsky L.P., Statistical Physics. Nauka, Moscow 1978.

20. Nicolis G., Prigogine I., Self Organization in Non Equilibrium Systems. Wiley, New York 1977; Mir, Moscow 1979.

21. Haken H., Advanced Synergetics. Springer, Berlin Heidelberg 1983;

22. Michailov A.S., Foundations of Synergetics I. Springer, Berlin Heidelberg 1990.

23. Michailov A.S., Loskutov A.Yu., Foundations of Synergetics II. Springer, Berlin Heidelberg 1991.

24. Murray G., Lectures on Nonlinear Differential Equations Models in Biology. Clarendon Press, Oxford 1977.

25. Vasiliev V.A., Romanovsky Yu.M., Yachno V.G. Autuwaves. Nauka, Moscow 1987.

26. Klimontovich Yu.L., Physica 179A, 471 (1991).

27. Klimontovich Yu.L., TMF92, 312 (1992).

28. Klimontovich Yu.L., Physics Let.A 170, 434(1992).

29. Van Kampen N.G., Stocastic Processes in Physics and Chemistry. North-Holland Amsterdam, 1983.

30. Risken H., The Fokker-Planck Equation. Springer Berlin, 1984.

31. Gardner C.W., Handbook of Stochastic Methodsfor Physics, Chemistry, and Natural Sciences.Springer, Berlin, Heidelberg, 1984.

32. Gantsevich S.V., Gurevich V.L., Katilus R., Rivista del Nuovo Cimento, 2,1 (1979).

33. Kogan Sh.M., Shul'man A.Ya., ZhETF, 56, 862 (1969).

34. Lifshits E.M., Pitaevsky L.P., Statistical Physics, (Nauka, Moscow 1978).

35. Keizer J., Statistical Thermodynamics of Nonequilibrium Processes.

(Springer, Berlin, Heidelberg, New York 1987).

36. Klimontovich Yu. L., Pis'ma v ZhTF, 9, 406 (1983).

37. Klimontovich Yu. L., Boon J. P., Europhys. Lett. 3(4) 395 (1987).

38. Klimontovich Yu. L., Pis'ma v ZhETF, 51(1), 43 (1990).

39. Kogan Sh. M., Usp. Fiz. Nauk 145, 285 (1985); Sov. Phys. Usp. 28, 171 (1985).

40. Voos R. F., Clarke J., J. Acoust. Soc Am. 643(1), 258 (1978).

41. Haken H. Synergetic Computers and Cognition. A Top-Down Approach to Neural Nets. Springer-Verlag, Berlin, Heidelberg, 1991.

42. Fuchs A., Haken H., in "Neural and Synergetic computers", ed. by H. Haken, Springer-Verlag, Berlin, 1988, p. 16.

43. Akhromeeva T. S., Kusdumov S. P., Malinetskii G. G., Samarskii A. A, Chaos and Sissipative Structures in "Reaction-Diffusion" Systems. Nauka, Moscow, 1992.

A MICROSCOPIC ANALYSIS OF A TWO-DIMENSIONAL CHEMICAL WAVE FRONT

A. Lemarchand, H. Lemarchand, A. Lesne

Laboratoire de Physique Théorique des Liquides, CNRS URA 765
Boîte 121, Université Pierre et Marie Curie
4 place Jussieu, 75252 Paris Cedex 05, France

M. Mareschal

Département de Chimie-Physique, Université Libre de Bruxelles
Campus Plaine, Bvd du Triomphe, 1050 Bruxelles, Belgium

ABSTRACT: A reactive lattice gas cellular automaton model is used to simulate a chemical wave front propagating in a two-dimensional (2D) medium. The corresponding macroscopic description is given by a reaction-diffusion equation first studied by Fisher and also by Kolmogorov, Petrovsky and Piskunov in the 1D case. The computed values of mean front properties such as the propagation velocity and the profile width agree with the 1D macroscopic values.

The microscopic interface between the two reactive species exhibits a self-similar character. Its fractal structure, described through several fractal dimensions, is shown to be independent of the reaction and diffusion parameter values. Moreover, it exhibits a two-scale organization.

A large variety of phenomena encountered in chemistry [1], biology [2] or materials science [3] are commonly described by non linear reaction-diffusion equations leading to complex space-time behaviors. Among these, wave front propagation has been subject of both theoretical and experimental interests. Much is known about propagating solutions in a 1D medium [4–7]. It is easy to check that in a 2D medium [8,9], proper 1D initial conditions will generate only effective 1D, macroscopic and purely deterministic wave solutions.

Departing from the pioneering approaches, we derive the present discussion from a microscopic simulation based on a lattice gas model. Such a microscopic analysis of a reactive wave front have already been performed on different chemical schemes [10,11]. The mean results we obtained agree with the macroscopic predictions; the simulation will also provide information on the microscopic mechanisms through the evidence of the fractal structure of the reactive interface and of its length scales.

1. SOME RESULTS OF THE MACROSCOPIC THEORY

We consider a two-dimensional (2D) system containing two chemical species A and B which may diffuse with an identical coefficient $\mathcal{D}$ and react according to the following autocatalytic reaction:

$$A + B \longrightarrow A + A \tag{1}$$

From a macroscopic point of view, $a = a(x,y,t)$ denotes the local mean fraction of particles A; imposing that the total local number density n does not vary with space (x,y) and time t, the deterministic evolution of the system[1] hence reduces to the following reaction-diffusion equation:

$$\frac{\partial a}{\partial t} = Ka(1-a) + \mathcal{D}\left(\frac{\partial^2 a}{\partial x^2} + \frac{\partial^2 a}{\partial y^2}\right) \tag{2}$$

where K is proportional to the kinetic constant of reaction Eq. (1). The constraint on the mean number density n is achieved by taking a spatially uniform initial number density $n_{(t=0)}$, since Eq. (1) does not modify the total local number of particles and since the two species have the same diffusion coefficient $\mathcal{D}$.

This equation was first studied by Fisher [4] and Kolmogorov, Petrovsky, and Piskunov [5] and will be further referred as FKPP equation. One may easily verify that any solution of Eq. (2), spatially analytical for any fixed time and with initial condition independent of y is analytical both in space and time and stays independent of y, so that the evolution reduces to those in a 1D medium. Hence one proves [4–6] that Eq. (2) admits uniformly translating solutions $a(x - vt)$, independent of y as soon as the initial (at time $t = 0$) condition does, moving in the direction x with a constant velocity v and replacing the unstable uniform and stationary state $a \equiv 0$ by the stable state $a \equiv 1$. An essential feature of Eq. (2) is the existence of solutions for any propagation velocity v greater than a minimum value equal to v_{min}:

$$v_{min} = 2\sqrt{K\mathcal{D}} \tag{3}$$

---

[1]Let us note that Eq. (2) is also convenient to describe any reaction:

$$A + B + G_1 \longrightarrow A + A + G_2$$

where G_1 and G_2 are gaseous species, which concentrations, or equivalently partial pressions, are kept equal to a prescribed value.

A large class of sufficiently steep profiles, including a step function, evolve to a uniformly translating solution $a(x - v_{min}t)$, propagating with the minimum allowed velocity.

The analytical expression of this asymptotic profile is not known but an approximate value s of the steepness at the inflexion point (which unicity can be proved) can be used to define [2] a front width as:

$$e_{th} = \frac{1}{s} \approx 8\sqrt{\frac{D}{K}} \tag{4}$$

In a 2D medium with initial conditions independent of y, the deterministic equation of evolution reduces to those of a 1D propagation front. This reduction to an effective dimension one is not straightforwardly possible for a 2D lattice gas cellular automaton. Our first objective is thus to compare the theoretical results of the 1D deterministic front with the corresponding results deduced for the 2D simulation. The links between this macroscopic theory and the mean results deduced from our microscopic simulation will be discussed in Section 3, after the description of the computational method in Section 2.

2. COMPUTATIONAL ASPECTS

We use a lattice gas cellular automaton model first introduced by Hardy, de Pazzis, and Pomeau known as the HPP model [12–14]. We consider a binary mixture with an equal number of particles A and B moving on a square lattice at integer times; note that the total number of particles remains constant under the reaction given in Eq. (1). We choose units such that lattice constant and velocity modulus equal one. No more than one particle is to be found at given time and node, moving in a given direction according to the so-called exclusion principle. Denoting by m the number of possible velocity directions at a node ($m = 4$ here), we define the reduced number density ρ as the average number of particles (of any species) per node divided by m. In order to ensure a step function initial profile, particle species is set to A in the left half of the lattice whereas it is set to B in the right half. A two-step cellular automaton updating rule is defined on the Boolean field of node states.

Velocities of the automaton particles can take only four values, of same modulus rescaled to 1, along each direction of the lattice. Step one is propagation along the direction of particle velocities.

Step two is collision. The collision rules respect microscopic laws, preserve (statistical) spatial homogeneity and isotropy, and leave invariant the total number of particles. All different configurations, including those obtained by permutation of the different chemical species, are considered for collisions between 2, 3 and 4 particles. When several configurations are accessible after a collision, the final state is randomly chosen among them. These two steps should then be considered as an effective tool designed to achieve the same statistical mixing as those present in the real medium. This mixing, hence the associated step, is essential to ensure the statistical stationarity and spatial homogeneity of the evolution, and support all the hypothesis upon the uniformity properties of the automaton we shall make. Let us emphasize that we mimic the microscopic diffusion of "real" particles as in a gaseous phase.

This approach contrasts with the three steps reactive lattice gas model first introduced by Kapral et al. [12]. In order to mimic the action of a real solvent, their simulation involves an additional step in which the velocities of the "real" particles are randomized; this average over microscopic scales defines "mesoscopic" particles, which are then propagated.

The occurrence of a reaction is restricted to binary collisions, in order to simulate a quadratic kinetics. For every binary collision taking place between particles of different species, the collision is said to be reactive (which implies the transformation of the colliding particle B into a particle A) if $K_{micro} \geq R$, where R is a random number sampled from a uniform distribution between 0 and 1 and where K_{micro} is a constant ($0 \leq K_{micro} \leq 1$) related to the activation energy of reaction (1). This process reproduces the Boolean identical and independent random variables of probability K_{micro} which are involved in the automaton collision steps.

In order to draw a parallel between the macroscopic and the microscopic dynamics, we need to determine the relations between the two macroscopic parameters K and $\mathcal{D}$ and the two quantities ρ and K_{micro} that we control in the simulation.

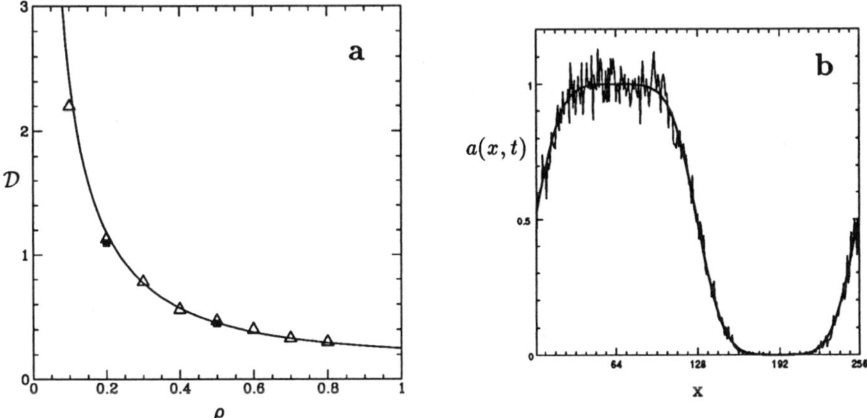

Figure 1: Diffusion coefficient $\mathcal{D}$ versus the reduced density ρ.
In Fig. 1a, the solid line (——) corresponds to the analytical result of Eq. (7)
obtained in the Boltzmann approximation. The solid squares ■ are simulation
results deduced from the mean square displacement of a tagged particle.
The open triangles $\triangle$ are simulation results deduced from the mixing of two
non reactive species; Fig. 1b illustrates the determination of $\mathcal{D}$ for the triangle
at $\rho = 0.2$. The method is to compare at a given time the simulated profile
(noisy line) with a family of theoretical profiles depending on $\mathcal{D}$. The best fit
(——), according to a least squares method, leads to the desired value of $\mathcal{D}$.

We first perform an analytical calculation of the diffusion coefficient as a function of
ρ in the framework of the Boltzmann approximation. The Green-Kubo formula [14]
relates $\mathcal{D}$ to the velocity autocorrelation function $\psi(t)$ of a tagged particle according
to

$$\mathcal{D} = \frac{1}{d} \lim_{t \to \infty} \int_0^t \psi(\tau)d\tau \tag{5}$$

where d is the dimensionality of the system ($d = 2$ here). Switching to discrete times
with time unit equal one time step and admitting that the ratio $\alpha = \psi(t+1)/\psi(t)$
is independent of t, we find

$$\mathcal{D} = \frac{1}{2} \left(\frac{1}{1 - \alpha} - \frac{1}{2} \right) \tag{6}$$

204

which leads for the HPP model to the following analytical form:

$$\mathcal{D} = \frac{1}{4\rho - \frac{8}{3}\rho^2 + \frac{2}{3}\rho^3} - \frac{1}{4} \tag{7}$$

We then numerically compute the effective diffusion coefficient in the automaton; by implementing two different methods as shown in Fig. 1. The first one lays on the computation of the mean square displacement of a tagged particle. The second one is based on the mixing of two non reactive species [15]. The simulation is stopped at a time t_0 for which the profile, obtained by averaging over y, looks like a sine. In absence of a chemical reaction, the macroscopic FKPP equation (2) for the fraction a of particles A reduces to the diffusion term. Starting from a step function initial condition, the analytical profile appears as a function $a(t, \mathcal{D})$ of the time t and the diffusion coefficient $\mathcal{D}$. Knowing $t = t_0$, we fit the simulated profile into the continuous family of theoretical profiles $a(t_0, \mathcal{D})$ obtained by varying $\mathcal{D}$. The best fit leads to the desired value of $\mathcal{D}$. The analytical expression of Eq. (7) and the above numerical determinations are in very good agreement.

After this analysis of $\mathcal{D}$ we turn to the other macroscopic parameter K that appears in the FKPP equation (2). We have to relate it to the microscopic control parameters ρ and K_{micro} of the simulation. Focusing on the chemical kinetics, we reduce the FKPP equation to its purely chemical term; the solution of this ordinary differential equation is

$$a(t) = \frac{a(0)}{a(0) + [1 - a(0)]e^{-Kt}} \tag{8}$$

to be compared to the simulated evolution of $a(t)$. The proper initial condition is now a homogeneous mixture of A and B species. Fig. 2 shows the "gauging curve" giving the relation between K, K_{micro} and ρ. Moreover, Fig. 2b proves that the simulation rules reproduce the desired kinetics.

We are now able to mimic any macroscopic evolution on the microscopic lattice through the proper choice of the simulation parameters.

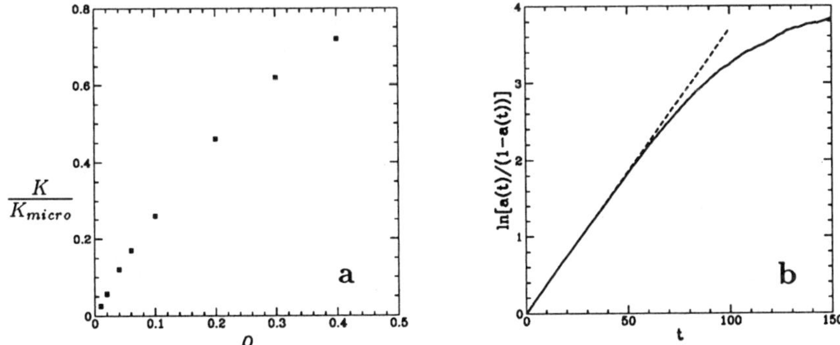

Figure 2: Chemical rate constant K versus the reduced density ρ.
Our numerical results are given in Fig. 2a. The method of determination is
illustrated in Fig. 2b for the point at $\rho = 0.4$. The initial condition of the
simulation is a homogeneous mixture of particles A and B reacting with a
microscopic rate constant $K_{micro} = 0.05$; we then plot $\ln[a(t)/(1 - a(t)]$ as a
function (———) of time t, for which the theory (...) predicts a slope K.

3. MEAN PROPERTIES OF THE SIMULATED WAVE FRONT

In our simulation model, the propagation of the front is taken into account in
the following manner. Classical periodic boundary conditions are imposed only in
the y direction . Defining a lattice column by the set of nodes with an identical
abscissa x, we couple the first and the last columns but keep them separated by a
permeable wall. If a particle A from the first column crosses the wall, it changes
into a particle B —- reciprocally for a particle B moving from the last column into
the first one.

In order to mimic an infinite medium in the x direction, it is necessary to coun-
terbalance the consumption of particles B due to the reaction given by Eq. (1).
When the number of particles A becomes greater than half the total number of particles, the wall is translated to the left and the all lattice is updated: the particles
A between the previous and the new wall positions are transformed into particles
B so that the numbers of particles A and B become approximately equal as they
were initially. The wall motion contains global information about the entire front

propagation and we define the velocity of the simulated front as the wall velocity. Its time averaged value $\bar{v}_s$ is computed for different values of K and $\mathcal{D}$. In Table I we compare $\bar{v}_s$ with the macroscopic value v_{min} given in Eq. (3).

K	0.0062	0.0069	0.02	0.023	0.046	0.026
$\mathcal{D}$	0.78	1.1	0.78	1.1	1.1	2.2
v_{min} Theory	0.14	0.17	0.25	0.32	0.45	0.47
$\bar{v}_s$ Simulation	0.15	0.19	0.24	0.32	0.45	0.44

Table I: It presents a comparison between the mean propagation velocity $\bar{v}_s$ deduced from the simulation and the value v_{min} predicted by the deterministic theory. We choose to order the results with v_{min} increasing. We find that $\bar{v}_s \approx v_{min} = 2\sqrt{K\mathcal{D}}$.

The agreement is satisfactory indicating that, in a 2D microscopic and stochastic medium, an initial step function profile evolves to a front propagating with the minimum velocity as in 1D macroscopic [6] or microscopic [7] media.

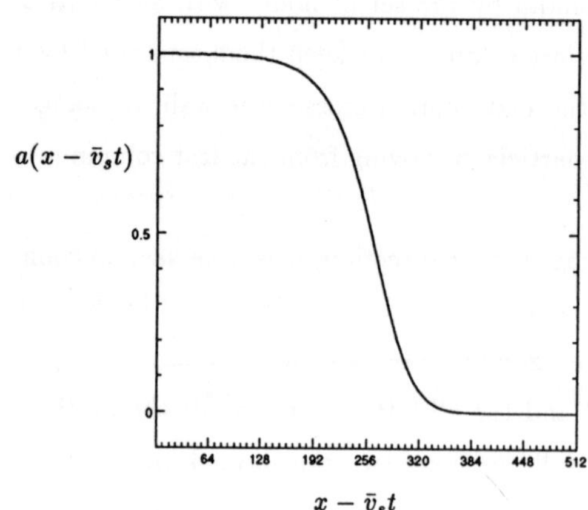

Figure 3: Mean profile.
It is obtained in the wall frame, propagating at the velocity $\bar{v}_s$, through averaging over y and t; the result is rescaled to the local mean fraction a of particles A. The numbers from 0 to 512 refer to the position on the discrete square lattice on which the simulation takes place. The parameter values used in the simulation correspond to $\mathcal{D} = 1.1$ and $K = 0.0069$.

Figure 3 represents the spatial variations of the mean fraction a of particles A in the wall frame after averaging over y and t. Hereafter, "mean" will refer to this average over y and t, besides equivalent to a statistical average at fixed time and position.

The shape of the mean profile depicted in Fig. 3 is characterized by its width $\bar{e}_s$ defined from the tangent at the inflection point. In Table II the mean front width $\bar{e}_s$ is compared to the macroscopic value e_{th} given in Eq. (4).

K	0.092	0.046	0.023	0.014	0.0062	0.0069
$\mathcal{D}$	1.1	1.1	1.1	1.1	0.78	1.1
e_{th} Theory	28	39	55	71	90	101
$\bar{e}_s$ Simulation	29	43	57	76	95	103

Table II: It presents a comparison between the mean profile width $\bar{e}_s$ deduced from the simulation and the value e_{th} predicted by the deterministic theory. We choose to order the results with e_{th} increasing.
We find that $\bar{e}_s \approx e_{th} \simeq 8\sqrt{\frac{\mathcal{D}}{K}}$.

As for the velocity, the agreement is satisfactory. These agreements between the macroscopic theory and the mean results derived from our microscopic simulations is relevant from both theoretical and computational points of view. On the one hand, it proves that the macroscopic predictions are not affected by the local microscopic and stochastic fluctuations. On the other hand, it supports the reliability of the simulation method. We are thus inclined to base on it further investigations on wave front properties, in particular at the microscopic level.

4. FRACTAL ANALYSIS OF THE REACTIVE INTERFACE

Although this numerical study is restricted to the FKPP model, the methods of analysis of the front dynamics and of its local structure are far more general, and could be applied in a large class of situations involving fronts or reactive boundaries. In this aim, we shall try to briefly expose the theoretical background which leads us to the definition of the front characteristics and which supports their physical

208

meaning.

We define this interface as the set of lattice sites which are occupied at the same time by two different species. An instantaneous representation is displayed in Fig. 4. A mere observation immediately reveals a spatial extension in the x direction and an asymmetrical distribution of points: the 2D interface presents a long tail on the left exactly as the 1D front profile [8].

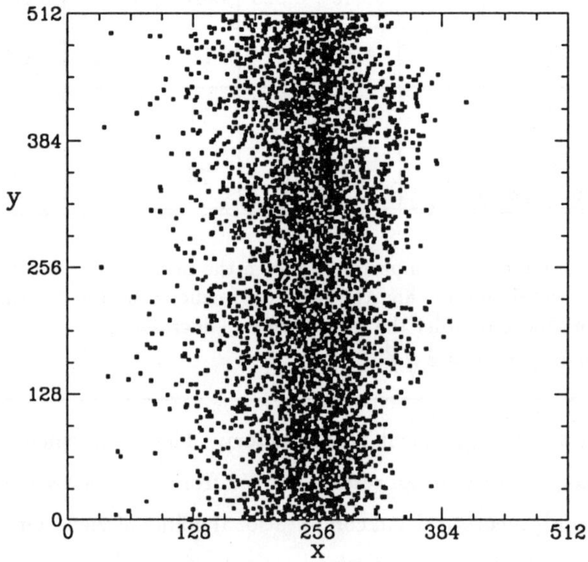

Figure 4: Reactive interface.
In this instantaneous plot for the conditions in Fig. 3, black points correspond to the lattice nodes where both A and B species are present.
Its obvious fractal character will be quantified through its capacity and its local similarity dimension; its scale invariance will be proved.

A natural approach is to investigate the seemingly fractal character of this interface, in the hope to elucidate the microscopic mechanisms and more specifically to describe the competition between diffusion and reaction.

To be more quantitative, we implemented two different operational definitions leading to a fractal dimension. They are based upon different theoretical characteristics of fractal sets of points, extended, in a computational form, to "real" experimental or numerical fractals which are defined, rather, as finite sets of cells, of linear size λ; λ thus appears as the smallest spatial scale of the fractal, which we call "resolution". It is then of the utmost importance to perform a self-similarity test, which will support the fractal character of the interface. Let us note $\mathcal{F}_\lambda$ the interface observed with the resolution λ. Two essential notions in a fractal characterization are the capacity $D_{\text{BC}}(\mathcal{F}_\lambda)$ and the similarity dimension $D_{\text{GP}}(\mathcal{F}_\lambda)$ theoretically defined as:

$$D_{\text{BC}}(\mathcal{F}_\lambda) = -\liminf_{r \to 0} \frac{\log_{10} N(\mathcal{F}_\lambda, r)}{\log_{10}(r)} \qquad D_{\text{GP}}(\mathcal{F}_\lambda) = \lim_{r \to 0} \frac{\log_{10} C(\mathcal{F}_\lambda, r)}{\log_{10}(r)}$$

(9)

where $N(\mathcal{F}_\lambda, r)$ is the number of squares of side r necessary to cover the interface and $C(\mathcal{F}_\lambda, r)$ the number of interfacial points in a disc of radius r.

The first step of an implementation for real fractals is to evidence the existence of these fractal dimensions. It is in practice impossible to perform the limit $r \to 0$ and the scaling laws associated to the definitions (9) are experimentally revealed by a linear region in the log log plots, as shown in Fig. 5. The dimension $D_{\text{BC}}(\mathcal{F}_\lambda)$ is thus defined as the slope of the linear part in $\log_{10}(r) \mapsto -\log_{10} N(\mathcal{F}_\lambda, r)$; this numerical calculation is known as the *box-counting method* [16] although it is a mere adaptation of the mathematical concept of capacity. D_{GP} is defined as the slope of the linear part in $\log_{10}(r) \mapsto \log_{10} C(\mathcal{F}_\lambda, r)$; this numerical calculation is known as the *Grassberger and Procaccia method* [17] although this method was initially designed in a much different context to compute the dimension of an invariant ergodic measure from a generic trajectory.

The second step is to test if these dimensions are independent of λ. Actually one can prove that the existence of any of these two dimensions and complementary its independence with respect to λ imply the self-similarity of the structure. This property means that the set $k.\mathcal{F}_\lambda$ deduced from the initial structure $\mathcal{F}_\lambda$ by an isotropic dilation of ratio k is identical to the set $\mathcal{F}_{k\lambda}$ observed with a resolution $k\lambda$. Such a structure $\mathcal{F}$ is then said to be *scale-invariant*.

To these purposes, we have computed the dimension of the structures deduced from the interface after contraction of the lattice parameter by a factor 4 (respectively 8). A cell of the contracted lattice is considered as an interfacial point as soon

210

as the corresponding 4×4 (8×8) square of the initial lattice contains at least 1 (2, in order to reproduce a decrease of sensitivity) interfacial point(s). In Figs. 5a and 5b, the curves corresponding to the initial lattice and to the contracted lattice have identical slopes, which allows us to relate this slope to a well-defined fractal dimension of the interface, independent of the chosen lattice parameter, thus having an intrinsic, hence relevant, meaning. Fig. 5a illustrates the box-counting method.

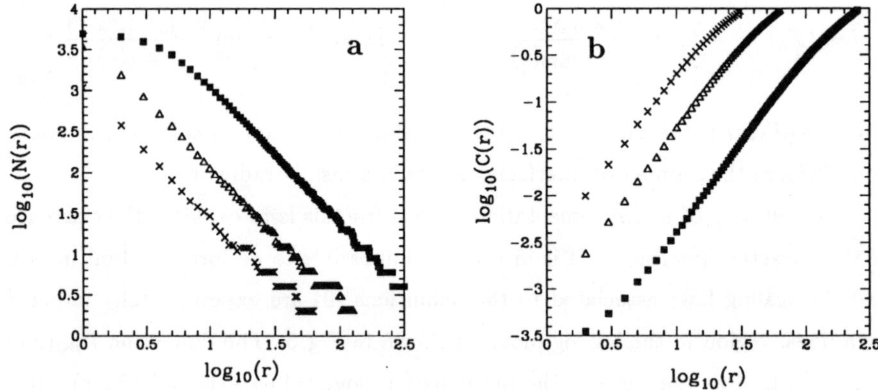

Figure 5: Self-similarity and fractal dimensions of the interface .
In Fig. 5a, we implement the *box-counting method*; defining $N(r)$ as the number of square boxes of side r required to cover the interface, we plot $\log_{10} N(r)$ versus $\log_{10} r$; the modulus of the slope of the linear region is the *capacity* D_{BC} of the reactive interface.

In Fig. 5b, we implement the *Grassberger and Procaccia method*; defining $C(r)$ as the spatial mean number of interface points in a disc of radius r around an interface point, we plot $\log_{10} C(r)$ versus $\log_{10} r$; the slope of the linear region is the similarity dimension D_{GP} revealing the scaling properties of the mean local structure.

Moreover, as explained in the text, the fact that these dimensions do not depend upon the lattice constant proves the self-similarity of the structure. The curves plotted with solid squares ■ correspond to the initial lattice. The other curves are obtained when defining the interface as a set of "macrocells" of respectively 4×4 nodes for the open triangles $\triangle$ and 8×8 nodes for the crosses $\times$.

The value found, denoted by $D_{BC} = 1.63 \pm 0.02$ appears to be independent of

K and $\mathcal{D}$ as can be seen in Table III. Very close results has been obtained using the alternative co-dimension method [16]. Both provide good numerical approach of the capacity of the interface and are upper bound for its Hausdorff dimension; moreover they do not depend upon the resolution chosen to observe the structure. A somewhat larger value $D_{GP} = 1.90 \pm 0.02$ is found using the Grassberger and Procaccia method illustrated in Fig. 5b.

K	0.0069	0.014	0.023	0.046
D_{BC}	1.63	1.64	1.63	1.62
D_{GP}	1.91	1.91	1.89	1.80 (?)
S	0.681	0.6810	0.675	0.678

$\mathcal{D}$	0.78	1.1	2.2
D_{BC}	1.38 (?)	1.63	1.61
D_{GP}	1.91	1.89	1.85
S	0.673	0.681	0.687

Table III: It presents the values of the fractal dimensions D_{BC} and D_{GP} and of the statistical entropy S for different chemical rate constants K and diffusion coefficients $\mathcal{D}$. It proves the independence of these quantities D_{BC}, D_{GP} and S with respect to the parameters K and $\mathcal{D}$. The discrepancy between the value of the two dimensions is elucidated below.

We have performed an other test, shown in Table III: the statistical entropy S of the interface is seen to be K and $\mathcal{D}$ independent, allowing us to conclude in favour of the universality of the fractal properties.

The discrepancy between the values of D_{BC} and D_{GP} is explained by the different length scales at which these dimensions are obtained. In fact the scaling law leading to the value of D_{GP} is valid at smaller scales than the law leading to D_{BC}. This conjecture is confirmed by a simulation on a larger lattice 1024×1024. The Grassberger and Procaccia method, whose validity was previously restricted to small scales, can be here extended to larger scales without saturation effects. The scale domains associated with $D_{GP} \approx 1.9$ and $D_{BC} \approx 1.7$ appear clearly separate as shown in Fig. 6.

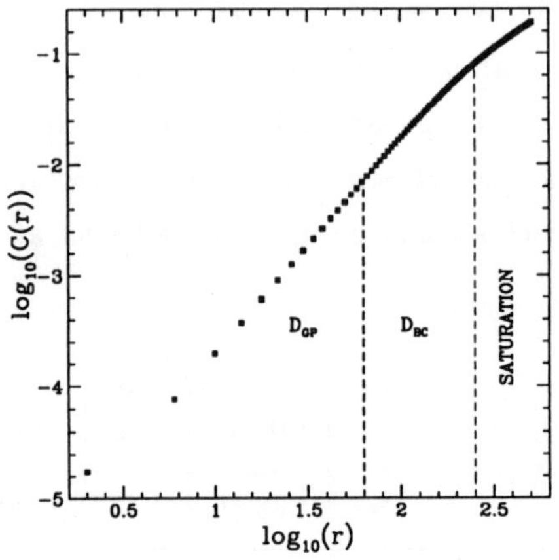
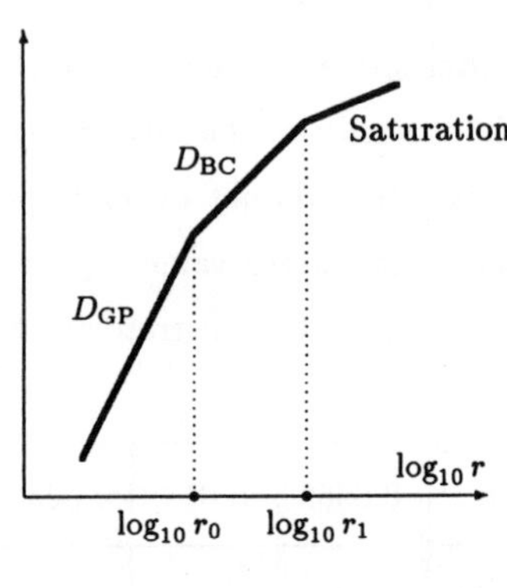

Figure 6: Bi-fractality of the reactive interface.
This figure evidences the two-scale structure of the interface. When observed with a fine resolution $r < r_0$, it exhibits a fractal character of dimension $D_{GP} \approx 1.9$ revealing reactive aggregates. At a larger scale $r_0 < r < r_1$, the relevant dimension is $D_{BC} \approx 1.7$, characteristic of a diffusive behaviour. For $r > r_1$, finite-size effects lead to a dimension ≈ 1.1, since at such a scale the interface appears as a fine strip. The scheme on the right side, where the relevant features are emphasized, helps the explanation.
Simulation have been here performed on a larger lattice 1024×1024 and for microscopic parameter values corresponding to $\mathcal{D} = 1.1$ and $K = 0.00184$.

We propose an explanation for the dependence of the fractal dimension with respect to the observation scale. At small scale, chemical effets overwhelm the diffusion consequences so that the dominant structure consists in compact reactive clusters; at a larger scale, these reactive aggregates are organized in the lacunary structure due to the diffusion process. We actually approximately obtain at large scale the fractal dimension $D = 7/4$ that Sapoval et al. [18] obtained in the purely diffusive case.

Such a fractal analysis not only evidences scale invariance properties, but also provides quantitative estimates of qualitative properties as the intricate, fingerlike or diffuse character of a structure; accurate comparisons with theoretical models and numerical or experimental results from other phenomena are then possible.

Acknowledgements: This work was partially supported by NATO CRG 920083.

REFERENCES:

1. *Nonequilibrium Chemical Dynamics: From Experiment to Microscopic Simulations*, edited by F. Baras and D. Walgraef, Physica A **188** (1-3) (1992).

2. J.D. Murray, *Mathematical Biology*, (Springer, Berlin, 1989).

3. *Non Linear Phenomena in Materials Science II*, Solid State Phenomena, vol 23/24, edited by G. Martin and L. Kubin (Trans Tech Publications, Zurich, 1992).

4. R.A. Fisher, Ann. Eugenics **7**, 335 (1937).

5. A. Kolmogorov, I. Petrovsky and N. Piskunov, Bull. Univ. Moscow. Ser. Intl. Sec A **1** (6), 1 (1937).

6. D.G. Aronson and II.F. Weinberger, Adv. Math. **30**, 33 (1978)

7. A. Lemarchand, H. Lemarchand and M. Mareschal, *Microscopic Simulation of a Chemical Wave Front Propagating into an Unstable State* submitted to J. Stat. Phys. (1992).

8. M. Bramson, P. Calderoni, A. de Masi, P. Ferrari, J. Lebowitz and R.H. Schonnmann, J. Stat. Phys. **45**, 905 (1986).

9. P. Clavin, P. Lallemand, Y. Pomeau and G. Searby, J. Fluid Mech. **188**, 437 (1988).

10. S. Cornell, M. Droz and B. Chopard, Phys. Rev. A **44**, 4826 (1991).

11. A. Lawniczak, D. Dab, R. Kapral and J.P. Boon, Physica D **47** 132 (1991).

12. J. Hardy, O. de Pazzis and Y. Pomeau, Phys. Rev. A **13**, 1949 (1976).

13. U. Frisch, D. d'Humières, B. Hasslacher, P. Lallemand, Y. Pomeau and J.P. Rivet, Complex Systems **1**, 649 (1987).

14. M. Ernst in *Liquids, Freezing and Glass Transition* edited by J.P. Hansen, D. Levesque and J. Zinn-Justin, North-Holland, Amsterdam (1991).

15. A. Noullez *Automates de gaz sur réseaux: aspects théoriques et simulations.* PhD Thesis, Bruxelles (1990).

16. K.R. Sreenivasan, Annu. Rev. Fluid Mech. **23**, 539 (1991).

17. P. Grassberger and I. Procaccia, Phys. Rev. Lett **50**, 347 (1983).

18. B. Sapoval, M. Rosso and J.F. Gouyet, J. Phys. Lett. **46**, L149 (1985).

Molecular Dynamics Simulations
of Nonequilibrium Effects in Chemical Systems

J. Gorecki

Institute of Physical Chemistry, Polish Academy of Sciences,
Kasprzaka 44/52, PL-01-224 Warsaw, Poland
    and
Institute for Molecular Science
Myodaiji, Okazaki 444, Japan

Abstract

A new technique for large scale computer simulations of systems with chemical reactions is proposed. It allows to extend a size of simulations, which are performed with the use of the molecular dynamics method for reactive hard spheres. Two applications of the new technique are presented. In the first of them the results of simulations for a model chemical system in which, due to the Hopf bifurcation, a limit cycle appears from a stable stationary state are discussed. The second example shows that the method may be successfully applied to obtain nonequilibrium spatial correlations between concentrations of reactants which may appear if the detailed balance condition is not satisfied. In this case a quantitative analysis of results is given.

I. Introduction

The growing interest in microscopic simulations of chemical systems far from an equilibrium state can be noticed in the recent years [1,2]. Experiments tell us that the influence of internal fluctuations may be very important in strongly nonlinear chemical systems. Thus, one needs a theory which goes beyond the standard phenomenological chemical kinetics based on the mass action law. Such a theory should take into account the presence of fluctuations. Computer simulations, performed at the microscopic level, allow us to test different theoretical approaches on idealized models and they play the role of "experiments" in which, contrary to the real experiments, all the elementary chemical processes are well known.

One of the most important factors is a scale of simulations. Of course, our computers do not allow us to deal with numbers of molecules, which are usually involved in real systems. On the other hand, the simulations, which are performed for systems with a small number of molecules are usually very affected by the internal fluctuations and the analysis of obtained data is very difficult [3,4]. Moreover, typical reactions involve many different chemical components. If a system considered in simulations is small then local fluctuations of reagents concentrations may lead to large spatial inhomogeneities, which may significantly affect their time evolution. To reduce this effect one needs to enlarge the scale of simulations as much as it is possible.

There is no doubt that molecular dynamics technique is one of the best methods for microscopic simulations [5]. It uses "realistic" potentials of interactions between atoms. The motion of particles is described by corresponding Newton's equations. On the other hand, the method is the most time consuming in computing. To save computer time in large scale simulations of nonequilibrium effects in chemical systems techniques, in which interactions between molecules are independent of their chemical features, are used. In practice it means that the parameter, which marks the chemical identity of a particle does not have any influence on its motion. The methods, such like the Bird algorithm [6], or reactive lattice gas cellular automaton model [7,8] are based on this idea. Both methods allow us to simulate systems composed of more than 1,000,000 objects on a small computer. However, both of them contain serious simplifications. The lattice gas cellular automaton model assumes that "molecules" may occupy the lattice nodes only. Moreover, the motion of molecules is replaced by jumps to neighboring lattice nodes within a single time step. Thus a spatial distribution of particles is influenced by the geometry of lattice used and, because all speeds are equal, the method does not allow to simulate thermal effects. On the other hand, Bird's algorithm assumes that all particles within a single cell are potential collision partners, regardless of their exact positions. Thus the algorithm is more suitable for simulations of low density systems, than of dense ones. In this paper I would like to present a new method, which is free of these shortcomings. It assumes the real space geometry and it takes into account correlations between positions of particles in order to check if they may collide.

II. Periodically Expanded Molecular Dynamics

The subject of this paper is to show that molecular dynamics simulations of a system with chemical reactions may, in some cases, be very efficient and

216

take into account many-millions of particles. The first step towards highly accurate, large scale, simulations was suggested in [9] and then successfully used to calculate the influence of nonequilibrium effects on a rate constant in a system with a thermally activated reaction [10]. The method was based on an observation that if the mechanical motion of objects is separated from chemical processes and if no energy is released or consumed in reactive collisions then the mechanical motion in a system with chemical reactions is identical as in an equilibrium system composed of the same geometrical objects. Thus, many different reaction paths may be created from a prerecorded equilibrium trajectory by taking a point on it, assigning the chemical identity parameter to all particles and then checking if successive collisions are reactive or not. Of course the parameter describing the chemical identity of a molecule have to be changed after each reactive collisions in which the molecule is involved.

The original method generates reaction paths which describe a chemical system with the same number of molecules as the one for which the equilibrium trajectory is recorded. However if the equilibrium simulation is performed for the periodic boundary conditions, then spatial extension of it is possible. The periodic boundary conditions mean that the simulated system is regarded as an elementary cell in an infinite system, which is invariant with respect to translations by the vectors of the side length. The most popular shape of a cell is square in the case of 2D systems and a cube in 3D ones. Knowing the evolution within a single (elementary) cell we have information about positions and velocities for corresponding (by symmetry) objects in all other cells. Thus, from a prerecorded trajectory one can easily obtain the evolution of a system which is extended periodically by a number of cells in each direction. Of course the periodic boundary conditions remain satisfied for the extended system too. If a chemical identity of molecules is neglected than such expansion does not bring us any new information because the evolution within all sub-cells of the extended system is identical. Moreover, it may lead to wrong conclusions because the correlations extending over a single cell are affected by artificially introduced periodicity. However, for a multicomponent chemical system in which the motion of particles is not related to chemical identity the situation is different. First, different chemical compositions may be initialized in various cells by marking the equivalent (by periodicity) particles in a different way. Secondly, a steric factor (reaction cross section), if it is not equal to unity, differentiate the time evolution in various cells, because a collision between isomorphic objects may be reactive in one cell and nonreactive in another. Because of the periodic boundary conditions a free flow of molecules, between the neighboring cells is ensured. Finally, one obtains the evolution of a system much larger than the original one.

Let us notice that the method described allows us to use computer facilities in an extremely efficient way. In the case of chemically reacting hard spheres, I am concerned in the following, only times of collisions and identities of colliding objects have to be recorded in order to restore a trajectory. Next, in order to create an expanded system one needs only one large array in a computer memory to store the actual chemical identities of all objects (the use of logical or integer*1 variables seems the most natural), whereas all the other quantities as velocities, positions and moments of collisions are periodic in space and they do not require a large memory. In practice, using personal computer with 16Mb of RAM I may easily handle 10 000 000 objects. I believe that molecular dynamics simulations for a similar scale of a system are too complex for the largest existing computers if the "orthodox" technique is applied. Reading the parameters of a single collision time we know that a collision occurs in all elementary cells and the corresponding particles are involved. Thus, the number of collisions per an object is the same as in the original simulation. Therefore the space on a computer disk required to record a long trajectory is much reduced if compared with the system of the same number of molecules in which the collisions in different regions of space are uncorrelated.

The periodically expanded molecular dynamics benefits us if the spatial correlations between different reagents are considered. Of course, the method is limited to correlations which range is restricted to a single cell only, as otherwise they will be affected by the introduced periodic structure. But, on the other hand, the periodic expansion simplifies calculations of spatial correlations because it is sufficient to calculate interesting intermolecular distances between particles within a single cell. The contributions to correlations of various types are obtained depending on the chemical identities of objects in a given cell .

III. Oscillations in a Simple Chemical Model

As the first example let us consider a very simple oscillating system. Three reagents A,B and C are assumed to react according to the following scheme:

$$A + A + A \underset{k_{-1}}{\overset{k_1}{\rightleftharpoons}} A + A + B$$

$$k_2$$
$$B + B + B \rightleftharpoons B + B + C$$
$$k_{-2}$$

$$k_3$$
$$C + C + C \rightleftharpoons C + C + A \qquad (1)$$
$$k_{-3}$$

The proposed model is highly artificial, as the detailed balance condition is violated and all the processes are described by three body interactions, which seem to be far less realistic than a two body processes. Nevertheless, the proposed model has some interesting features. For simplicity let us assume that the rate constants for forward reactions for all three processes (1) are the same and moreover all the backward reactions are characterized by the same rate constant:

$$k_1 = k_2 = k_3$$

and

$$k_{-1} = k_{-2} = k_{-3}$$

Now the system is characterized by a single parameter - the ratio of the rate constants for the forward and the backward reactions: $\lambda = k_{-1}/k_1$. The kinetic equations describing the time evolution of concentrations read:

$$\frac{da}{d\tau} = -a^3 + c^3 - \lambda ac^2 + \lambda(1 - a - c)a^2$$

$$\frac{dc}{d\tau} = -c^3 + (1 - a - c)^3 - \lambda(1 - a - c)^2 c + \lambda c^2 a \qquad (2)$$

where a and c denote concentrations of A and C respectively and $\tau = k_1 t$. These two equations describe the system completely, as $a+b+c$ remains constant during the reaction (I use the normalization of concentrations in which $a + b + c = 1$). It may be shown easily that 3 is the critical value of the parameter λ. If $\lambda < 3$ then the system has a single stable stationary state for which $a = b = c = 1/3$. On the other hand if $\frac{k_f}{k_b} > 3$ then this state becomes unstable and a stable limit cycle appears.

There rises a question if one needs so unrealistic model as the presented above to get a limit cycle. It is known that the models of chemical oscillations in a closed system characterized by two independent reactants, in which only the binary reactions are considered, are isomorphic with the Lotka-Voltera model [11]. Thus, we need at least a single reaction involving a triple collision in order to introduce a nonlinearity, which is strong enough to give a limit cycle. Let us notice that the most popular model showing the Hopf bifurcation (Brusselator [12]) does include a single tri-molecular reaction. The model considered above may be also easily simplified to a reaction scheme containing a single triple reaction. However, after such simplification, the nice symmetry with respect to concentrations of species is lost.

A comment on how to simulate triple collisions in a hard sphere system is needed. Of course, the probability that three hard objects (as spheres in my simulations) collide at the same time is just zero. Thus, as triple encounters, I regard these binary collisions for which there exists a sphere which is closer than r_0 with respect to centers of both colliding spheres, where r_0 characterizes the considered reaction. Moreover, I check that there is no other sphere which is closer to the colliding one. It may be shown that such definition is a fair representation of tri-molecular reaction. The proper choice of r_0 depends on density of a system. Here I used r_0 equal to 1.25σ, where σ denotes the sphere diameter.

The equilibrium simulations have been done for $N = 500$ hard spheres placed in a cubic box with the side length $d = 14.7\sigma$. The packing fraction is $\eta = 0.0824$. The masses and diameters of spheres representing A, B and C are the same (32 a.u. and 0.5 nm, respectively).

The time evolution of a small system (composed of 500 spheres only) is presented in figures 1 and 2. The behavior for λ below the critical point is shown in figure 1a,b. Both the plot of concentrations of A and C as functions of time and the plot of trajectory on the phase space $a \times c$ clearly indicate that the system is just fluctuating around the stable state.

The behavior of system (1) above the critical point ($\lambda = 5.4$) is shown in figure 2 a,b. One may notice that the state $a = c = 1/3$ is now unstable and the system tends to the limit cycle.

The behavior around the critical point is the most interesting. Figures 3 and 4 show the time evolution for two different values of λ. Two evolutions starting from the same concentrations of A, B and C but different spatial distributions of spheres are shown. The dashed line describes a small system composed of 4000 spheres. The solid line shows results obtained for a large system composed of 1,372,000 spheres. It may be noticed that a smaller system develops "oscillatory" behavior for the values of critical parameter much below the critical point. On the other hand for the large system oscillations are clearly damped for $\lambda = 2.8$. These results indicate that the oscillatory behavior may appear below the critical point if internal fluctuations are strong enough (compare [13]).

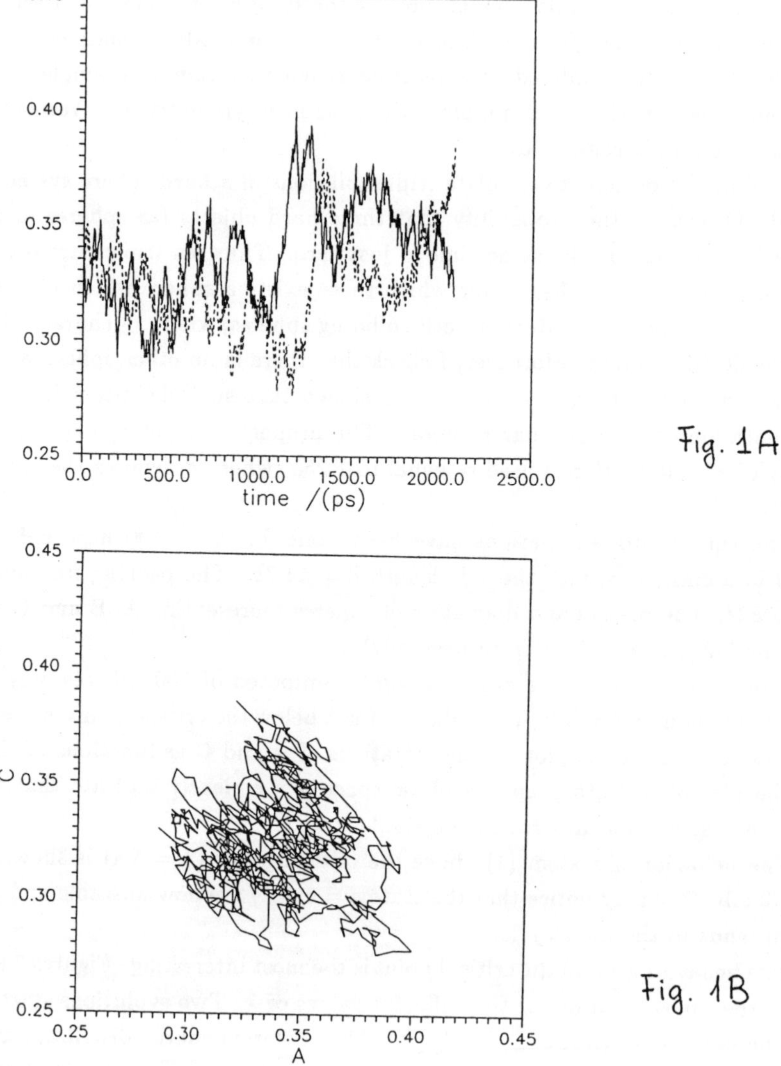

Fig. 1. The behavior of system (1) for a small value of the critical parameter λ ($\lambda = 2$). Fig. 1A shows the time evolution of a and c. The solid line represents the concentration of **A**, the dashed line - the concentration of **C**. Fig. 1B gives the trajectory on the phase space $a \times c$.

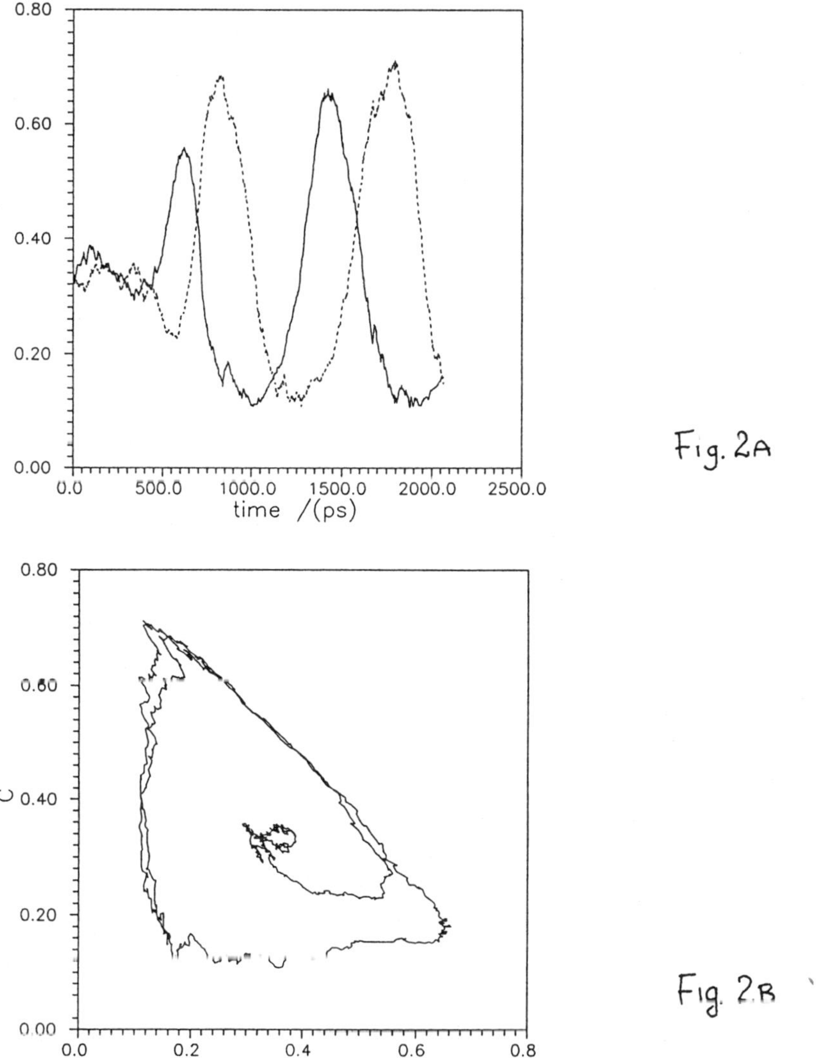

Fig. 2. The same as in fig. 1., but for a large value of the critical parameter λ ($\lambda = 5.4$).

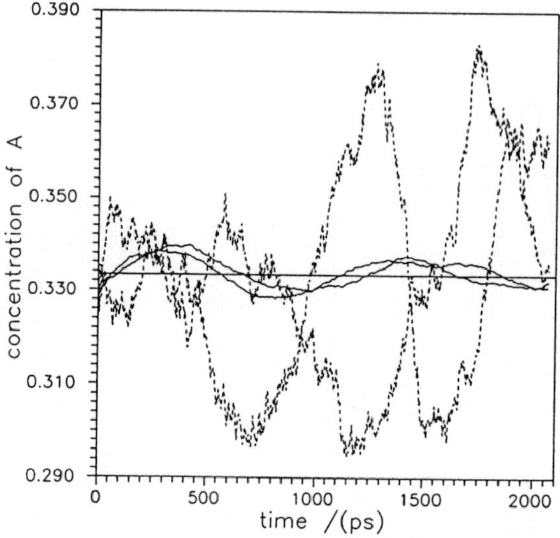

Fig. 3. Time evolution of a in system (1) for $\lambda = 2.8$. Two trajectories are created for 4000 spheres (dashed line) and for 1,372,000 spheres (solid line).

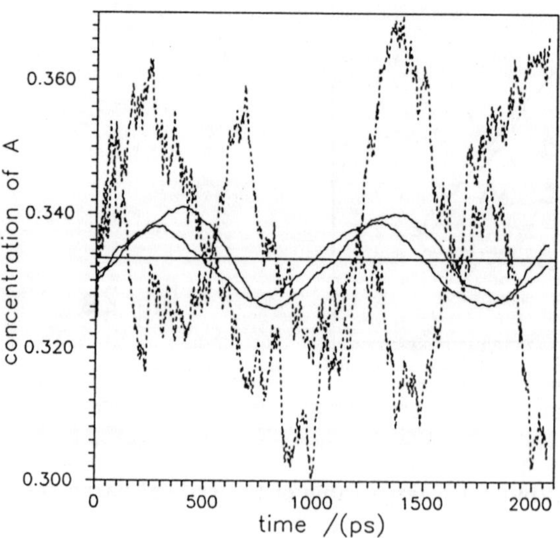

Fig. 4. Time evolution of a in system (1) for the critical value of λ ($\lambda = 3.0$). Notation as in fig. 3.

IV. Nonequilibrium Spatial Correlations of Concentration [14]

Let us suppose that several reactions takes place in a system at the same time:

$$A_i + B_i \underset{k_{-i}}{\overset{k_i}{\rightleftharpoons}} C_i + D_i \qquad \text{for} \quad i=1,2 \dots N \qquad (3)$$

and let k_i, k_{-i} denote the rates of the forward and the backward reactions respectively. The detailed balance condition requires that any of the forward reactions is balanced by the corresponding backward process, thus for each pair of reactions (3):

$$k_i a_i b_i = k_{-i} c_i d_i \qquad \text{for} \quad i=1,2 \dots N \qquad (4)$$

where a_i, b_i, c_i and d_i are the concentrations of chemical species involved in the reaction i. However, if we have input and output fluxes of the chemical species, then we may have a steady state of a system for which the detailed balance condition is not satisfied.

If reactions are accompanied with diffusion of chemical species then the nonequilibrium spatial correlations may appear in the system, as it was pointed by Kuramoto[15]. The appearance of spatial correlations is directly related to the breaking of the detailed balance condition in chemical kinetics, as it was shown for the Schlögl model by Nicolis and Malek Mansour[16] and later studied in a general case by Kitahara et. al.[17] .

The first microscopic simulations of such nonequilibrium correlations were done by Nicolis et. al. [18], who were using the Bird technique of direct simulation of tho Boltzmann equation for a system with chemical reactions. In these simulations, in order to keep stationary concentrations of reagents, at which the detailed balance condition is not satisfied, the molecules of reactants are continuously added to a system and the molecules of products are subsequently removed. Here, instead of keeping the stationary concentrations in a state for which the detailed balance condition is not satisfied by adding and removing molecules [18, 19], a closed system is considered, but with thc ratc conctants different from those, which satisfy Eq.(4).

The model which is considered below consists of the following reactions:

$$X + X \underset{k_{-1}}{\overset{k_1}{\rightleftharpoons}} X + Y$$

$$X + Y \underset{k_{-2}}{\overset{k_2}{\rightleftharpoons}} Y + Y \tag{5}$$

The reactions proceed in a closed system so the sum of concentrations of X and Y (denoted as x and y respectively) is constant. The kinetic equations for system (5) read:

$$\frac{dx}{dt} = -k_1 x^2 + k_{-1} xy - k_2 xy + k_{-2} y^2$$

$$\frac{dy}{dt} = k_1 x^2 - k_{-1} xy + k_2 xy - k_{-2} y^2 \tag{6}$$

It is clear that $x + y = c$ is a constraint of these kinetic equations. Equations (6) admit a single stationary state $x_s, y_s; 0 \leq x_s, y_s \leq c$, which is stable.

The detailed balance condition for reactions (5) is satisfied if and only if the following relationship between the rate constants is valid:

$$\frac{k_1}{k_2} = \frac{k_{-1}}{k_{-2}} = \frac{c - x_s}{x_s} \tag{7}$$

Now let us consider the situation in which the rate constants k_i, k_{-i} do not satisfy the detailed balance condition, i.e.:

$$\frac{k_1}{k_2} \neq \frac{k_{-1}}{k_{-2}}$$

which can be realized in the presence of nonequilibrium constraints on the system. For example rate constants may depend on concentrations of the other species which concentrations are kept constant by the exchange with an external source. Another example is given by photochemical reactions for which the rate constants are sensitive with respect to the intensity of light [20]. Therefore, if the values of rate constants can be controlled by external parameters one may create a stationary state for which detailed balance condition is not held.

If the reactions (5) are coupled with diffusion, the kinetic equations read:

$$\frac{\partial x}{\partial t} = -k_1 x^2 + k_{-1} xy - k_2 xy + k_{-2} y^2 + D\nabla^2 x$$

$$\frac{\partial y}{\partial t} = k_1 x^2 - k_{-1} xy + k_2 xy - k_{-2} y^2 + D\nabla^2 y \tag{8}$$

Writing equations (8) it is assumed that the diffusion constant D is the same for the both reagents X and Y.

The existence of nonequilibrium spatial correlations in a steady state may be proven if one considers a master equation, which treats diffusion as a jump process and a chemical reaction as a birth-and-death process. The full analysis of the problem is presented in the reference [21]. It may be shown that the nonequilibrium spatial correlations have the form:

$$< n(\mathbf{r})m(\mathbf{r}') >_{chem} = n_s\delta_{NM}\delta(\mathbf{r} - \mathbf{r}') + \frac{C_{NM}}{8\pi D}\frac{1}{|\mathbf{r} - \mathbf{r}'|}\exp(-\kappa\,|\,\mathbf{r} - \mathbf{r}'\,|) \quad (9)$$

where n and m denote the concentrations of the reagents N and M. The constants C_{NM} and κ are related to the chemical dynamics and diffusion. For the reactions (5) they read:

$$C_{XX} = -2(k_1 x_s^2 - k_{-1}x_s y_s)$$

$$C_{XY} = 2(k_1 x_s^2 - k_{-1}x_s y_s)$$

$$C_{YY} = -2(k_1 x_s^2 - k_{-1}x_s y_s) \quad (10)$$

and

$$\kappa^2 = \frac{1}{D}(-2k_1 x_s + (k_{-1} - k_2)(c - 2x_s) - 2k_{-2}(c - x_s)) \quad (11)$$

As it has been mentioned for a system with the detailed balance condition satisfied all constants C_{NM} are equal to zero and the nonequilibrium spatial correlations should be absent. These correlations may appear in a stationary state only if the detailed balance condition is not hold. For the model (5) the absolute values of all constants C_{NM} are the same, which indicates the equal strength of correlations between different reagents. Moreover the constant κ, describing the range of correlations, does not depend on the choice of reactants which correlations are considered.

In the simulations both reactants X and Y are represented by spheres with the same mass ($m = 32a.u.$) and diameter ($\sigma = 5A$). Their chemical identity is described by an additional parameter which has no influence on the mechanical motion of a sphere.

In order to control the values of rate constants I have introduced steric factors [$s_i, s_{-i}, i = 1,2$ for reactions (5)], which describe what fraction of all collisions between given reagents is reactive. In the following I show the results which has been obtained by periodic extension of three systems:

(I). a low density system. The original simulations were done for $N = 500$ hard spheres placed in a cubic box with the side length $d = 14.7\sigma$. The packing fraction is $\eta = 0.0824$. This system was expanded by 7 box lengths in each direction so the total number of molecules is 171500.

(II). a middle density system ($N = 1200$ spheres, $d = 12.9\sigma, \eta = 0.29$). This system was expanded by 8 box lengths in each direction leading to 614400 spheres.

(III). a high density system ($N = 1331$ spheres, $d = 11.1\sigma, \eta = 0.51$). This system was also expanded by 8 box lengths in each direction and the total number of spheres is 681472.

Three different sets of steric factors for reactions (5) have been considered in simulations:

(a) a system with the detailed balance condition satisfied:

$$s_1 = 0.4, \quad s_{-1} = 0.2, \quad s_2 = 0.2, \quad s_{-2} = 0.1.$$

The concentrations at the stable state are: $x_s = c/3$ and $y_s = 2c/3$.

(b) a system close to a detailed balance:

$$s_1 = 0.4, \quad s_{-1} = 0.2, \quad s_2 = 0.2, \quad s_{-2} = 0.2.$$

for which the stationary concentration of X equals to: $x_s = (\sqrt{2} - 1)c$.

(c) a system far away from a detailed balance:

$$s_1 = 0.8, \quad s_{-1} = 0.4, \quad s_2 = 0.2, \quad s_{-2} = 0.8.$$

for which the concentration of X at its stationary state equals to: $x_s = \frac{(\sqrt{65}-7)}{2}c$.

At the beginning of simulations the chemical identities of spheres are assigned such that the concentrations of X and Y correspond to their stationary values. The initial velocities are assigned according to a Maxwellian distribution which corresponds to 300K. Then the system is allowed to relax for more than 400 collisions per sphere before the spatial correlations of density are measured. Of course, during the relaxation reactive collisions are permitted.

One may expect (compare Ref.[17]) that the nonequilibrium spatial correlations have a limited range, which is of the order of the mean free path. Therefore a cutoff (R) in the range of intermolecular distances, up to which the nonequilibrium correlations are expected to be important, may be introduced. The interval of interesting intermolecular distances $[\sigma, R]$ is divided into a number of subintervals of the same length Δ. The simulations give us the average number of spheres representing reactant N which distance from a sphere representing reactant M lies within $[\sigma + i\Delta, \sigma + (i + 1)\Delta]$. Let us denote this quantity as

$G_{NM}(i)$. The radial distribution function $(g(r))$ for the system as a whole (i.e. with neglected chemical identity of spheres) can be obtained by scaling $G(i)$ by the value, calculated assuming a uniform distribution of spheres:

$$g([\sigma + i\Delta, \sigma + (i+1)\Delta]) = \frac{G(i)}{\frac{4}{3}\pi((\sigma + (i+1)\Delta)^3 - (\sigma + i\Delta)^3)c} \tag{12}$$

where $G(i)$ denote the average number of spheres within the interval $[\sigma + i\Delta, \sigma + (i+1)\Delta]$, without consideration of their chemical identity.

Similarly we can calculate the partial radial distribution functions g_{XX}, g_{XY} and g_{YY}:

$$g_{XX}([\sigma + i\Delta, \sigma + (i+1)\Delta]) = \frac{G_{XX}(i)}{\frac{4}{3}\pi((\sigma + (i+1)\Delta)^3 - (\sigma + i\Delta)^3)x_s} \tag{13}$$

$$g_{XY}([\sigma + i\Delta, \sigma + (i+1)\Delta]) = \frac{G_{XY}(i)}{\frac{4}{3}\pi((\sigma + (i+1)\Delta)^3 - (\sigma + i\Delta)^3)x_s} \tag{14}$$

$$g_{YY}([\sigma + i\Delta, \sigma + (i+1)\Delta]) = \frac{G_{YY}(i)}{\frac{4}{3}\pi((\sigma + (i+1)\Delta)^3 - (\sigma + i\Delta)^3)y_s} \tag{15}$$

Typical radial distribution functions (they were calculated for a high density system (III) with reaction rates (c)) are plotted in Fig. 5. It can be noticed that all the partial radial distribution functions (shown by the dashed lines) are distinctly different from the equilibrium radial distribution function g which is drawn using the solid line.

In order to compare the nonequilibrium spatial correlations observed in simulations with the theory, one needs to separate the equilibrium correlations introduced by the hard sphere potential from the total partial distribution function:

$$< n(\mathbf{r})m(\mathbf{r}') >_{total} - < n(\mathbf{r})m(\mathbf{r}') >_{eq} + < n(\mathbf{r})m(\mathbf{r}') >_{chem} \tag{16}$$

Having in mind that the accuracy of g_{NM}/g obtained in simulations is higher than g_{NM} itself, it is convenient to calculate the chemical contribution to correlations of concentration in the following way:

$$< n(\mathbf{r})m(\mathbf{r}') >_{chem} = < n(\mathbf{r})m(\mathbf{r}') >_{total} - < n(\mathbf{r})m(\mathbf{r}') >_{eq}$$

$$= n_s m_s g(|\mathbf{r} - \mathbf{r}'|)(\frac{g_{NM}(|\mathbf{r} - \mathbf{r}'|)}{g(|\mathbf{r} - \mathbf{r}'|)} - 1)$$

$$= [\frac{n_s}{c}\frac{m_s}{c}g(|\ \mathbf{r} - \mathbf{r}'\ |)(\frac{g_{NM}(|\ \mathbf{r} - \mathbf{r}'\ |)}{g_{00}(|\ \mathbf{r} - \mathbf{r}'\ |)} - 1)]c^2 \tag{17}$$

The correlations given by Eq.(17) with the total density factor c^2 excluded are drawn using a solid line in Figs. 6 and 7.

The nonequilibrium spatial correlations (or rather the lack of them) for a detailed balance case in a system characterized by the low density are shown on Figures $6A$ and $6B$. Similarly to the results obtained for the other densities, $< nm >_{chem}$ for the set (a) is fluctuating around zero, showing no indication for the presence of correlations.

On the other hand, for all the cases in which the detailed balance condition was not satisfied, we observed the existence of nonequilibrium correlations. For a given set of steric factors their range seems to be independent on the reagents involved. The range of nonequilibrium correlations is approximately equal to 1.5σ for $\eta = 0.0824$, 0.4σ for $\eta = 0.29$ and 0.15σ for $\eta = .51$. The mean free paths l of a sphere in these hard sphere systems, estimated from the average velocity and the collision frequency with the use of the formula:

$$l \cong (6\eta g(\sigma^+))^{-1} \tag{18}$$

are equal to: 1.63σ , 0.24σ and 0.05σ, respectively so they are similar to the range of nonequilibrium correlations.

The results obtained in molecular dynamics simulations are accurate enough to allow us for a quantitative comparison with the theory. Equation (10) says that correlations between concentrations of various reagents in the model (5) differ by their signs only. This conclusion is fully supported by the results of molecular dynamics. The absolute values of $< xx >_{chem}$, $< xy >_{chem}$ and $< yy >_{chem}$ observed in simulations are almost the same (compare Fig.6A with Fig.6B and Fig.7A with Fig.7B) and small deviations between these functions may be regarded as a statistical error. The accuracy achieved allows us to check if the functional form of $< nm >$ predicted by the theory (Eq.(9)) describes the observed correlations. On the basis of Eq.(9) we may expect that the data of simulations may be approximated with the use of two constants: B and κ, which are related with $< nm >$ in the following way:

$$< n(\mathbf{r})m(\mathbf{r}') >= \frac{B \exp(-\kappa |\mathbf{r} - \mathbf{r}'|)}{|\mathbf{r} - \mathbf{r}'|} \quad \text{for} \quad \mathbf{r} \neq \mathbf{r}' , \tag{19a}$$

or

$$\ln(< n(\mathbf{r})m(\mathbf{r}') > * |\mathbf{r} - \mathbf{r}'|) = \ln(B) - \kappa |\mathbf{r} - \mathbf{r}'| \tag{19b}$$

The best fit for the simulation data based on Eq.(18) is plotted using the short dashed line on Figs. 6 and 7. It is clear that the functional form given by Eq.(19) describes the decay of nonequilibrium spatial correlations of concentration in a correct way. Equations (10) and (11) relate both B and κ with the diffusion and the rates of reactions. For a hard sphere system all these quantities can be easily found as the rate of collisions ν may be approximated using:

$$\nu = 4\sigma \sqrt{\frac{\pi k_B T}{m}} g(\sigma^+) \tag{20}$$

and all the reaction rates come from scaling ν by corresponding steric factors. The diffusion constant is equal to:

$$D = \frac{1}{16} \frac{\sigma}{\eta g(\sigma^+)} \sqrt{\frac{\pi k_B T}{m}}, \tag{21}$$

and the value of the radial distribution function at the point of contact of spheres $g(\sigma^+)$ may be obtained from the equation:

$$1 + 4\eta g(\sigma^+) = \frac{(1 + \eta + \eta^2 + \eta^3)}{(1 - \eta)^3} \tag{22}$$

The values of κ obtained from Eqs.(11, 20-22) are compared with the best numerical fit of the simulation data in the Table 1.

Table 1.
The values of κ (in σ^{-1} units) ; comparison of Eq.(11) with the numerical fit of simulation data.

system	I	I	II	II	III	III
reaction	b	c	b	c	b	c
κ_{theory}	0.60	1.01	4.04	6.82	18.8	31.7
κ_{fit}	1.73	2.57	6.25	9.58	25.9	39.5

It may be noticed that the values of κ, which describe the decay of observed correlations, are higher than those predicted by the theory what means that the nonequilibrium correlations are decreasing faster than it might be expected.

This effect is illustrated in Fig.8. The short dashed line shows the best numerical fit for the simulation data (and thus it approximates the short distance part of correlations in the most accurate way), whereas the long dashed curve uses κ coming from Eq.(11). It can be seen that κ obtained from the theory gives a better fit for the long distance decay of correlations than the numerical fit. The similar effect (faster decay of correlations at short distances than at long distances) was also observed in simulations based on Bird's method[17]. We think that the origin of discrepancy may be explained in the following way. Let us notice that to calculate κ from Eq.(11) we use the equilibrium values of the diffusion constant and the phenomenological rate constants. These values can be applied to describe large scale effects, whereas at very short distances the motion of spheres is correlated and multi-collision processes may have an important influence on the motion. Thus the decay of correlations at very short distances may be different than predicted by equilibrium quantities. This change in the rate of the decay can be nicely seen in Fig.8.

Next let us notice that if two different sets of the steric factors are considered then the ratio of $\kappa's$ obtained for these sets should not depend on systems densities. For the steric factors (c) and (b) this ratio is equal to 1.69. Although the values of κ obtained from the fit of simulation data differ by more than 50% from those predicted by the theory their ratios are equal to 1.49, 1.53 and 1.52 for the systems I, II and III respectively. At this point the agreement with the theory is remarkably good.

It is more difficult to compare the amplitudes of nonequilibrium correlations with the theory. First let us notice that the theory does not take the effect of excluded volume into account. Treating Eq.(9) literally and calculating the ratio of correlations corresponding to the sets of steric factors (b) and (c) at the distance equal to $|\mathbf{r} - \mathbf{r'}| = \sigma$ one obtains:

$$\frac{< n(\mathbf{r})m(\mathbf{r'}) >_{chem,(c)}}{< n(\mathbf{r})m(\mathbf{r'}) >_{chem,(b)}} = \frac{C_{NM,(c)}}{C_{NM,(b)}} \exp(-(\kappa_{(c)} - \kappa_{(b)})\sigma) \tag{23}$$

The ratio $C_{NM,(c)}/C_{NM,(b)}$ does not depend on density and it is equal to 6.27. The second factor in Eq.(23) decreases with the increasing density because the difference of $\kappa's$ is increasing. Thus, according to the theory, the ratio:

$$\frac{< n(\mathbf{r})m(\mathbf{r'}) >_{chem,(c)}}{< n(\mathbf{r})m(\mathbf{r'}) >_{chem,(b)}}$$

should decrease with the density. This conclusion is not supported by the results of simulations. The ratios between nonequilibrium correlations is equal to: 5.5,

4.7 and 4.5 for systems I, II and III, respectively. These values are similar to those, predicted by the theory for $|\mathbf{r} - \mathbf{r}'| \to 0$. Moreover, according to Eq.(9), if $|\mathbf{r} - \mathbf{r}'| \to 0$ than the ratio of the amplitudes of nonequilibrium correlations corresponding to two different densities but to the same set of steric factors is equal to the ratio of $g(\sigma^+)$ (if one excludes the factor c^2). For the densities considered this ratio is equal to:1.9 for (II/I) and 2.6 for (III/II). If we compare the amplitudes at the point of contact of spheres we obtain: 2.17 and 2.77 for the set (b) of steric factors and 1.85 and 2.63 for the set (c). Thus, for a hard sphere system $|\mathbf{r} - \mathbf{r}'|$ in Eqs.(9) and (23) should correspond to the distance between spheres rather than to the distance between their centers. The fact that the decay of correlations is different from the estimation based on the equilibrium diffusion coefficient, indicates that the nonequilibrium correlations may influence a short scale motion of molecules. We believe that the results obtained in our molecular dynamics simulations may be used as further tests of new theoretical approaches to this problem.

V. Conclusions

In this paper I have shown two simple applications of periodically expanded molecular dynamics technique. In the both cases (oscillations and nonequilibrium spatial correlations) the technique allowed to complete simulations, which seems to be extremely difficult if the standard methods are applied. The author believes that the periodically expanded molecular dynamics will find a lot of new applications in studies on nonequilibrium chemical systems.

Acknowledgments

The author is grateful to Professor Kazuo Kitahara for introducing him to the problem of nonequilibrium spatial correlations of reactant's concentrations. The author would like to thank the Computer Center of the Okazaki National Research Institutes where calculations were done. The final part of this work was supported by the grant no. 2 P303 018 05 provided by the Polish National Science Foundation (KBN).

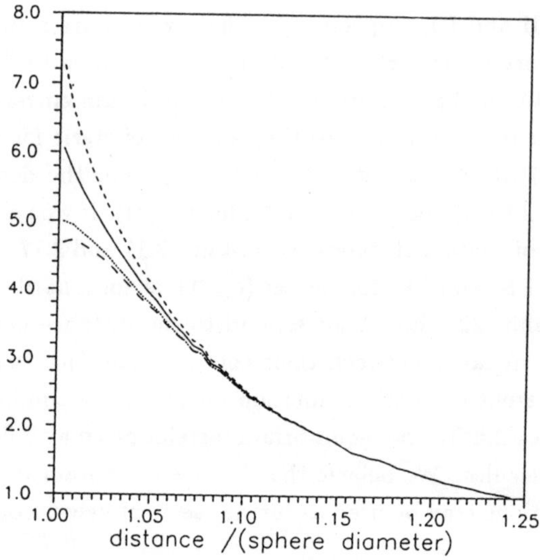

Fig. 5. The spatial correlations obtained in molecular dynamics simulations of the high density system (III) with the fast reaction (c). The average is calculated for 20000 consecutive collisions. The comparison between the radial distribution function for the system as a whole g (solid line) and the partial radial distribution functions: g_{XX} (dotted line, second curve from the bottom), g_{XY} (short dashed, the upper curve) and g_{YY} (long dashed, the bottom curve) is shown.

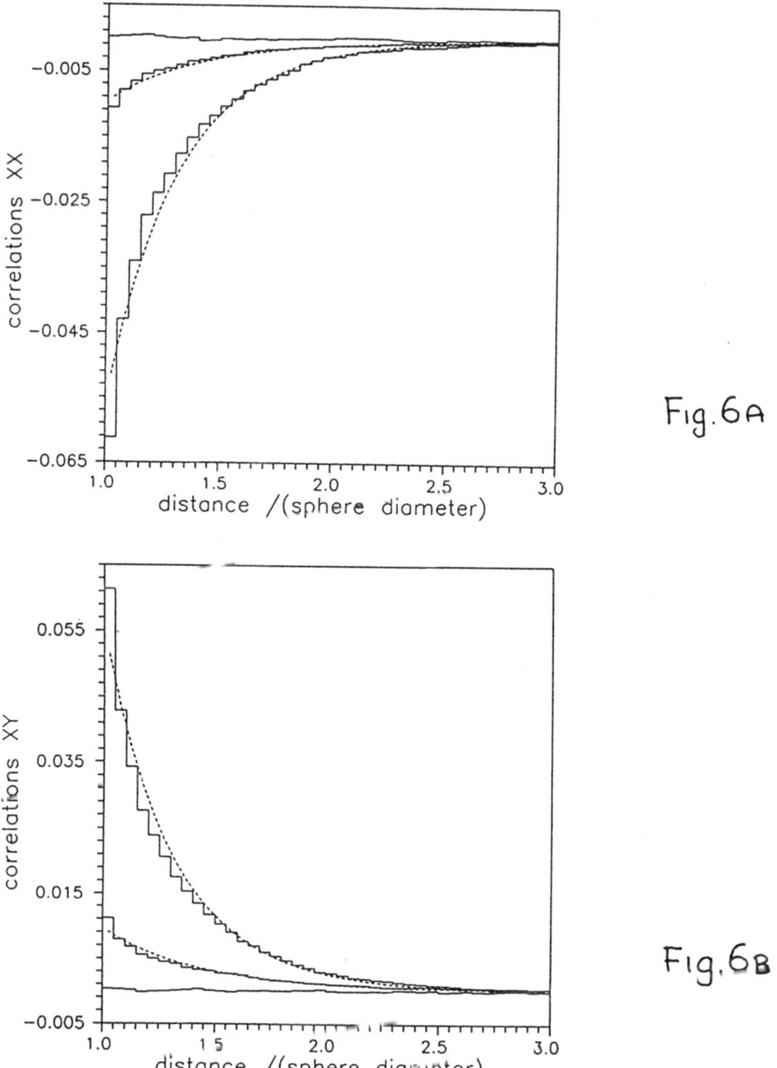

Fig. 6. The nonequilibrium spatial correlations in the system at low density (I, $\eta = 0.082$); the solid line shows the data obtained in simulations, whereas the dashed line is the fit for $< nm >_{chem}$ based on Eq.(19a). Fig. 6A: correlations between X and X, Upper curve - the system at the detailed balance (a), middle curve - slow reaction (b), lower curve - fast reaction (c); Fig. 6B: correlations between X and Y, lower curve - the system at the detailed balance (a), middle curve - slow reaction (b), upper curve - fast reaction (c).

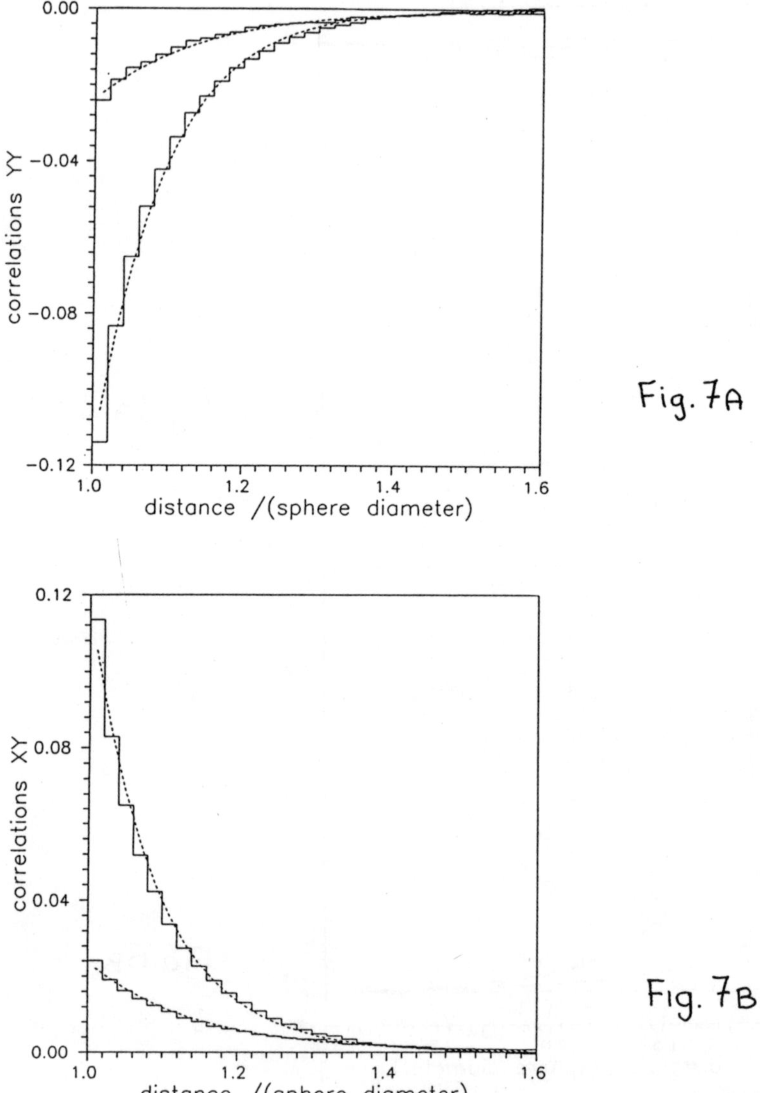

Fig. 7. The nonequilibrium spatial correlations $< nm >_{chem}$ in the system characterized by a middle density (II, $\eta = 0.29$); notation as in Fig. 6. Fig. 7A: correlations between Y and Y; the upper pair of curves - slow reaction(b), the lower pair - fast reaction(c); Fig. 7B: Correlations between X and Y; the upper pair of curves - fast reaction(c), the lower pair - slow reaction(b).

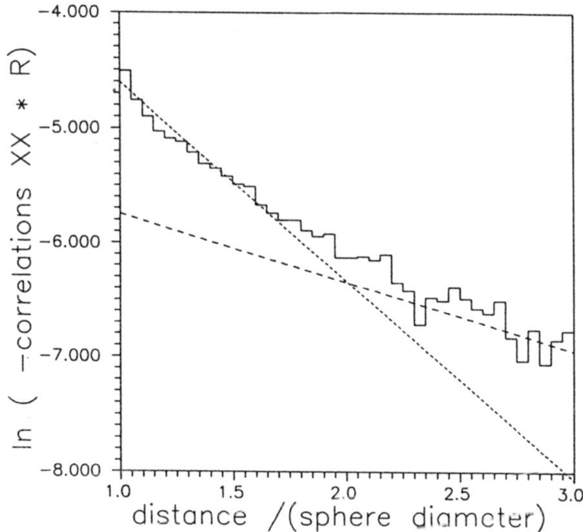

Fig. 8. A test for $< xx >_{chem}$ given by Eq.(18a) for a system with the set of steric factors (c). Solid line - molecular dynamics simulations; short dashed line - the best fit for B and κ; long dashed line - the fit with κ calculated from Eqs.(11,19).

References

[1] many interesting articles on microscopic simulations of nonequilibrium systems may be found in M. Mareschal and B.L. Holian, eds., *Microscopic Simulations of Complex Hydrodynamic Phenomena*, Plenum Press, New York 1992.

[2] M. Malek Mansour and F. Baras, Physica A **188**, 253 (1992).

[3] J. Gorecki and A.L. Kawczynski, J. Chem. Phys. **92** 7546 (1990).

[4] A.L. Kawczynski and J. Gorecki, J. Phys. Chem. **96** 1060 (1992).

[5] M.P. Allen and D.J. Tildesley, *Computer Simulation of Liquids*, Clarendon Press, Oxford 1987; W. G. Hoover, *Computational Statistical Mechanics*, Elsevier, Amsterdam 1991.

[6] G. A. Bird, *Molecular Gas Dynamics*, Clarendon, Oxford (1976).

[7] R. Kapral, J. Math. Chem. **6**, 113 (1991).

[8] R. Kapral, A. Lawniczak and P. Masiar, Phys. Rev. Lett. **66**, 2539 (1991); J. Chem. Phys. 96, 2762 (1992).

[9] R. Der and S. Fritzsche, Chem. Phys. Lett. **121**, 177 (1985).

[10] J. Gorecki, J. Popielawski and A.S. Cukrowski, Phys. Rev. **44** 3791 (1991).

[11] Gy. Pota, J. Chem. Phys. **78**, 1621 (1983).

[12] I. Prigogine and R. Lefever, J. Chem. Phys. **48** , 1695 (1969)

[13] A.L. Kawczynski and J. Gorecki, J. Phys. Chem. **97** 10358 (1993).

[14] the detailed analysis of spatial correlations in the system discussed in this chapter is presented in the manuscript: J. Gorecki, K. Kitahara, K. Yoshikawa and I. Hanazaki, *Molecular Dynamics Simulations of Nonequilibrium Spatial Correlations in Reaction Diffusion System*, prepared for publication.

[15] Y. Kuramoto, Prog. Theor. Phys. 52 (1974) 711.

[16] G. Nicolis and M. Malek Mansour, Phys. Rev. 29A (1984) 2845.

[17] K. Kitahara, K. Seki and S. Suzuki, J. Phys. Soc. Japan 59 (1990) 2309.

[18] G. Nicolis, A. Amellal, G. Dupont and M. Mareschal, Journal of Molecular Liquids, 41, (1989) 5.

[19] F. Baras, J. E. Pearson and M. Malek Mansour, J. Chem.Phys. 93 (1990) 5747; M. Mareschal and A. De Wit, J. Chem. Phys. 96 (1992) 2000.

[20] Y. Mori, P.K. Srivastava and I. Hanazaki, Physica D 50 (1991) 59; P.K. Srivastava, Y. Mori and I. Hanazaki, Chem. Phys. Lett. 177 (1991) 213.

[21] K. Kitahara, K. Seki and S. Suzuki in *Proceedings of Molecular Dynamical Processes ...* , ed. by A. Blumen et. al., World Scientific, Singapore (1990) p403.

Recent results on theory of Brownian motion

Akio Morita

Department of Chemistry, College of Arts and Sciences,

The University of Tokyo, Komaba, Meguro-ku, Tokyo 153, Japan

ABSTRACT

It is shown that the formulation of the generalized Langevin equation with the introduction of a projector leads to a serious difficulty that the correlation function of the position of a particle at time 0 and the random force at time t is constant which directly contradicts the fundamental requirements of time-dependent statistical mechanics. Also, starting with the Fokker-Planck-Kramers equation and the Boltzmann equation with the collision term by Bhatnagar, Gross and Krook, we have obtained the exact stationary solution for an open system with parabolic potentials possessing positive and negative force constants. And these results have been used for improvement of Kramers' theory on chemical kinetics.

1. Critical investigation of the generalized Langevin equation

In order to show our point clearly and simply, we take rather unconventional approach. As usual, we define the projector $\hat{P}$,

$$PX(t) = \frac{<v, X(t)>}{<v, v>} v \tag{1}$$

where $X(t)$ is a dynamic variable at time t, v is the velocity of a Brownian particle at time 0, and the angular brackets represent the ensemble average. We also employ the operator $\hat{Q}$ through the relation,

$$\hat{Q} = 1 - \hat{P} \tag{2}$$

and the Louville operator $\hat{L}$

$$\frac{dv(t)}{dt} = i\hat{L}v(t) \tag{3}$$

Now, we introduce the random acceleration $R(t)$ by the relation,

$$R(t) = e^{\hat{Q}i\hat{L}t}\dot{v} \tag{4}$$

which leads immediately to

$$\dot{R}(t) = \hat{Q}i\hat{L}R(t) = i\hat{L}R(t) - \frac{<v,i\hat{L}R(t)>}{<v,v>}v = i\hat{L}R(t) + \frac{<\dot{v},R(t)>}{<v,v>}v = i\hat{L}R(t) + \gamma(t)v \tag{5}$$

where the memory function $\gamma(t)$ is given by

$$\gamma(t) = \frac{<R,R(t)>}{<v,v>} \tag{6}$$

Because $\dot{R}(t) = i\hat{L}R(t) + \gamma(t)v$ in Eq. (5), we can write

$$R(t) = e^{i\hat{L}t}R(t) + \int_0^t \gamma(t')e^{i\hat{L}(t-t')}v dt' = \dot{v}(t) + \int_0^t \gamma(t-t')v(t')dt' \tag{7}$$

that is nothing but the well-known generalized Langevin equation (GLE). We have seen that $R(t)$ in Eq.(4) is equivalent to GLE. In view of Eqs. (1) and (2), it immediately follows that

$$\hat{Q}v = 0 \tag{8}$$

which leads to

$$<v,\hat{Q}i\hat{L}Y(t)> = <\hat{Q}v,i\hat{L}Y(t)> = 0 \tag{9}$$

where $Y(t)$ is a dynamic variable like $X(t)$ in Eq.(1). This equation together with $<v,\dot{v}> = 0$ that follows from the requirement of stationary condition leads to

$$<v,R(t)> = <v,e^{\hat{Q}i\hat{L}t}\dot{v}> = 0 \tag{10}$$

This relation is also well-known stating that $R(t)$ is always orthogonal to v. Now, let us consider

$$<x,\dot{R}(t)> = -<i\hat{L}\hat{Q}x,R(t)> \tag{11}$$

where $\dot{x}=v$ in which x is the position of the particle. Since $<v,x> = 0$, it is seen that

$$\hat{Q}x = x \tag{12}$$

which gives

$$<x,\dot{R}(t)> = -<v,R(t)> = 0 \tag{13}$$

This result is our main point. Equation (13) states that the correlation function, $<x,R(t)>$ should not depend on time, namely,

$$\langle x, R(t) \rangle = \langle x, R \rangle = \langle x, \dot{v} \rangle \tag{14}$$

Note that although $R(t)$ is always orthogonal to v, it is also always normal to x, furthermore, since x and v are orthogonal in view of $\langle x, v \rangle = 0$, $R(t)$ must be parallel with the direction of x with the time-invariant magnitude of R, which is quite strange. After all, the result in Eq. (14) contradicts with one of the most fundamental requirements in irreversible, non-equilibrium statistical mechanics that is

$$\lim_{t \to \infty} \langle x, R(t) \rangle = \langle x \rangle \langle R(t) \rangle = 0 \tag{15}$$

which follows from the fact that the random acceleration should not be correlated with anything else as time goes infinity. This is indeed a serious difficulty in the formulation of GLE. In the ordinary Langevin equation, the random acceleration correlates only with itself, thus allowing us to write

$$\langle x, R(t) \rangle = \langle x \rangle \langle R(t) \rangle = 0 \quad \text{for } white \ noise \text{ and } t > 0 \tag{16}$$

We see that GLE cannot produce the relation in Eq.(16), which is believed to hold as t becomes large. Hence a large amount of studies based on GLE not taking into account the present difficulty must be viewed very cautiously.

One of the well-known facts that cannot be interpreted by the ordinary Langevin equation is the velocity correlation functions of liquids, typically, helium, for example [1]. Morita has recently reconsidered this problem based on a fluctuating potential described by the following equation of motion [2]

$$\frac{d^2}{dt^2} x(t) + \beta \frac{d}{dt} x(t) + [\omega^2 + \lambda(t)] x(t) = w(t) \tag{17}$$

where $w(t)$ is white noise, $\langle w(t) \rangle = 0$ and $\langle w(t_1) w(t_2) \rangle = 2\varepsilon\delta(t_1 - t_2)$ and $\lambda(t)$ is dichotomous noise, $\langle \lambda(t) \rangle = 0$ and $\langle \lambda(t_1)\lambda(t_2) \rangle = E^2 exp(-\Gamma|t_1 - t_2|)$. In Fig. 1, the computer simulation result represented by circles is shown together with the theoretical curve obtained from Eq.(17). Note that the fitting is satisfactory when $\beta = 0$ as well as $w(t) = 0$ in which case irreversiblity is introduced through $\lambda(t)$ rather than $w(t)$. It is emphasized here that we can take into account the non-exponentially decaying velocity correlation function in terms of the fluctuating potential within the frame of the ordinary Langevin function.

240

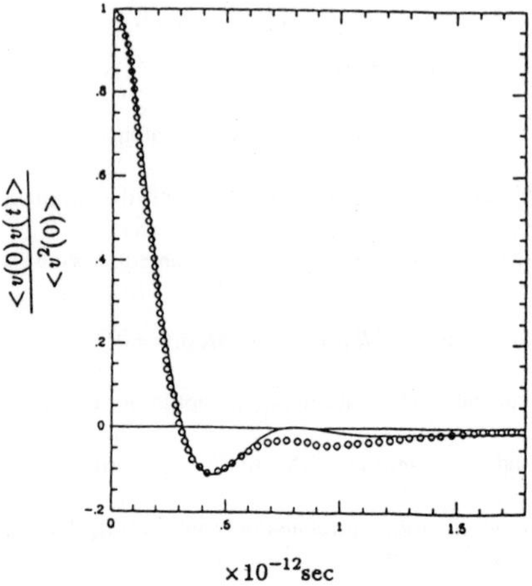

Fig. 1 Comparison of velocity correlation function.

2. Exact stationary solution of FPK and BBGK for an open system and its application to chemical kinetics theory

Another part of the report is concerned with the exact non-equilibrium stationary solutions obtained from FPK and BBGK, which are

$$\frac{\partial P(x,v,t)}{\partial t} + v\frac{\partial P}{\partial x} + \frac{f(x)}{m}\frac{\partial P}{\partial v} = \beta\frac{\partial(vP)}{\partial v} + \beta\frac{k_B T}{m}\frac{\partial^2 P}{\partial v^2} \tag{18}$$

and

$$\frac{\partial P(x,v,t)}{\partial t} + v\frac{\partial P}{\partial x} + \frac{f(x)}{m}\frac{\partial P}{\partial v} = \beta\sqrt{\frac{m}{2\pi k_B T}}\exp\left[-\frac{mv^2}{2k_B T}\right]\int_{-\infty}^{\infty} P(x,v,t)dv - \beta P \tag{19}$$

respectively. To this end, on setting $[\partial P(x,v,t)/\partial t] = 0$, we should find the expression for the distribution function, $\rho(x)$ by eliminating the velocity dependence from the relation,

$$\rho(x) = \int_{-\infty}^{\infty} P(x,v)dv$$

The solutions were successively obtained using computer assisted symbolic manipulation software for parabolic potentials which are given by

$$V_c(x) = V_0 - \frac{1}{2}m\,\omega_c^2(x-x_c)^2 \tag{20}$$

and

$$V_a(x) = \frac{1}{2}m\omega_a{}^2 x^2 \tag{21}$$

Since the detailed derivation is lengthy and described elsewhere [3], we just describe here the final results in the following. If the stationary process for an open system is considered, the flux J should not depend on x in view of the continuity equation. Our results were obtained assuming the Maxwell distribution for the velocity.

$$\phi_c J = -\frac{k_B T}{m} e^{-V_c(x)/k_B T} \frac{d}{dx}\left[e^{V_c(x)/k_B T} \rho(x) \right] \quad (for\ FPK) \tag{22}$$

and

$$\phi_a J = -\frac{k_B T}{m} e^{-V_a(x)/k_B T} \frac{d}{dx}\left[e^{V_a(x)/k_B T} \rho(x) \right] \quad (for\ FPK) \tag{23}$$

where

$$\phi_c = \frac{\beta + \sqrt{\beta^2 + 4\omega_c{}^2}}{2} \tag{24}$$

and

$$\phi_a = \frac{\beta - \sqrt{\beta^2 - 4\omega_a{}^2}}{2} \tag{25}$$

We see that ϕ_c can change from $2\omega_c$ to ∞ as β from 0 to ∞, while ϕ_a decreases from $2\omega_a$ to 0 as β goes from $2\omega_a$ to ∞. It is important to point out that there exists the lower limit for β in order to keep the system stationary. This value corresponds to the critical damped case for an ordinary damping oscillator. Under the value, the system starts oscillating which breaks up the stationary motion. It is also apparent that by regarding the left hand sides of Eqs. (22) and (23) as apparent flux as a function of β, the large value of β induces the barrier crossing of the potential top whereas it decreases the movement of the particle in the potential well. In other words, the large value of β leads to the far from equilibrium distribution for potential in Eq. (20) while it gives rise to the near equilibrium behaviour described by the Boltzmann distribution for potential in Eq. (21). Whereas for BBGK, we find that

$$\phi_c{}^{(b)} J = -\frac{k_B T}{m} e^{-V_c(x)/k_B T} \frac{d}{dx}\left[e^{V_c(x)/k_B T} \rho(x) \right] \quad (for\ BBGK) \tag{26}$$

and

$$\phi_a{}^{(b)} J = -\frac{k_B T}{m} e^{-V_a(x)/k_B T} \frac{d}{dx}\left[e^{V_a(x)/k_B T} \rho(x) \right] \quad (for\ BBGK) \tag{27}$$

where

242

$$\frac{\omega_c}{\phi_c^{(b)}} = \cfrac{1}{y_c^{-1}+\cfrac{1\cdot2}{y_c^{-1}+\cfrac{2\cdot3}{y_c^{-1}+\cfrac{3\cdot4}{y_c^{-1}+\cdots}}}} = \frac{1}{2}y_c^{-1}\left[\psi\left[\frac{y_c^{-1}}{4}+\frac{1}{2}\right]-\psi\left[\frac{y_c^{-1}}{4}\right]\right]-1 \tag{28}$$

$$\frac{\omega_a}{\phi_a^{(b)}} = \int_0^\infty sech^2\,u\;e^{iy_a^{-1}u}\,du = -\frac{1}{2}iy_a^{-1}\left[\psi\left[-i\frac{y_a^{-1}}{4}+\frac{1}{2}\right]-\psi\left[-i\frac{y_a^{-1}}{4}\right]\right]-1 \tag{29}$$

In the above equations,

$$y_c^{-1} = \frac{\beta}{\omega_c} \qquad y_a^{-1} = \frac{\beta}{\omega_a}$$

and

$$\psi(z)= \frac{d ln\;\Gamma(z)}{dz}$$

where $\psi(z)$ and $\Gamma(z)$ are the psi and gamma functions, respectively. As seen from Fig. 2, the behaviour of ϕ_c and $\phi_c^{(b)}$ is very similar, whereas $\phi_a^{(b)}$ is always complex for an arbitrary β, which does not leads to the stationary solution for potential in Eq. (21). In BBGK collisions are so strong that they do not allow for system's stationarity.

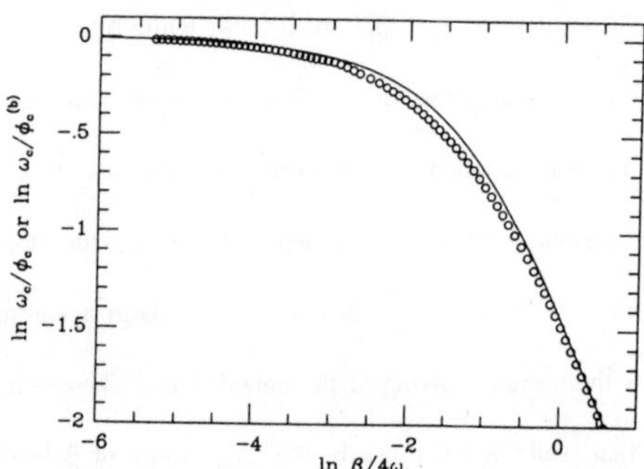

Fig. 2 Plots of $ln\,(\omega_c/\phi_c)$ (full curve) and $ln\,[\omega_c/\phi_c^{(b)}]$ (dotted curve) vs $ln\,(\beta/4\omega_c)$.

With the above results, we could improve Kramers' treatments of the reaction rate process by avoiding his assumption of FPK reducing to an ordinary differential equation. We see that we have the lower limit for β and the Kramers' turn-over shows up when the motion of the particle at the potential well cannot be neglected. In Fig. 3, experimental data [4] for isomerization reaction of stilbene between the cis and trans forms are shown together with theoretical ones. The same experiments are considered by the author who introduced random force of dichotomous colored noise. The result is shown in Fig. 4 [5].

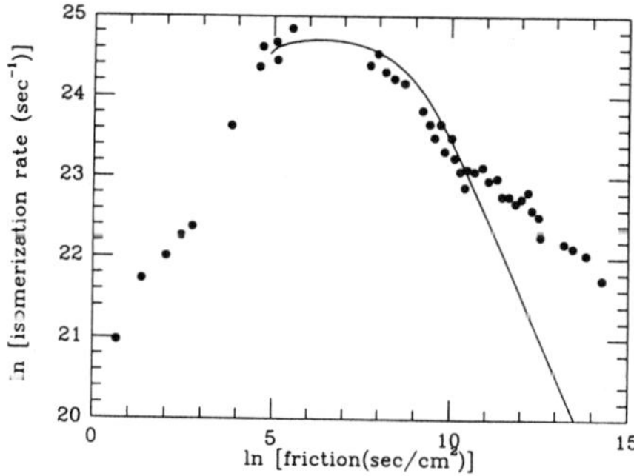

Fig. 3 Comparison of experimental data with the result from improved Kramers' theory.

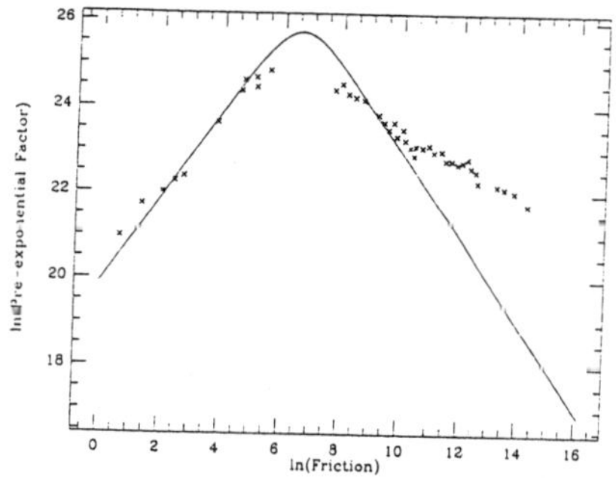

Fig. 4 Comparison of experimental data with the result given in ref. 5.

References

1. I. Prigogine, "*Spatial inhomogeneities and transient behaviour in chemical kinetics*", Edited by P. Gray, G. Nicolis, F. Baras, P. Borckmans and S. K. Scott, Manchester Univ. Press, U. K. (1990)

2. A. Morita, *Molecular Simulations*, **6**, 221, (1991)

3. A. Morita, *J. Chem. Phys.*, **96**, 3678 (1992)

4. G. R. Fleming and P. G. Wolynes, *Phys. Today*, **43**, 2715 (1990)

5. A. Morita, *J. Math. Chem.*, in press

STOCHASTIC MODELING OF BRANCHED-CHAIN REACTIONS ON THE SURFACE IN HETEROGENEOUS CATALYTIC PROCESSES OF OXIDATION OF GASES

E.L. Pechatnikov [1], V.V. Barelko [1], R. Danielak [2] and M. Frankowicz [2]

[1] *Institute of Chemical Physics of the Russian Academy of Sciences Chernogolovka, Russia*
[2] *Department of Theoretical Chemistry, Jagiellonian University Kraków, Poland*

Abstract

We consider a stochastic model of heterogeneous catalytic oxidation of gases on a surface. This model can be derived under certain assumptions both from Langmuir-Hinshelwood and Eley-Rideal mechanisms. The characteristic feature of the model is the kinetics of active centers on the surface; their diffusion and the branched-chain mechanism of their multiplication. The model reproduces experimentally observed phenomena: hysteresis in temperature dependence of reaction rate and 'memory' effects. It is shown that in the limit of strong diffusion results of the Monte Carlo simulations agree with the deterministic description based on the law of mass action. However, the decrease of the rate of the surface diffusion results in the deviation of the behavior of the stochastic model from the deterministic one (the breakdown of the law of mass action due to local effects like clusterization). In particular, the narrowing (and even disappearance) of the region with multiple steady states as well as the jumpwise changes of the critical initial conditions for the initiation of a new phase are obtained. The wavelike propagation of a new phase in the absence of diffusion is observed. The distribution of the induction time as a function of diffusion and temperature is also investigated.

1   Introduction

Whereas branched-chain mechanisms for gas phase reactions are well known and have been investigated during several decades [1], ideas that such mechanisms are also present in the field of heterogeneous catalysis are relatively new [2]. A branched-chain model was applied to describe critical effects of non-thermal origin in oxidation of NH_3, H_2, CO and hydrocarbons at Pt surface. The above model was applied to explain: the mechanism of surface recrystallization during reaction (so-called catalytic etching) [3, 4], non-thermal autowave phenomena in catalysis and the effect of spatial stratification [5]. It was also confirmed by data on temperature-stimulated phase transitions [6].

In the analysis of dynamic behavior of a catalytic surface the model [2] introduces the stages of birth, branching and death of active centers known from the theory of gas phase branched-chain reactions [1]. It is based on physical ideas developed by Frenkel [7] and generally accepted in crystal science as TLK (terraces, ledges, kinks) theory. According to this theory, a two-dimensional quasi-gas of atoms torn out from the lattice and adsorbed on the surface (so-called adatoms) is in equilibrium with the crystal lattice. In the terms of branched-chain theory adatoms may be interpreted as active centers[1], their emerging from the lattice as the stage of 'birth', and their return back to the lattice as 'death'. The original notions of Frenkel were supplemented with the idea that the energy released in a catalytic act may lead to excitation of the surface and creation of excited active centers, which in turn may activate other atoms of the lattice (the stage of branching).

An adequate and perspective tool for analysis of such problems are stochastic simulations [9]. These methods were successfully applied to model the nonlinear behavior in oxidation processes on the surface [10]. It was found that even for simple systems stochastic models may show specific critical effects. The purpose of the present work is to demonstrate capabilities of a stochastic branched-chain model to simulate such experimental phenomena as hysteresis in temperature dependence of reaction rate, memory effects, statistical character of induction time near critical points. The last example demonstrates considerable deviations from the law of mass action in the case of a finite value of diffusion rate.

2 Kinetic scheme of branched-chain process on the catalytic surface

Following [2] we assume that a surface without reaction consists of active centers (adatoms, Z) and passive centers (Z_0), being in dynamic equilibrium; Z can appear e.g. by thermal excitation of the lattice or disappear (e.g. by returning to the lattice or by sublimation).

$$Z_0 \longrightarrow Z \tag{1}$$

$$Z \longrightarrow Z_0 \tag{2}$$

The equilibrium may be shifted by temperature (i.e. lattice energy) changes or by chemical reactions occurring at the surface. In the following we will describe

---

[1]The reasoning will be the same if instead of adatoms one treats vacancies as active centers [8].

246

the whole catalytic process (adsorption-desorption, reactions between adsorbed species) only by considering states of catalytic centers. Of course, the behavior of real systems is a superposition of chemical interactions and possible 'reactions' between centers. Nevertheless under certain conditions we can neglect chemical reactions and still get phenomena observed for the full process.

Let us take as an example the process of gas oxidation on a surface of a catalyst. We will first consider Langmuir-Hinshelwood mechanism, and next Eley-Rideal mechanism [11]. In the case of the Langmuir-Hinshelwood mechanism we assume that the rates of processes of adsorption-desorption

$$O_2 + 2Z \; \rightleftarrows \; 2OZ \tag{3}$$

$$X + Z \; \rightleftarrows \; XZ \tag{4}$$

are large in comparison with the process (1-2), the reaction (leading to the appearance of excited centers, Z^*)

$$OZ + XZ \longrightarrow 2Z^* + XO \tag{5}$$

and with processes of energy dissipation

$$Z^* + Z_0 \longrightarrow Z + Z \tag{6}$$

$$Z^* \longrightarrow Z \tag{7}$$

In the above formulas OZ is oxygen atom adsorbed on an active center Z, X - other reactant (CO, H_2 or else), XZ is adsorbed reactant. We assume that the deexcitation of Z^* may lead to the creation of the new active center from a passive one (stage 6) or to energy dissipation to the lattice (stage 7). Stage (6) represents the branching of the chain. We assume that O_2 and X concentrations in the gaseous phase are constant. With above assumptions we can take the ratio $OZ : XZ : Z = $ const and identify these species as Z. Thus we may replace stages (3-5) by

$$Z + Z \longrightarrow 2Z^* \tag{8}$$

The Eley-Rideal scheme:

$$
\begin{array}{lcll}
O_2 + 2Z & \rightleftarrows & 2OZ & (a) \\
X + OZ & \longrightarrow & XOZ & (b) \\
XOZ & \longrightarrow & XO + Z^* & (c)
\end{array}
\tag{9}
$$

in the case of large adsorption rate of X (stage 9b) and desorption rate of product (9c) may be reduced to the summary process (8) too.

So the final form of kinetic scheme in both cases is:

$$
\begin{array}{rcll}
Z_0 & \xrightarrow{k_1} & Z & (a) \\
Z & \xrightarrow{k_2} & Z_0 & (b) \\
Z + Z & \xrightarrow{k_3} & 2Z^* & (c) \\
Z^* + Z_0 & \xrightarrow{k_4} & 2Z & (d) \\
Z^* & \xrightarrow{k_5} & Z & (e) \\
Z_i + Z_j & \longrightarrow & Z_j + Z_i & (f)
\end{array}
\tag{10}
$$

where $k_1 - k_5$ are rate constants of corresponding stages. In addition to chemical-like transitions we have included diffusion (10f), Z_i and Z_j stand for Z, Z_0 or Z^*.

3 The algorithm for Monte Carlo modeling

We take a lattice of $n \times n$ square cells with periodic boundary conditions. A cell may be in one of three possible states: Z_0, Z and Z^*. We have $n_{Z_0} + n_Z + n_{Z^*} = n^2$.

Every transition in scheme (10) is connected with number p_i , $0 < p_i < 1$ ($i = 1, ..., 5$) – probability of transitions (a-e) respectively.

All variants of possible transitions: monomolecular (10a,b or e) and these involving interactions with one of m neighbors [2] (10c,d) - are considered for a randomly chosen cell. We calculate the probability for a cell to keep its state as

$$
P_j^0 = (1 - q_j) \prod_{i=1}^{m} (1 - q_{ij}), \; j = 1, 2, ..., n^2.
\tag{11}
$$

Here j denotes a cell, q_j - probability of corresponding monomolecular transition (one of p_1, p_2, p_5) for the species in the cell, q_{ij} - the probability of reaction with i-th neighbor (p_3, p_4).

The probability that some reaction occurs

$$
P_j^r = 1 - P_j^0
\tag{12}
$$

is subdivided proportionally to $q_j, q_{1j}, ..., q_{mj}$ and the choice of the process is performed by random number generator.

Diffusion is simulated by exchange of two randomly chosen adjacent cells[3]. The diffusion rate D is given by the average number of diffusive exchanges for a cell in a unit of time (see Section 5.2 below).

---

[2] Numerical experiments were carried out with m=4 and n in the range from 40 to 100. We have not found any essential n-dependence in our results for n from this range.

[3] Note that when the states of both cells are identical, nothing changes.

4   The deterministic model

The deterministic model of a well mixed system with homogeneous initial conditions is described by the following set of equations

$$\dot{x} = -W_1 + W_2 - W_4,$$
$$\dot{z} = 2W_3 - W_4 - W_5, \tag{13}$$
$$x + y + z = 1.$$

Here W_i denote reaction rates defined in accordance with the law of mass action:

$$W_1 = k_1 x, \ W_2 = k_2 y, \ W_3 = k_3 y^2, \ W_4 = k_4 xz, \ W_5 = k_5 z,$$

and

$$x = n_{Z_0}/n^2, \ y = n_Z/n^2, \ z = n_{Z^*}/n^2,$$

are concentrations of Z_0, Z and Z^* respectively, $k_i = p_i$ for monomolecular stages (10a,b,e), $k_i = mp_i$ where m is a number of neighbors for bimolecular stages with the same reagents (10c) and $k_i = 2mp_i$ for bimolecular stages with different reagents (10d). The last equation in (13) is a conservation law for this system.

5   Numerical results and discussion

We will first compare the model (13) and MC results. Then we will give stochastic interpretation of some experimental critical phenomena, mentioned in the Introduction. Finally, we will consider some special properties of the stochastic model.

5.1   Assumptions for modeling

The stationary points of system (13) may be defined from the algebraic equation of fourth degree, but the result is too complicated for the analysis. However, some usual physical assumptions are helpful in deriving simple expressions for roots and the condition for the existence of three stationary points.

First, we should assume that the concentration of excited centers is small and consider the conservation law

$$x + y = 1 \tag{14}$$

instead of (13.3). Further, let us consider steady states with low and high activity separately.

The state with low activity exists, if $W_1 > W_4$. In the critical point they are equal; it means that

$$k_1 = k_4 z, \tag{15}$$

i.e., the state with low activity becomes unstable (disappears) when the contribution to the active centers concentration due to the process of branching becomes equal to the contribution from the process of spontaneous birth. Thus, from (13.1), (14), and equality (15), concentrations of centers in this critical point are

$$x^{ign} = \frac{k_2}{2k_1 + k_2}, \ y^{ign} = \frac{2k_1}{2k_1 + k_2}, \ z^{ign} = \frac{k_1}{k_4}, \tag{16}$$

and the condition for the validity of the 'reduced' conservation law (14) for the lower branch is $k_4 \gg k_1$. The substitution of (16) to stationary equation (13.2) yields an equation for the rate constants in the critical point for which the state with low activity disappears

$$2k_3 \left(\frac{2k_1}{2k_1 + k_2} \right)^2 - k_4 \frac{k_1}{k_4} \frac{k_2}{2k_1 + k_2} - k_5 \frac{k_1}{k_4} = 0. \tag{17}$$

To analyze the state with high activity we may neglect the contribution to the process from the rate of spontaneous birth of active centers W_1. In this case the equation (13.1) yields

$$k_2 y - k_4 x z = 0. \tag{18}$$

The system of stationary equation (13.2) and equations (14), (18) reduces to a quadratic equation. The corresponding determinant is non-negative and in the critical point it equals to zero

$$\Delta \equiv (k_2 + 2k_3)^2 - 8k_2 k_3 (1 + \frac{k_5}{k_4}) \geq 0. \tag{19}$$

Equality (19) imposes restrictions on rate constants in the critical point, where steady state with high activity dissappears.

Formulae (17) and (19) approximately describe the region of coexistence of multiple steady states for the model (13). This region has rather complicated shape in the space of parameters $(k_1, ..., k_5)$. The following assumptions are made to simplify the analysis and to reduce our problem to the one with a single temperature-dependent parameter.

First, we may fix the rate constants of monomolecular stages (10a,b,e). It means that we fix the life-times of active centers Z and Z^* and thus neglect their temperature dependence. This dependence should be taken into account in more detailed considerations, as the main cause of increase of the catalytic activity with temperature is evidently the increase of the number of active centers. Second assumption concerns the character of branching process. We assume that the probability for an excited center to react with a single neighboring passive center (step (10d)) equals to the probability of deexcitation (step (10e)) and it is much larger than the rates of other processes. We take the rate constant k_5 of monomolecular stage (10e) equal to 1 choosing the appropriate time scale. To provide an equality between p_4 and p_5 we should take $k_4 = 8$.

These assumptions allow us to concentrate attention on the role of the most important stage from the point of view of the mechanism of branching, namely

250

the birth of excited centers Z^* in process (10c). So we keep only one temperature-dependent rate constant k_3 and treat its variation as the variation of temperature. The interval of non-uniqueness of steady states is

$$k_3^{ext} < k_3 < k_3^{ign}, \qquad (20)$$

where $k_3^{ext} = k_2(k_a + (k_a^2 - 1)^{1/2})/2$ (from (19));
$\quad k_3^{ign} = k_b(k_2 k_4 + k_b k_5)/(8 k_1 k_4)$ (from (17));
$\quad k_a = 1 + 2k_5/k_4$; $k_b = 2k_1 + k_2$.

Numerical experiments were carried out with $k_1 = 1 \cdot 10^{-3}$, $k_2 = 1.5 \cdot 10^{-2}$, $k_4 = 8$ and $k_5 = 1$ in (13). For this set of the rate constants the region of two steady states is in the range $1.476 \cdot 10^{-2} < k_3 < 3.658 \cdot 10^{-2}$. The estimation (20) yields $k_3^{ext} = 1.500 \cdot 10^{-2}$, $k_3^{ign} = 3.639 \cdot 10^{-2}$.

5.2 Comparison of deterministic model (13) and MC simulations.

Fig.1 and 2 show transitions to a single stationary state for model (13) and MC modeling. The equilibrium state without chemical reaction $x = k_1/(k_1 + k_2)$, $y = k_2/(k_1 + k_2)$, $z = 0$ is used as an initial conditions. The Monte Carlo time step is Δt; it is proportional to probabilities p_i, (i=1,...,5). The model (13) is valid in the limit of large D and small Δt.

In the absence of diffusion ($D = 0$) we have significant deviations from the homogeneous phenomenological kinetics (Fig.1, curves 1 and 2). For large D and small values of rate constants k_1, k_2 and k_3, even if Δt is relatively large deterministic and stochastic results practically coincide (Fig.1, curves 1 and 3).

The role of Δt's value becomes significant, if the rate constants of stages (10a-e) are of the same order. Results presented in Fig.2 demonstrate a significant discrepancy between stochastic and deterministic results (curves 1, 2). Note that differences appear not only in dynamics, but also for steady states. If Δt decreases the stochastic results approach to the deterministic ones even in absence of diffusion (curves 1 and 3). It is rather obvious: if we solve differential equations we take small time increments. Nevertheless for some fast chemical reactions, apparently, we cannot neglect the time of chemical conversion in comparison with the characteristic time of process. In this case the results obtained from a model with finite increment of time may be more physically justified than these obtained from model with continuous time.

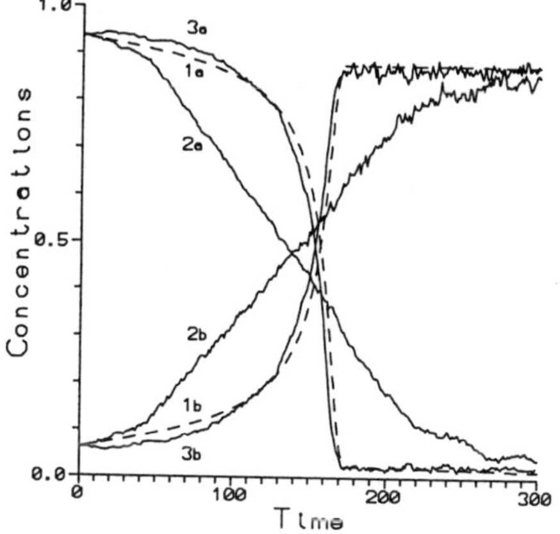

Figure 1: The influence of the rate of diffusion D on the dynamic behavior of MC-model in comparison with model (13), $k_1 = 10^{-3}$, $k_2 = 1.5 \cdot 10^{-2}$, $k_3 = 8 \cdot 10^{-2}$, $k_4 = 8$, $k_5 = 1$,
  1 (dashed) — results of (13);
  2, 3 (solid) — MC-model;
  2 — $D = 0$, $\Delta t = 0.1$;
  3 — $D = 10$, $\Delta t = 1$;
  a — concentration of Z_0,
  b — concentration of Z.

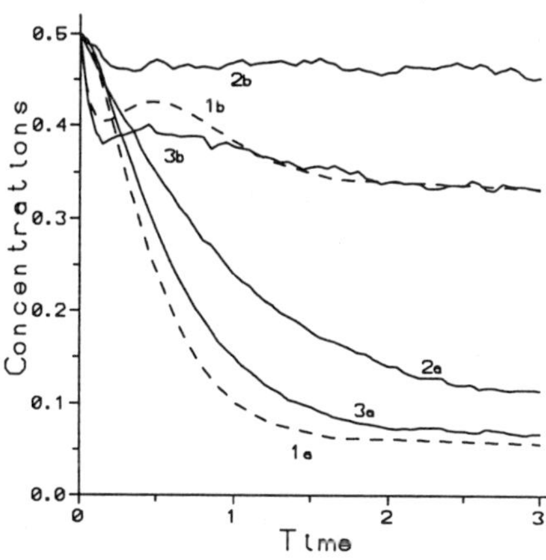

Figure 2: The influence of Δt value on the dynamic behavior and steady state of MC-model compared with model (13), $k_1 = 1$, $k_2 = 1$, $k_3 = 4$, $k_4 = 8$, $k_5 = 1$.
  1 (dashed) — results of (13);
  2, 3 (solid) — MC-model;
  2 — $D = 30$, $\Delta t = 1$;
  3 — $D = 0$, $\Delta t = 0.025$;
  a — concentration of Z_0,
  b — concentration of Z.

5.3 Critical phenomena caused by branched-chain birth of active centers

The dependencies of concentration of active centers [4] $Z_a = Z + Z^*$ on k_3 ('temperature') are shown in Fig.3. Solid lines show values of steady states as functions of k_3 for the model (13), points marked by crosses are steady states calculated by Monte-Carlo simulations with $D = 0.5$. Slow increase of k_3 leads to a jump in stationary activity from lower to upper branch at k_3^{ign}. If k_3 is slowly decreased, the jump from upper to lower branch occurs at k_3^{ext}, $k_3^{ext} < k_3^{ign}$. These results qualitatively reproduce experimental data on hysteretic dependence of reaction rate on temperature[5]. The cause of these phenomena is a nonlinear law for birth and a linear one for death of Z. Notice that the region of non-uniqueness of steady state is narrower if diffusion is smaller. In the limit case $D = 0$ this region is extremely small, if exists. Corresponding values of the steady states are marked by circles in

---

[4] We use the sum of Z and Z^* as a measure of lattice activity, because of their leading role in activation process.

[5] k_3^{ign} and k_3^{ext} correspond to the temperature of ignition, T^{ign} and of extinction, T^{ext} [2].

252

Fig.3 . The system demonstrates non-periodic oscillations of concentrations with significant amplitude in the narrow range of k_3 values. We may consider this range as the region of non-uniqueness of steady state.

Before the discussion of simulations of the phenomenon of 'memory' of a catalyst, we describe briefly corresponding experiments [2]. The catalyst working in the regime of high activity is quickly cooled down to the temperature lower than T^{ext}. After some period of time temperature is increased again. The shorter the period of keeping the catalyst at low temperature is, the lower is the temperature of a following ignition. As the catalyst 'remembers' its previous active state, this phenomenon is called the *memory effect*.

In the terms of branched-chain model the memory effect is interpreted as keeping high active centers concentration on the surface during some period after cooling. In our numerical experiments we imitate increasing of cooling period by decreasing the initial concentration of active centers. Then we determine whether this concentration is sufficient to reach a state with high activity. Dashed curve in Fig.3 gives dependence of unstable stationary states on k_3 for the model (13). Error bars show the critical initial concentration of active centers for several k_3 values for $D = 0.5$. For large values of D these critical values should be near to dashed line.

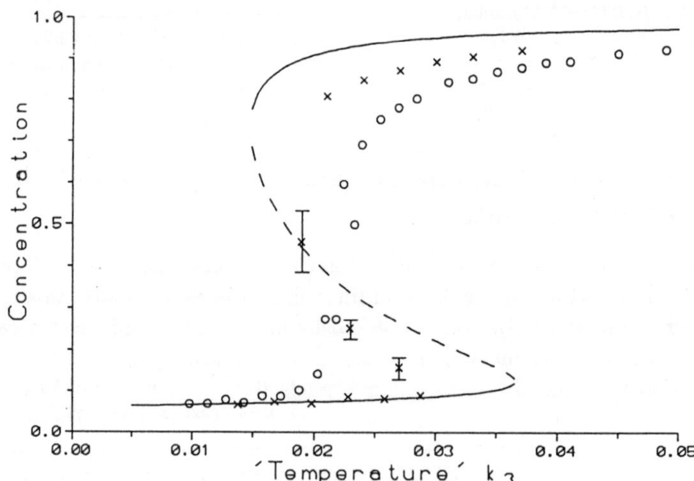

Figure 3: Dependence of stable (solid lines) and unstable (dashed line) stationary Z_a concentration on k_3 for model (13), other rate constants as for Fig.1, $k_3^{ext} = 1.476 \cdot 10^{-2}, k_3^{ign} = 3.658 \cdot 10^{-2}$;

    $\times$ — average values on stable branches for MC-model, $D = 0.5$, $k_3^{ext} = 1.79 \pm 1 \cdot 10^{-2}$, $k_3^{ign} = 2.89 \pm 1 \cdot 10^{-2}$;

    $\circ$ — $D = 0$, $k_3^{ext} = 2.1 \pm 1 \cdot 10^{-2}$, $k_3^{ign} = 2.3 \pm 1 \cdot 10^{-2}$;

    $\mathbf{I}$ — critical initial active centers concentration with error bars (for 50 runs), $D = 0.5$.

5.4 Some phenomena of stochastic nature

Up to now we have discussed phenomena appearing both in the deterministic and in the stochastic models. However, capabilities of stochastic models for description of kinetic experiments are in general far richer. It is connected with the fundamental discreteness and stochastic nature of surface processes, especially nonlinear ones. The characteristic parameters of nonlinear processes, such as critical temperature, induction time etc. obtained experimentally are in fact random variables. The comparison of experimental and calculated distributions may give an additional tool for identification of proper reaction mechanisms.

Fig.4 reproduces typical distributions of induction time. We take k_3 slightly larger than k_3^{ign} for the model (13) and begin simulations from lattice completely filled with passive centers Z_0. We define the induction time τ as the time of achievement of 50% of stationary Z_a concentration. As one could expect, the average value and dispersion of induction times decrease with increase of the distance between k_3 and k_3^{ign} (distributions 1 and 2 on Fig.4). The increase of diffusion results in the enlargement of induction time (Fig.4, distributions 1 and 3).

Another reason for differences between stochastic models and models of type (13) comes from the breakdown of the law of mass action for surface processes. This law cannot describe phenomena connected with clusterization of species on a surface. A bright example of complex behavior of simple surface system which disobeys the law of mass action is described in [11]. Critical phenomena in such systems are difficult for theoretical studies, even by such methods as the mean-field approach [12] or the transfer matrix method [13].

As an illustration of the role of clusterization let us consider a process of transition to the state of high activity, in particular the dependence of critical conditions of such transition on diffusion rate. The numerical procedure is organized as follows. The rate k_3 is chosen in the range of existence of two steady states for model (13). The lattice is filled randomly with Z_0 and Z cells in given ratio. The initial ratio is classified as *subcritical*, if the process leads to the lower steady state and *supercritical* in the opposite case.

Corresponding dependencies are shown in Fig.5 . The most interesting result is that the dependence of Z^{crit} on D has a jumpwise characteristics. For low values of diffusion rate the system attains the active state even if the lattice is fully covered by Z_0 at the beginning. It means that if diffusion is absent or relatively small the passive steady state loses its stability. The spontaneous local appearance of the active phase gives rise to its wavelike propagation on the lattice. The explanation of this fact is that diffusion hinders clusterization of Z and decreases the branching rate because of reaction (10c). As diffusion does not influence the death rate of Z, the resulting probability of Z cluster growth is larger for smaller diffusion rates. The symmetric situation has place in the case of extinction. So the system under conditions of small diffusion evolves in essentially inhomogeneous way. The propagation of the activity wave is caused only

254

by chemical reaction and it is not connected with a transfer of active centers.

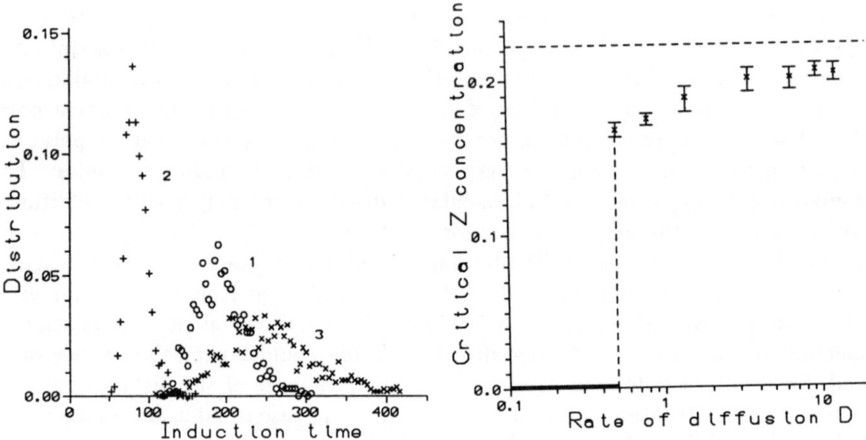

Figure.4: Distribution of induction times for MC-model. k_1, k_2, k_4, k_5 as for Fig.1.
 $1 - k_3 = 5 \cdot 10^{-2}, D = 0$, average time $t_{av} = 193.7$, variance $\sigma = 32.0$;
 $2 - k_3 = 10 \cdot 10^{-2}, D = 0, t_{av} = 85.4$, $\sigma = 13.5$;
 $3 - k_3 = 5 \cdot 10^{-2}, D = 0.5, t_{av} = 263.4$, $\sigma = 68.6$.

Figure 5: Jumpwise dependence of critical initial concentration of Z centers, necessary for ignition, $k_3 = 0.03$, other rate constants as for Fig.1. $D^{crit} = 0.50 \pm 2$. Horizontal dashed line — the limit case of strong diffusion.

The mechanism of autowave propagation, described above, may be characterized as inductive one. It differs fundamentally from classical autowaves in systems with heat and/or mass transfer, described by reaction-diffusion equations. The difference is that inductive autowaves can be observed only for low diffusion rates, whereas classical autowaves exist only in diffusive media and disappear if transfer processes are suppressed (wave velocity tends to zero) [5]. The inductive mechanism presented above is close to the one introduced in [14] (so-called axiomatic model for excited media in biology) and it needs a special consideration.

References

[1] N.N.Semenov, On some problems of chemical kinetics and reactivity, AN SSSR Press, 1958 (in Russian).

[2] Yu.E. Volodin, V.V. Barelko and P.I. Khalzov, Chem. Eng. Commun. 18 (1982) 271.

[3] K.T. Baker and R.B.Thomas, Platinum Metals Rev. 18 (1974) 130.

[4] E.S. Gen'kin, Ya.A. Gal'chenko, V.V. Barelko, Yu.E. Volodin and V.I. Chernyshev, Doklady AN SSSR 316, 658 (English translation: Plenum Publishing Corp. 1991, p. 84-86).

[5] Yu.E. Volodin, V.N. Zvyagin, A.N. Ivanova and V.V. Barelko, in Adv. Chem. Phys.,Vol. 77, ed. by I. Prigogine and S.A. Rice (1990), p. 551.

[6] E.L. Pechatnikov and V.V. Barelko, Sov. J. Chem. Phys. 9 (1992)1345-1363.

[7] Ya.I. Frenkel, Zhurnal eksper. i teor. fiziki 16 (1946) 13 (in Russian); Ya.I. Frenkel, in Ya.E. Geguzin (ed.), Surface diffusion and spreading, Moscow, Nauka, 1969 (in Russian).

[8] S.Kishimoto, Bull. Chem.Soc. of Japan 48 (1975) 1937.

[9] See e.g. X.-G. Wu and R. Kapral, Physica A 188 (1992) 284 and references therein.

[10] R.M. Ziff, E. Gulari and Y. Barshad, Phys. Rev. Lett. 56 (1986) 2553-2556.

[11] R.P.H. Gasser, An introduction to chemisorption and catalysis by metals, Clarendon Press, Oxford, 1985.

[12] A.V. Myshlyavtsev, J.L. Sales, G. Zgrablich and V.P. Zhdanov, J. Stat. Phys. 58 (1990) 1029.

[13] R. Dickman, Phys. Rev. A34 (1986) 4246.

[14] N. Wiener and A. Rosenblueth, Arch. Inst. Cardiologia de Mexico 16 (1946) N 3, p.4.

256

KRAMERS-MOYAL TYPE EXPANSION FOR SYSTEMS
DRIVEN BY NONLINEAR NOISE: AN EXAMPLE

Jerzy Łuczka[*] and Adam Gadomski[**]

[*] Department of Theoretical Physics, Silesian University,

40-007 Katowice, Poland

[**] Department of Polymer Physics, Silesian University,

41-200 Sosnowiec, Poland

ABSTRACT. *A linear system driven by quadratic Ornstein-Uhlenbeck noise is considered. An exact evolution equation for a probability density is presented. It has the form of a generalized non-local diffusion equation with a defined integral kernel. The evolution equations is transformed to a Kramers-Moyal type equation. The problem of the truncation of this equation to the Fokker-Planck one is discussed, in particular the limit of a small correlation time of the noise is studied.*

1. INTRODUCTION

The role of random processes in physics, chemistry and biology has become more and more important (see, e. g., [1] and refs. therein). Stochastic processes (noises) appear in modeling of evolution of a system with a large number of degrees of freedom as a result of elimination of irrelevant variables. The system is effectively described by a few dynamic equations for relevant variables in which random parameters are involved. Here we consider a one-dimensional case for which such an equation has the form

$$\dot{x} = F(x, \xi(t)), \quad \dot{x} = dx/dt \qquad (1)$$

$x \in \mathbb{U}$ - a phase space of the system, $\xi(t)$ - a random perturbation.

Eq. (1) defines certain stochastic process $x(t)$ and the problem of determination of its probabilistic characteristics is fundamental. Some simple characteristics can be inferred from a one-dimensional probability distribution $p(x, t)$ of the process $x(t)$. Its evolution is determined by an infinitesimal generator L of the process $x(t)$ defined by the equation

$$\frac{\partial}{\partial t} p(x, t) = (Lp)(x, t) \tag{2}$$

If an infinitesimal generator $\mathscr{L}$ of the noise $\xi(t)$ is known,

$$\frac{\partial}{\partial t} \rho(\xi, t) = (\mathscr{L}\rho)(\xi, t) \tag{3}$$

where $\rho(\xi, t)$ is a probability density for $\xi(t)$ and $\xi \in A$ (A - a phase space of the process $\xi(t)$), then for a joint process $\{x(t), \xi(t)\}$ one gets [2]

$$\frac{\partial}{\partial t} \mathbb{P}(x, \xi, t) = (\mathbb{L}\mathbb{P})(x, \xi, t)$$

$$= -\frac{\partial}{\partial x} [F(x, \xi)\mathbb{P}(x, \xi, t)] + (\mathscr{L}\mathbb{P})(x, \xi, t) \tag{4}$$

where $\mathbb{P}(x, \xi, t)$ is a probability distribution for a joint process $\{x(t), \xi(t)\}$ and $\mathbb{L}$ is its infinitesimal generator.
We are interested not in $\mathbb{P}(x, \xi, t)$ but rather in $p(x, t)$. How to get $p(x, t)$ from known Eq. (4)? By the use of the equation

$$p(x, t) = \int_A \mathbb{P}(x, \xi, t)d\xi \tag{5}$$

(if ξ belongs to a discrete set then an integral is replaced by a sum), we get a relation between (4) and (2). This is a typical contraction problem in dynamics that is based on reduction of unobserved ("fast") variables (here, the noise $\xi(t)$ in (4)) by their elimination. In some cases it is a trivial task, but generally it is a nontrivial problem.

2. LINEAR NOISE

This is the case described by

$$\dot{x} = f(x) + g(x)\xi(t) \tag{6}$$

As an example, let us the consider Ornstein-Uhlenbeck (correlated) noise [3],

$$\langle\xi(t)\rangle = 0, \quad \langle\xi(t)\xi(s)\rangle = \alpha D \exp(-\alpha|t-s|) \tag{7}$$

258

and $\xi(t)$ is the (stationary) Gaussian process.

Now, Eq. (4) has the form [3]

$$\frac{\partial}{\partial t} \, \mathbb{P}(x, \, \xi, \, t) = - \, \frac{\partial}{\partial x} \, [f(x) + g(x)\xi]\mathbb{P}(x, \, \xi, \, t)$$

$$+ \, \alpha \, \frac{\partial}{\partial \xi} \, \mathbb{P}(x, \, \xi, \, t) + \alpha^2 D \, \frac{\partial^2}{\partial \xi^2} \, \mathbb{P}(x, \, \xi, \, t) \qquad (8)$$

The exact reduction of Eq. (8) to Eq. (2) is possible only for linear models,

$$\dot{x} = -ax + \xi(t), \qquad x \in \mathbb{R} \qquad (9)$$

$$\dot{x} = - \, [a + \xi(t)]x, \qquad x \in \mathbb{R}^+ \qquad (10)$$

For example, for the process (9), Eq. (2) has the form [4]

$$\frac{\partial}{\partial t} \, p(x, \, t) = a \, \frac{\partial}{\partial x} \, xp(x, \, t) + D(t) \, \frac{\partial^2}{\partial x^2} \, p(x, \, t) \qquad (11)$$

where

$$D(t) = \frac{\alpha D}{\alpha + a} \, [1 - e^{-(\alpha + a)t} \,] \qquad (12)$$

The process $x(t)$ defined by Eq. (9) is non-Markovian! Its probability distribution satisfies the Fokker-Planck equation (11) with the diffusion function $D(t)$ of (12). The same equation (11) describes a Markovian process driven by non-stationary Gaussian white noise (non-correlated noise),

$$\dot{x} = -ax + \Gamma(t), \qquad x \in \mathbb{R} \qquad (13)$$

$$<\Gamma(t)> = 0, \qquad <\Gamma(t)\Gamma(s)> = 2D(t)\delta(t-s) \qquad (14)$$

Because no analytical theory exists for the general case (6), many approximate methods have been elaborated. Part of them is based on approximate Fokker-Planck equations [3, 5, 6]

$$\frac{\partial}{\partial t} \, p(x, \, t) = - \, \frac{\partial}{\partial x} \, f(x)p(x, \, t)$$

$$+ \, \frac{\partial}{\partial x} \, g(x) \, \frac{\partial}{\partial x} \, g(x)\mathbb{D}(x, \, t)p(x, \, t) \qquad (15)$$

with modified diffusion functions $\mathbb{D}(x, t)$. A critical analysis of this approach is carried out in Ref. [7].

3. NONLINEAR NOISE

If in Eq. (1), the function $F(x, \xi(t))$ is **nonlinear** in $\xi(t)$, then $\xi(t)$ cannot be modeled by the white noise. The process $\xi(t)$ should be correlated (with non-zero correlation time τ_c). Then $x(t)$ is non-Markovian and the white-noise limit **does not exist** in a general case. Is it now possible to approximate an exact evolution equation for $p(x, t)$ by a diffusion-type equation of the form (15)? In Ref. [8] a general theorem is proved but it is difficult to check assumptions involved in it for the given model. In Refs. [9, 10] some results concerning a model with quadratic noise are presented. This model is given by

$$\dot{x} = -ax + \xi^2(t), \qquad x \in \mathbb{R} \tag{16}$$

with the Ornstein-Uhlenbeck noise $\xi(t)$ of Eq. (7).

As a possible application of Eq. (16) one can mention input – output systems [11], thermoelectrical instruments [12], turbulent fluid flows [13] and quantum optics [14].

In a series of papers [15   ], one of us considered the problem (16) and some special cases (for specific values of the ratio α/a) were analyzed. Here we present for the first time exact results valid for all values of parameters of the model (16) and (7). Using the same method as in Ref. [15] one can show that the reduced probability distribution (5) satisfies the equation

$$\frac{\partial}{\partial t} p(x, t) = \frac{\partial}{\partial x} [ax - \alpha D]p(x, t)$$

$$+ \alpha D \frac{\partial^2}{\partial x^2} \int_{-\infty}^{\infty} \mathbb{H}(x-y; t)p(y, t)dy \tag{17}$$

where

$$\mathbb{H}(x; t) = (1/2\pi)\int_{-\infty}^{\infty} e^{-i\omega x} H(\omega, t)d\omega \tag{18}$$

$$H(\omega,\ t)\ =$$

$$\frac{1}{i\omega}\ \frac{J_{\nu-1}(s_\omega)J_{-\nu+1}(s_\omega e^{-at/2})-J_{-\nu+1}(s_\omega)J_{\nu-1}(s_\omega e^{-at/2})}{J_{\nu+1}(s_\omega)J_{-\nu+1}(s_\omega e^{-at/2})-J_{-\nu-1}(s_\omega)J_{\nu-1}(s_\omega e^{-at/2})}$$

$$(19)$$

and

$$s_\omega = 2\nu(iD\omega)^{1/2},\qquad \nu = -\ 2\alpha/a \qquad\qquad (20)$$

and J_ν stands for the Bessel function.

Eq. (17) is an integro-differential equation and has an interesting structure and interpretation. The first term of the right-hand side of (17) is a drift with a noise-induced part αD. It comes from the averaged equation (16). If the kernel $\mathbb{H}(x;\ t)$ was proportional to the Dirac delta, $\mathbb{H}(x;\ t)\ \propto\ \delta(x)$, then the second term would be purely diffusive. Therefore $\mathbb{H}(x;\ t)$ can be named a space-nonlocal diffusion function and Eq. (17) can be interpreted as **a space-nonlocal diffusion equation.**

It is argued [8-10] that if the correlation time $\tau_c = 1/\alpha$ of the noise $\xi(t)$ is small enough then $\xi(t)$ is close to white noise and $x(t)$ differs little from the process described by the corresponding Fokker-Planck equation. Such approximated equations are obtained in Refs. [9, 10]. From our Eq. (17) we can derive an exact Kramers-Moyal series of the form

$$\frac{\partial}{\partial t}\ p(x,\ t) = \frac{\partial}{\partial x}\ [ax - \alpha D]p(x,\ t)$$

$$+ \sum_{n=2}^{\infty} \varepsilon_n(a,\alpha,D,t)\ \frac{\partial^n}{\partial x^n}\ p(x,\ t) \qquad\qquad (21)$$

The coefficients $\varepsilon_n(a,\alpha,D,t)$ can be calculated explicitly, but here we present ε_2 and ε_3 in the long-time limit, $t\to\infty$,

$$\varepsilon_2(a,\alpha,D) = \frac{2(\alpha D)^2}{a+2\alpha}\ ,\qquad \varepsilon_3(a,\alpha,D) = \frac{-4(\alpha D)^3}{(a+2\alpha)(a+\alpha)} \qquad (22)$$

The structure of ε_n $(n \geq 4)$ can be inferred from (22).

We cannot truncate the series (21) at an arbitrary order. According to the Pawula theorem [16], $p(x, t)$ as a solution of an equation of the type (21) is non-negative, $p(x, t) \geq 0$, for three cases only:

(a) if $\varepsilon_n = 0$ for $n \geq 2$ or

(b) if $\varepsilon_n = 0$ for $n \geq 3$ or

(c) one should keep all terms .

The limit of short correlation time of the noise does not generally exist: if $\alpha \to \infty$ then ε_n are of orders α, noise induced drift term is of order α as well as the deterministic part of drift is $(-ax)$. So, the series (21) diverges. But if $a \equiv 0$ or $a \propto \alpha$ then the limit $\alpha \to \infty$ exists, the white noise limit exists!. Nevertheless, Eq. (21) cannot be reduced to the Fokker-Planck form. A diffusion-type approximation is rather artificial (we formally truncate the series putting $\varepsilon_n \equiv 0$ for $n \geq 3$). This approximation leads to the Gaussian distribution which is symmetrical and does not disappear for all (x, t) [15]. As it follows from (16), the relation

$$x(t) \geq x(0) \exp(-at) \tag{23}$$

holds for all realizations of $x(t)$ and therefore $p(x, t) \equiv 0$ for $x < x(0)\exp(-at)$. Hence, $x(t)$ cannot be described by the Gaussian probability and ordinary truncation of Eq. (21) is incorrect.

ACKNOWLEDGMENTS

The authors thank the Organizers of the Symposium "Far-from-Equilibrium Dynamics of Chemical Systems" (September 6-10, 1993, Borki, Poland) for financial support.

The work supported in part by the KBN (Poland).

REFERENCES

1. P. Hänggi, P. Talkner and M. Borkovec, Rev. Mod. Phys. 62 (1990) 251.

2. N. G. van Kampen, Stochastic processes in physics and chemistry (North-Holland, Amsterdam, 1987), Chap. XIV.

3. J. Łuczka, Physica A 153 (1988) 619;

4. R. F. Fox, Phys. Rev. A 34 (1986) 4525.

5. J. Masoliver, B. J. West, Phys. Rev. A 35 (1987) 3086.

6. R. F. Fox, J. Stat. Phys. 46 (1987) 1145.

7. J. Łuczka, Phys. Lett. A **139** (1989) 29; Acta Phys. Polon. A **77** (1990) 427.

8. R. L. Stratonovich, Conditional Markov processes and their application to optimal control (Elsevier, Amsterdam, 1968).

9. V. I. Tikhonov and M. A. Mironov, Markovian processes (Sovietskoe Radio, Moscow, 1977) [in Russian].

10. M. San Miguel and J. M. Sancho, Z. Phys. B - Condensed Matter **43** (1981) 351.

11. W. B. Davenport, jr., R. A. Johnson and D. Middleton, J. Appl. Phys. **23** (1952) 377.

12. V. I. Tikhonov, Zh. Tekh. Fiz. **XXV** (1955) 817.

13. H. H. Shen, Physica A **159** (1989) 273.

14. M. Mizuno and K. Ikeda, Physica D **36** (1989) 327.

15. J. Łuczka, J. Stat. Phys. **42** (1986) 1009; **45** (1986) 309; **47** (1987) 505.

16. H. Risken, The Fokker-Planck equation (Springer, Berlin, 1989).

THE ROLE OF SPECIES TEMPERATURES IN NONEQUILIBRIUM REACTIVE EFFECTS IN MIXTURES

Duncan Napier and Bernie D. Shizgal[†]

Department of Chemistry, University of British Columbia, Vancouver, British Columbia, Canada
[†] also with the Department of Geophysics and Astronomy

Abstract

The nonequilibrium effects for a model reactive system, $A + B \rightarrow$ products, that arise from the perturbation of the distribution functions of both components from Maxwellian are studied. The main objectives of these calculations is the evaluation of the fractional decrease of the nonequilibrium rate coefficient from the equilibrium value. The reactive process causes the temperatures of the two species to differ from the system temperature and this effect can play an important role in the determination of the departure of the rate coefficient from the equilibrium value. These effects are often examined with the Chapman Enskog method of solution the Boltzmann equation, which we refer to as the Weak Non-Equilibrium (WNE) approach. A second Chapman-Enskog type method, involves the expansion of the distribution functions about Maxwellians at different temperatures and is referred to as Strong Non-Equilibrium (SNE) approach. The present paper considers the departure of the species temperatures in a reactive system in terms of the WNE and SNE methods.

I. Introduction

The departure from equilibrium of chemically reactive systems has been of concern for over four decades [1]-[10] and continues to be of considerable interest as a fundamental kinetic theory problem, as well as of practical importance. The nonequilibrium effects associated with the reentry of space vehicles in the terrestrial atmosphere, with shock wave phenomena, plasma processing of materials, and other systems play an important role in the understanding of these chemically reactive processes for real systems. The description of the transport processes for molecular systems coupled to chemical nonequilibrium remains an important problem [11]. The main objective is the theoretical description of physical situations that may be far from equilibrium.

The main purpose of this paper is to reexamine the nonequilibrium effects associated with a simple reactive system of the type,

$$A + B \rightarrow \text{products},$$

where the reactive process perturbs the distribution functions of both species from the equilibrium Maxwellian distributions and the rate coefficient differs from the equilibrium rate coefficient. If $k^{(0)}$ and k are, the equilibrium and nonequilibrium rate coefficients, respectively, the fractional decrease of the nonequilibrium rate coefficient from the equilibrium rate coefficient defined by,

$$\eta = [k - k^{(0)}]/k^{(0)}, \tag{1}$$

is the main objective of these studies. The determination of these nonequilibrium effects in a model system has been studied in detail in the previous papers [1]-[10]. Additional references to earlier papers are to be found in the papers referenced. Shizgal and Karplus [6] showed that in two component reactive systems there is an important effect associated with the departure of the species temperatures from the system temperature. The main objective of the present paper is to reconsider these nonequilibrium effects in light of the discussions by Cukrowski et al [9]. The work is also motivated by the general methods of solution of the Boltzmann equation introduced by Pascal and Brun [11] in their study of transport processes in molecular systems.

The usual approach to this problem is based on the treatment of the reactive process as a small perturbation of the elastic collision processes that bring the system to equilibrium. This is equivalent to the recognition that reactive cross sections are typically much smaller than elastic cross sections. The distribution functions are expressed in terms of Maxwellians at the *same temperature* plus a small correction that arises because of the reactive process. The departure of the distribution functions from Maxwellian is then determined with a Chapman-Enskog type approach [12] that has been described at length in previous papers. We briefly review this approach in Section II of the paper. This approach has been referred to as weak nonequilibrium (WNE) by Pascal and Brun [11] in their recent study of transport processes in molecular systems. In this case the distribution function for the molecular system is expanded about a single temperature even though the translational and vibrational temperatures may differ. The different species temperatures in the present paper are somewhat analogous to different translational and vibrational temperatures in molecular systems.

In Section III, we consider the expansion of the distribution functions about Maxwellians at *different temperatures* and consider a Chapman-Enskog type procedure to determine the departure from these local Maxwellians. Pascal and Brun refer to this approach as strong nonequilibrium (SNE) and in their application expand the distribution function about equilibrium Boltzmann distributions at different translational and vibrational temperatures. As discussed by Pascal and Brun, the results with SNE do not coincide with WNE when the vibrational and translational temperatures coincide.

The present paper is a preliminary report of the formalism involved with the WNE and SNE approaches. Further details of the calculations with extensive results will be presented in a forthcoming publication. In the present work, we carry out a numerical integration of the lower order hydrodynamic equations for the temperatures of two reactive species and compare the long time values with the Chapman-Enskog results, that is, with the WNE results. We show that the two approaches coincide when the requisite separation of time scales exists.

II. The Chapman-Enskog Approach - Weak Nonequilibrium (WNE)

The methodology for the study of the nonequilibrium effects in the reactive process has been presented in earlier papers [5]-[8]. We here present a brief overview of the approach. The distribution functions for species $\gamma = 1$ and 2, for this spatially homogenous system in the absence of external forces, are assumed to be given by two coupled Boltzmann equations of the form,

$$\frac{\partial f_1}{\partial t} = \frac{1}{\epsilon}[I_{11} + I_{12}] - R_{12} \tag{2}$$

$$\frac{\partial f_2}{\partial t} = \frac{1}{\epsilon}[I_{22} + I_{21}] - R_{21}, \tag{3}$$

where $I_{\gamma\eta}$ are the like $(\gamma = \eta)$ and unlike $(\gamma \neq \eta)$ elastic collision operators, parameterized by elastic cross section $\sigma_{\gamma\eta}$ and defined explicitly elsewhere ([5] - [8],[12]). The quantities $R_{\gamma\eta}$ are the reactive collision terms that depend on the reactive cross section σ_{12}^*. These time dependent equations are solved with the assumption that the reactive cross sections are small relative to the elastic cross sections. The parameter $\epsilon \approx \sigma/\sigma^*$ has been inserted into the Boltzmann equations to take this ordering of collision terms into account. We thus set,

$$f_\gamma = f_\gamma^{(0)}[1 + \epsilon\psi_\gamma] \tag{4}$$

and the equations of order $1/\epsilon$ obtained with Eq. (4) in Eqs. (2) and (3) determined so that the zero order distribution functions, are the Maxwellians characterized by time dependent number densities, $n_\gamma(t)$, and the single temperature, $T(t)$. The term, ψ_γ, is the perturbation from the Maxwellian.

The Chapman-Enskog approach [12] involves the assumption that a steady state occurs in a time short on the reactive time scale and that the time dependence is implicit through the variation with time of the number densities and the temperature, that is,

$$\frac{\partial f_\gamma}{\partial t} = \frac{\partial f_\gamma^{(0)}}{\partial n_\gamma}\left(\frac{dn_\gamma}{dt}\right)^{(0)} + \frac{\partial f_\gamma^{(0)}}{\partial T}\left(\frac{dT}{dt}\right)^{(0)},$$

where the variation with time is determined from the Boltzmann equations.

With the substitution of the Eq. (4) into the Boltzmann equations, Eqs. (2) and

(3), we find the Chapman-Enskog equations for the perturbations ψ_γ, by equating terms of zero order in ϵ,

$$J_{11}(\psi_1) + J_{12}(\psi_1) + J_{12}(\psi_2) = f_1^{(0)} H_1(c_1) \tag{5}$$

$$J_{22}(\psi_2) + J_{21}(\psi_2) + J_{21}(\psi_1) = f_2^{(0)} H_2(c_2) . \tag{6}$$

The first collision operators on the LHS of Eqs. (5) and (6) are the *self* collision terms for A-A and B-B elastic collisions. The second and third collision operators take account of A-B collisions and couple the two equations. The inhomogeneous terms, H_γ are determined by the strength of the reactive process and involve the reactive collision term, R_{12}, and the terms that result from $\partial f_\gamma / \partial t$, preceeding Eq. (5).

The solution of these coupled linear integral equations is obtained with the expansion of the unknown functions ψ_γ in Sonine (Laguerre) polynomials, $S_\gamma^{(i)}(x_\gamma^2)$ defined in previous papers [5], [6], and we get the set of algebraic equations for the coefficients $a_j^{(\gamma)}$ given by,

$$\sum_{j=1}^{N} \left[\left(n_1^2 [S_1^{(i)}, S_1^{(j)}] + n_1 n_2 \{ S_1^{(i)}, S_1^{(j)} \} \right) a_j^{(1)} + n_1 n_2 \{ S_1^{(i)}, S_2^{(j)} \} a_j^{(2)} \right] = n_1 n_2 \alpha_i^{(1)}, \tag{7}$$

and

$$\sum_{j=1}^{N} \left[\left(n_2^2 [S_2^{(i)}, S_2^{(j)}] + n_1 n_2 \{ S_2^{(i)}, S_2^{(j)} \} \right) a_j^{(2)} + n_1 n_2 \{ S_2^{(i)}, S_1^{(j)} \} a_j^{(1)} \right] = n_1 n_2 \alpha_i^{(2)}, \tag{8}$$

where the bracket matrix elements $[S_\gamma^{(i)}, S_\gamma^{(j)}]$, the brace matrix elements $\{ S_\gamma^{(i)}, S_\eta^{(j)} \}$ and the coefficients $\alpha_i^{(\gamma)}$ are defined elsewhere [6], [7]. Since the density is defined in terms of $f_\gamma^{(0)}$, $\int f_\gamma^{(0)} \psi_\gamma d c_\gamma = 0$, and hence $a_0^{(\gamma)} = 0$. An important aspect of this WNE approach is that the *single* temperature is defined by

$$\tfrac{3}{2} n k T = \int f_1^{(0)} \tfrac{1}{2} m_1 c_1^2 d c_1 + \int f_2^{(0)} \tfrac{1}{2} m_2 c_2^2 d c_2 \tag{9}$$

where $n = n_1 + n_2$. As a consequence of this definition we have

$$n_1 a_1^{(1)} + n_2 a_1^{(2)} = 0. \tag{10}$$

Consistent with this definition of the temperature is that the two equations in Eqs. (7) and (8) with $i = 1$ are the negative of one another; their sum being equal to zero is a reflection of conservation of energy when both species are taken into account. Consequently, the set of equations that are solved is the set with Eq. (10) replacing either of the two equations with $i = 1$ in Eqs. (7) and (8).

The solution of Eqs. (7) and (8) together with Eq. (10) as discussed above yields the expansion coefficients and the fractional decrease in the equilibrium rate of reaction is given by,

$$\eta = -\sum_{\gamma=1}^{2}\sum_{i=1}^{N} a_i^{(\gamma)} A_i^{(\gamma)}/A_0, \tag{11}$$

where the $A_i^{(\gamma)}$ integrals are defined by

$$A_i^{(\gamma)} = \int\int\int f_1^{(0)} f_2^{(0)} S_\gamma^{(i)} \sigma^* g\, d\Omega\, dc_1\, dc_2.$$

An important aspect of this treatment of the nonequilibrium effects is that the temperatures of each species differ from the common temperature of the system, that is,

$$T_\gamma = T^{(0)}[1 - a_1^{(\gamma)}] \tag{12}$$

III. Strong Nonequilibrium (SNE)

Since for some instances, it appears that the species temperatures play an important role [6], [7], [9], [10], it is useful to consider the expansion of the distribution functions about local Maxwellians characterized by densities n_γ and *different* species temperatures, T_γ, that remain to be specified. This is equivalent to the assumption that the rate of collisions between unlike species is very slow relative to the self collision rate. This is an approach that is considered in plasma systems, in ionospheric applications and is also employed in the determination of mobility of ions in neutral gases. The Boltzmann equations, Eqs. (2) and (3), are rewritten in the form,

$$\frac{\partial f_1}{\partial t} = \frac{1}{\epsilon} I_{11} + I_{12} - R_{12} \tag{13}$$

$$\frac{\partial f_2}{\partial t} = \frac{1}{\epsilon} I_{22} + I_{21} - R_{21}, \tag{14}$$

where the term in $1/\epsilon$ multiplies only the self-collision integral terms. The collision term between unlike species is considered to be much smaller than the like species collision terms. The exchange of energy between components is slow owing to a small σ_{12} cross section or a disparate mass ratio. The distribution function is now written in the form,

$$f_\gamma = F_\gamma^{(0)}[1 + \phi_\gamma], \tag{15}$$

where $F_\gamma^{(0)} = n_\gamma(t)[m_\gamma/2\pi k T_\gamma(t)]^{3/2} \exp[-m_\gamma c^2/2k T_\gamma(t)]$ is the local Maxwellian with a *different* temperature for each species. These Maxwellians are the solutions to the equation of order $1/\epsilon$ that results with the substitution of Eq. (15) into Eqs. (13) and (14). To lowest order, the time dependence of the species densities corresponds to the integral of R_{12} over all velocities and the time dependence of the species temperatures is obtained from the Boltzmann equations, Eqs. (13)

268

and (14), cf. [6], [7]. With the substitution of this expansion into the Boltzmann equations, Eqs. (2) and (3) and the evaluation of

$$\frac{\partial F_\gamma^{(0)}(n_\gamma, T_\gamma)}{\partial t} = \frac{\partial F_\gamma^{(0)}}{\partial n_\gamma}\left(\frac{dn_\gamma}{dt}\right)^{(0)} + \frac{\partial F_\gamma^{(0)}}{\partial T_\gamma}\left(\frac{dT_\gamma}{dt}\right)^{(0)},$$

in terms of the variation of the densities and the temperatures, yields the uncoupled set of integral equations,

$$J_{\gamma\gamma}(\phi_\gamma) = F_\gamma^{(0)}G_\gamma(c_\gamma) \quad \gamma = 1,\ 2. \tag{16}$$

The inhomogeneous terms G_γ are evaluated much the same way as H_γ except that they are parametrized by the different species temperatures, and the cross collision integral term is also include together with the reactive terms. We employ the expansions,

$$\phi_\gamma(y_\gamma) = \sum_{i=2}^{N} b_j^{(\gamma)} S_\gamma^{(j)}(y_\gamma^2), \tag{17}$$

and get the set of algebraic equations for the coefficients $b_j^{(\gamma)}$ given by,

$$\sum_{j=2}^{N} n_1^2[S_\gamma^{(i)}, S_\gamma^{(j)}]b_j^{(\gamma)} = n_1 n_2 \beta_i^{(\gamma)}, \quad \gamma = 1,\ 2 \tag{18}$$

where the square bracket matrix elements $[S_\gamma^{(i)}, S_\gamma^{(j)}]$ are the one-component matrix elements employed in an earlier paper [5]. The coefficients $\beta_i^{(\gamma)}$ depend on the two temperatures T_1 and T_2. The integrals in Eq. (18) involve the lowest order matrix elements of the unlike species collision operator [13]. Since $\int F_\gamma^{(0)}\phi_\gamma dc_\gamma = 0$ and $\int F_\gamma^{(0)} m_\gamma c_\gamma^2 \phi_\gamma dc_\gamma = 0$, we have that $b_0^{(\gamma)} = b_1^{(\gamma)} = 0$.

The set of equations, Eq. (18), is coupled to the time dependent equations for n_γ and T_γ. The solution of Eq. (18) would in principle require specifying an initial condition, integrating dn_γ/dt and dT_γ/dt and inverting Eq. (18) at each time.

When $T_1 = T_2$, which only occurs when the two components are identical, the results for the correction to the equilibrium rate of reaction calculated with the SNE formalism coincides with the results with the WNE. In the limit $A = B$, the expansion coefficients for both species are the same in both WNE and SNE. The unlike collision operators in the WNE add to give the self-collision operator so that Eqs. (7) and (8) reduce to Eqs. (16). (There is a factor of 2 introduced in the WNE owing to the counting of collisions twice.)

IV. Results and Discussion

The main purpose of the present note is to test the range of validity of the Chapman-Enskog approach in terms of the departure of the species temperature from a common temperature. The calculations are carried out for the model used in many previous applications. Elastic cross sections are assumed to be hard spheres and the reactive cross section is the line-of-centers cross section with a threshold energy E^*. An important parameter is the reduced threshold energy $\epsilon^* = E^*/kT$ [5]-[7]. In the WNE, the elastic collision cross sections are considered to be significantly larger than the reactive cross section. A steady state is assumed to be attained in a time of the order of the elastic collision time before much reaction has occurred. We have integrated the hydrodynamic equations obtained with $F_\gamma^{(0)}$ that give the time dependence of n_1, n_2, T_1 and T_2 with the SNE approach. The results of the integration of the hydrodynamic equations are shown in Figure 1, for different choices of system variables including the ratio of the elastic to reactive cross sections, $A = \sigma_{12}/\sigma^*$. The results shown in Fig. 1, are for an increasing separation in the elastic and reactive time scales. The solid curves are for equal number densities and masses and the *splitting* of the temperatures are equal. The dashed curves are for a number density ratio $n_1/n_2 = 4$ and an asymmetric splitting of the temperatures is shown. It is clear that steady state temperatures have been attained on a time scale short relative to the reactive time scale.

A comparison of the steady asymptotic temperatures in Figure 1 with the WNE results, that is, Eq. (12) is shown in Table 1. As the ratio of elastic to reactive time scales is sufficiently large, as controlled by ϵ^* and A, the two results are in complete agreement. This aspect of the Chapman-Enskog approach has been discussed previously [8]. The basic conclusion is that the separation of elastic and reactive time scales must be of the order of 10^{-3} to 10^{-4} for the Chapman-Enskog approach to be valid. For this situation, the departure from equilibrium will be correspondingly very small. Further details and results based on this formalism will be presented in a forthcoming publication [15].

Table 1.

Comparison of the species temperatures obtained by integration of the hydrodynamic equations (SNE) and the Chapman-Enskog method (WNE).[a]

ϵ^*/kT	A	$\left(\frac{T-T_1}{T}\right)^{(WNE)}$	$\left(\frac{T-T_1}{T}\right)^{(SNE)}$	$\left(\frac{T-T_1}{T}\right)^{(SNE)} / \left(\frac{T-T_1}{T}\right)^{(WNE)}$
1.0	1.0	5.173(-2)	5.983(-2)	0.865
2.0	1.0	3.172(-2)	3.418(-2)	0.928
4.0	1.0	7.727(-3)	7.820(-3)	0.988
1.0	100.0	5.173(-4)	5.180(-4)	0.999
2.0	100.0	3.172(-4)	3.174(-4)	0.999
4.0	100.0	7.727(-5)	7.727(-5)	1.000
1.0	10000.0	5.173(-6)	5.173(-6)	1.000
2.0	10000.0	3.172(-6)	3.172(-6)	1.000
4.0	10000.0	7.727(-7)	7.727(-7)	1.000

[a]$n_1/n_2 = 1, m_1/m_2 = 2; (-n) = 10^{-n}$

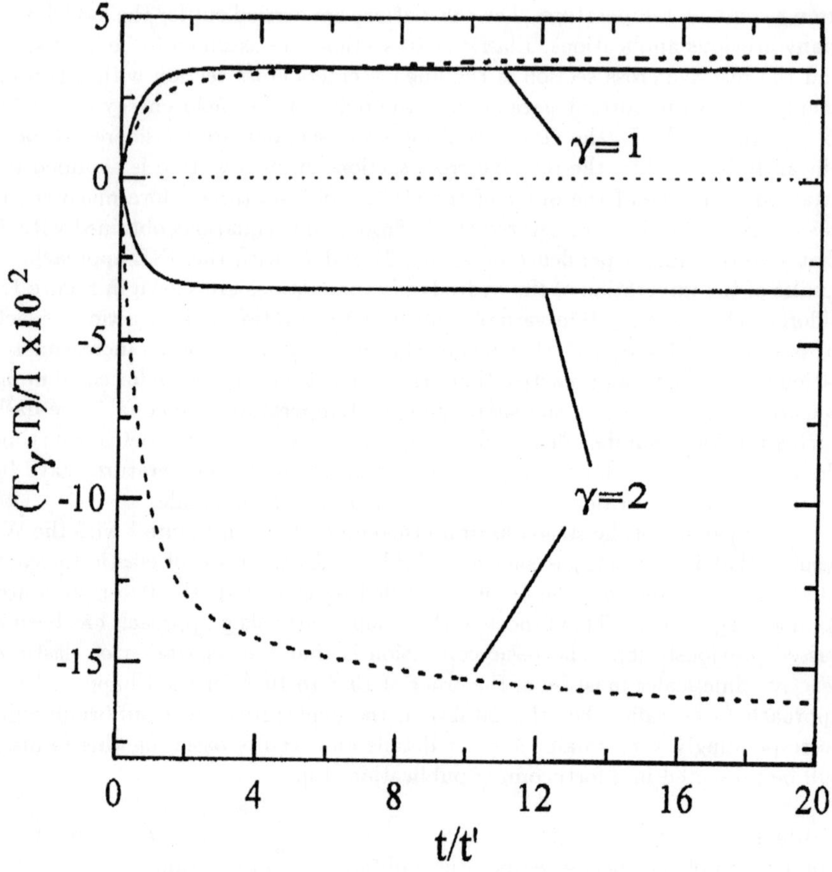

Figure 1

Time dependence of the species temperatures:
T_γ obtained from the numerical integration of hydrodynamic equations in the SNE formalism; $\epsilon^* = 2$, $A = 1$ $m_1/m_2 = 2$, $n_1/n_2 = 1$ (solid curve), and $n_1/n_2 = 4$ (dashed curve).

References

[1] I. Prigogine and E. Xhrouet, Physica **15**, 913 (1949); I. Prigogine and M. Mahieu, *ibid* **16**, 51 (1952).

[2] R. D. Present, J. Chem. Phys. **31**, 747 (1959).

[3] J. Ross and P. Mazur, J. Chem. Phys. **35**, 19 (1961).

[4] C. W. Pyun and J. Ross, J. Chem. Phys. **40**, 2572 (1964).

[5] B. Shizgal and M. Karplus, J. Chem. Phys. **52**, 4262 (1970).

[6] B. Shizgal and M. Karplus, J. Chem. Phys. **54**, 4345 (1971).

[7] B. Shizgal and M. Karplus, J. Chem. Phys. **54**, 4357 (1971).

[8] B. Shizgal, J. Chem. Phys. **55**, 76 (1971).

[9] A. S. Cukrowski, S. Fritzsche, and J. Popielawski, in *Proceedings of the International Symposium on Far-from-Equilibrium Dynamics of Chemical Systems*, Swidno, Poland 3-7 September, 1990, edited by J. Popielawski and J. Gorecki (World Scientific, Singapore, 1991).

[10] A. S. Cukrowski, J. Popielawski, L. Qin and J. S. Dahler, J. Chem. Phys. **97**, 9086 (1992).

[11] S. Pascal and R. Brun, Phys. Rev. E. **47**, 3251 (1993).

[12] S. Chapman and T. G. Cowling *The Mathematical Theory of Nonuniform Gases* (Cambridge University Press, Cambridge, 1970).

[13] B. Shizgal and J. M. Fitzpatrick, Chem. Phys. **6**, 54 (1974).

[14] B. Shizgal and M. J. Lindenfeld, Chem. Phys. **41**, 81 (1979).

[15] D. Napier and B. D. Shizgal, (to be published).

STATISTICAL TREATMENT OF CHEMICAL REACTIONS FAR FROM EQUILIBRIUM

R. Haberlandt

Research Group Statistical Theory of Non-Equilibrium Processes, WIP, KAI e.V.,Permoserstraße 15, 04303 Leipzig, Germany; Department of Physics, Leipzig University,Linnéstraße 5, 04103 Leipzig, Germany;

Abstract

To understand complex irreversible processes in detail one must study the equations of motion of these transport processes analytically or at least numerically. For chemical reactions and other transport processes - taking place simultaneously - observables are selected to construct a generalized canonical ensemble $\sigma(t)$. Equations of motion are given. Generalized non-Markovian rate equations for chemical reactions of inhomogeneous systems far from equilibrium with internal degrees of freedom and transport equations for other coupled transport processes are derived. Reciprocity relations between different generalized correlation coefficients are given. Finally, special cases are considered and remarks on application of this approach are made.

1 Introduction

The recent development of different methods, e.g. experimental techniques up to spectroscopic short time fs technique, theories of transport processes, computer simulations, as a bridge between analytical theory and experimental data gives the possibility to try to understand general complex chemical reactions including situations far from equilibrium as intended for instance by this [1] and former international symposia, e.g. [2],[3],[4],[5]. To understand complex irreversible processes in detail one must study the equations of motion of these transport processes analytically or at least numerically.

Firstly, a short review of a general derivation of equations of motion by use of generalized canonical ensembles $\sigma(t)$ is given. Secondly, generalized non-Markovian rate equations for chemical reactions of inhomogeneous systems far from equilibrium with internal degrees of freedom coupled with other transport processes are derived. Thirdly, some generalized reciprocity relations (between different generalized correlation coefficients) are given. Lastly, some remarks to apply this approach are made.

2 Statistical Description of Irreversible Processes far from Equilibrium

2.1 Physical background - selection of observables - preparation of the ensemble

In solving any physical problem experimentally or theoretically the first task is the selection of the observables describing the physical situation.
We consider at first a general case and choose the n observables

$$A_1(r), A_2(r), \cdots A_i(r), \cdots A_n(r) \quad \text{for short} \quad A(r). \tag{1}$$

Then we construct - following an information-theoretical approach - a generalized canonical ensemble [6],[7]

$$\sigma(t) = \exp\left\{ -\sum_{i=1}^{n} \lambda_i(t) A_i(r) \right\} \tag{2}$$

and investigate the time dependence by master equations. Here the $\lambda_i(t)$ are Lagrangian multipliers, ensuring that the expectation values of the observables $A_i(r)$ $i = 1, 2, \ldots n$ - describing a "A - density" at point r in the considered inhomogeneous system - can be calculated by

$$< A >_i = Tr(A_i \sigma(t)) \tag{3}$$

with

$$Tr(\sigma(t)) = 1 \tag{4}$$

in the same manner as by the usual canonical ensemble density

$$< A >_i = Tr(A_i \rho(t)) \tag{5}$$

with

$$Tr(\rho(t)) = 1. \tag{6}$$

At any time - in general for $t = 0$ - both the densities are chosen to be equal

$$\rho(0) = \sigma(0). \tag{7}$$

Near equilibrium the λ_i become the canonical conjugated quantities of the observables A_i, for instance

$$\text{for } A_1 = H, \ \lambda_1(t) \to 1/kT, \quad \text{for } A_2 = N, \ \lambda_2(t) \to -\mu/kT. \tag{8}$$

274

In the contrary to $\rho(t)$ the new introduced $\sigma(t)$ does not obey the Liouville equation with the Liouvillian $L(t)$ and the Hamiltonian $\mathcal{H}(t)$

$$\frac{\partial \rho(t)}{\partial t} = -iL(t)\rho(t) = -i\hbar[\mathcal{H}(t), \rho(t)]. \tag{9}$$

In the case of chemical reaction of the type

$$A + B \rightleftharpoons C + D \tag{10}$$

again the selection of the density-like variables just mentioned (compare eq.(1)) [8] will be specified to

$$H_t(r), A_n(r), N_i(r) \tag{11}$$

with the time dependent - due to time-dependent external fields - Hamiltonian density $H_t(r)$ (for simplicity we will drop the index t further on), with $A_n(r)$ standing for occurring additional transport processes and the particle number density for the reacting particles $N_i(r)$ (i = A,B,C,D).
To transform the chosen grand canonical form into a generalized canonical form again the $N_i(r)$ will be substituted by the deviations from their equilibrium values

$$\Delta N_i(r) = N_i(r) - < N_i >^\circ I \tag{12}$$

(I unity operator). One gets another useful simplification by taking into account the conservation of the total number of the interacting particles and introducing the reaction progress variables

$$N = \frac{N_i}{\nu_i} \quad \Delta N = \frac{\Delta N_i}{\nu_i} \quad (i = A, B, C, D). \tag{13}$$

Taking into account additionally internal degrees of freedom - described by subscript α (as abbreviation for the internal state j of the species $i = A, B, C, B \quad \alpha_i^j$) - one finally has the following set of observables

$$H(r), A_n(r), N_\alpha(r). \tag{14}$$

Introducing a generalized affinity

$$\mathcal{A}(r, t) = \sum_{\alpha_i^j} \mathcal{A}_{\alpha_i^j}(r, t) = -\sum_{\alpha^j} \sum_i \nu_i \mu_{i\alpha^j}(r, t) \tag{15}$$

with the additional Lagragian parameters $\mu_{i\alpha^j}(r, t)$ - becoming near equilibrium the common chemical potential μ_i - one gets the following generalized canonical distribution [8]

$$\sigma(t) = \exp\left\{-\lambda_o(t) - \int d^3r\, \lambda_1(r, t)[H(r) - \sum_\alpha \mathcal{A}_\alpha(r, t)\Delta N_\alpha(r)]\right\} \tag{16}$$

to describe far from equilibrium bimolecular reactions of the type

$$\nu_A A(\alpha_A^j) + \nu_B B(\alpha_B^j) \rightleftharpoons \nu_C C(\alpha_C^j) + \nu_D D(\alpha_D^j) \tag{17}$$

in inhomogeneous systems with internal degrees of freedoms and occurring transport processes simultaneously.

2.2 Derivation of Master equations

To derive the desired master equations we use a method of Robertson [7] and find

$$\begin{aligned}
\frac{\partial \sigma(t)}{\partial t} =\ & -iP(t)L(t)\sigma(t) \\
& + \int dt' P(t)L(t)T(t,t')[1 - P(t')]\sigma(t')\lambda_{n'}(r',t')
\end{aligned} \tag{18}$$

with a time-dependent projection operator

$$P(t)\mathcal{B} = \sum_n \int d^3 r \, \frac{\delta \sigma(t)}{\delta < A_n(r) >} Tr[A_n(r)\mathcal{B}] \tag{19}$$

which is defined to connect the time dependence between the usual distribution function $\rho(t)$ and the generalized canonical distribution function $\sigma(t)$

$$\dot{\sigma}(t) = P(t)\dot{\rho}(t) \tag{20}$$

and a generalized time development operator $T(t,t')$ with

$$\frac{\partial T(t,t')}{\partial t'} = iT(t,t')[1 - P(t')]L(t'). \tag{21}$$

2.3 Derivation of equations of motion

Multiplying the master equation by A_i and forming the trace we get the equation of motion for $< A_i >$:

$$\frac{\partial < A_n(r) >_t}{\partial t} = < \dot{A}_n(r,t) >_t + \sum_{n'} \int_0^\infty dt' \int d^3 r \, K_{nn'}(r,r',t,t') \tag{22}$$

with

$$< \dot{A}_n(r,t) >_t = iTr\{L(t)A_n(r)\sigma(t)\} \tag{23}$$

and the generalized correlation functions

$$K_{nn'}(r,r',t,t') = < \dot{A}_n(r,t)T(t,t')[1 - P(t')]\bar{\dot{A}}_{n'}(r',t') >_{t'} \tag{24}$$

276

and the Kubo-transform

$$\bar{A} = \int_0^1 dx\, \sigma(t)^x A \sigma(t)^{-x} - < A >_t I \tag{25}$$

(I - Unity operator). Thus, one gets for transport coefficients a generalization of the linear response theory of Kubo [9] with the correct time behaviour in the non-linear case.

Taking into account additionally fluctuations one can derive Fokker-Planck-equations, respectively Langevin equations. But this is beyond the scope of this lecture [10].

2.4 Chemical reactions and other transport processes simultaneously

For the interesting case of s coupled bimolecular chemical reaction and n transport processes

$$\nu_A A(\alpha_A^j) + \nu_B B(\alpha_B^j) \rightleftharpoons \nu_C C(\alpha_C^j) + \nu_D D(\alpha_D^j) \tag{26}$$

we find -with the selected variables Hamiltonian density $H(r)$, reaction progress variables $\Delta N_\alpha = \frac{\Delta N_{x\alpha}}{\nu_x}$ ($x = A, B, C, D$ and the internal states α) and variables for transport processes $A_k(r)$- the following generalized rate equation [8]

$$\frac{\partial < N_l(r) >_t}{\partial t} = < \dot{N}_l(r,t) >_t$$
$$+ \sum_{\alpha,l',\alpha'} \int dt' \int d^3r \, K_{l\alpha,l'\alpha'}(r,r',t,t') \mathcal{A}'_{l'\alpha'}(r',t')$$
$$+ \sum_{\alpha,k} \int dt' \int d^3r \, M_{l\alpha,k}(r,r',t,t') \lambda_k(r',t') \tag{27}$$

with the generalized correlation functions

$$K_{l\alpha,l\alpha'}(r,r',t,t') = < \dot{N}_{l\alpha}(r,t) T(t,t')[1 - P(t')]\bar{\bar{N}}_{l'\alpha'}(r',t') >_{t'} \tag{28}$$

$$M_{l\alpha,k}(r,r',t,t') = < \dot{N}_{l\alpha}(r,t) T(t,t')[1 - P(t')]\bar{\bar{A}}_k(r',t') >_{t'} \tag{29}$$

which depend on position r' of the system at the point r because of the inhomogenity and nonlocality assumed for the system.

Equation (27) shows three different reasons - corresponding to the different terms in this equation - to change the particle number density considered

1. streaming within the homogeneous system,

2. reactions between the different species including mutual interaction of different interactions l and l' characterized by $K_{l\alpha,l\alpha'}(r,r',t,t')$,

3. transport processes occurring simultaneously indicated by
$M_{l\alpha,k}(r,r',t,t')$.

Similarly, one gets the transport equation for the kth transport process of the other n coupled processes

$$\frac{\partial < A_k(r) >_t}{\partial t} = < \dot{A}_k(r,t) >_t$$
$$+ \sum_{k'=1}^{n} \int dt' \int d^3r \bar{K}_{k,k'}(r,r',t,t')\lambda_{k'}(r',t')$$
$$+ \sum_{l\alpha} \int dt' \int d^3r \bar{M}_{kl\alpha}(r,r',t,t')\mathcal{A}'_{l\alpha}(r',t')$$

$$(30)$$

for $k = 1, 2, \ldots, n$ with easily understandable slightly changed notion of the generalized correlation functions

$$\bar{K}_{kk'}(r,r',t,t') = < \dot{A}_k(r,t)T(t,t')[1 - P(t')]\bar{A}_{k'}(r',t') >_{t'} \qquad (31)$$

$$\bar{M}_{kl\alpha}(r,r',t,t') = < \dot{A}_k(r,t)T(t,t')[1 - P(t')]\bar{N}_{l\alpha}(r',t') >_{t'} . \qquad (32)$$

It is to be seen that $< \Delta\dot{N}_l >, < \dot{A}_k >$ as well as $K, M, \bar{K}, \bar{M}$ depend - and that in non-linear manner - on $\mathcal{A}_l, \lambda_k$ or $< \Delta N_l >, < A_k >$, respectively. Thus, the $2(s + n)$ equations (27),(30) can be considered as equations to determine the $2(s + n)$ unknown quantities $\mathcal{A}_l, \lambda_k$ or $< \Delta N_l >, < A_k >$, respectively ($l = 1, 2, \ldots, s; k = 1, 2, \ldots, n$).
With these results one gets for the irreversible processes far from equilibrium generalized Onsager reciprocity relations and for chemical reactions near equilibrium in homogeneous local systems with the usual correlation functions in the framework of the linear response theory the result of Yamamoto [11] (see below).

3 Specification, Application

The equations (27),(30) can be considered as starting point in giving

1. general relations for the generalized correlation functions

2. rate equations for the linear case of a single reaction of homogeneous systems near equilibrium [11]

3. a comparison of reactions in more dense systems with those in diluted systems

For these cases some summarizing remarks will be given at the end of this paper. For further possibilities as

- computer simulations of model reactions [4],[12]

- applications to plasma chemistry [13] and so on

one should look into the literature. The possibility to construct a fluctuation theory of chemical reactions using this formalism and the results of [10] should here be mentioned only. Investigations of coupled irreversible processes in porous media are in progress [14].

3.1 General relations

3.1.1 Generalized reciprocity relations

As a special application we will consider the mutual influence of the elementary reactions. The calculation of the different generalized correlation functions show the existence of generalized (Onsager-) reciprocity relations of the following kind

$$K_{i\alpha,i'\alpha'} = K_{i'\alpha',i\alpha} \tag{33}$$

only in the linear case. But from eqs.(27) written down for different species one finds some relations between different generalized correlation functions

$$K_{i,i'\alpha'} = \sum_{\alpha} K_{i\alpha,i'\alpha'} \qquad\qquad K_{i\alpha,i'} = \sum_{\alpha'} K_{i\alpha,i'\alpha'}$$

$$K_{i,i'} = \sum_{\alpha}\sum_{\alpha'} K_{i\alpha,i'\alpha'} = \sum_{\alpha} K_{i\alpha,i'} = \sum_{\alpha'} K_{i,i'\alpha'}$$

$$K = \frac{1}{\nu_i\nu_{i'}}K_{i,i'} = \frac{1}{\nu_i}K_i = \sum_{\alpha}\frac{1}{\nu_i}K_{i\alpha} = \frac{1}{\nu_i'}K_i' = \sum_{\alpha'}\frac{1}{\nu_i'}K_{i'\alpha'} \tag{34}$$

3.1.2 Deviation from the internal equilibrium

As another special question we discuss the influence of deviations from the "internal equilibrium" (intramolecular equilibrium) upon the chemical reactions. To this behalf we divide the coefficients $\mu_{i\alpha}(r,t)$ into two parts

$$\mu_{i\alpha} = \bar{\mu}_{i\alpha} + \delta\mu_{i\alpha}, \tag{35}$$

where $\bar{\mu}_{i\alpha}$ describes the contribution of the internal equilibrium and $\delta\mu_{i\alpha}$ the deviation from it. Using the conditions for the internal equilibrium

$$\bar{\mu}_{i\alpha j} = \bar{\mu}_{i\alpha j'} = \cdots = \mu_i \tag{36}$$

and - using the notion $\mu_{i\alpha}{}^0$ to describe the contribution of the thermal equilibrium - one gets

$$\begin{aligned}
\Delta\mu_{i\alpha} &= \mu_{i\alpha} - \mu_{i\alpha}{}^0 = \mu_{i\alpha} - \mu_i{}^0 \\
&= \mu_i - \mu_i{}^0 + \delta\mu_{i\alpha} = \Delta\mu_i + \delta\mu_{i\alpha}
\end{aligned} \tag{37}$$

and

$$\sum_i \nu_i \mu_i{}^0 = 0 \tag{38}$$

Finally one gets from eq.(27) neglecting the transport processes

$$\begin{aligned}
\frac{\partial < N(r) >_t}{\partial t} &= \; < \dot{N}(r,t) >_t \\
&+ \int dt' \int d^3r [K(r,r',t,t')\mathcal{A}'(r',t') + \\
&\sum_\alpha \sum_{l'} \sum_{\alpha'} K_{i\alpha,i'\alpha'}(r,r',t,t')\delta\mu_{i'\alpha'}(r',t')].
\end{aligned} \tag{39}$$

That means: the contribution of the deviations from the internal equilibrium to the change of the particle number density during the reaction is not only represented by the additional third term, but also by a cross effect due to the influence of the internal degrees of freedom upon the calculation of $K(r,r',t,t')$. This result is valid for inhomogeneous systems far from equilibrium. For homogeneous systems the first term vanishes - as mentioned above - and near equilibrium linearization yields decoupling into two additive terms only. The third term is due to deviation of the internal equilibrium only and will vanish performing the chemical reaction through states of internal equilibria only.

3.2   Rate equations near equilibrium

To discuss this special case - a single bimolecular reaction between homogeneous reactants near equilibrium - we specify equation (27) to

$$\begin{aligned}
\frac{\partial < N >_t}{\partial t} &= \; < \dot{N} >_t \\
&+ \int dt' K(t,t')\mathcal{A}'(t')
\end{aligned} \tag{40}$$

with

$$< \dot{N} >_t = Tr[NL\sigma] = 0 \tag{41}$$

for the homogeneous systems considered and with the correlation function

$$K(t,t') = < \dot{N}(t)T(t,t')[1 - P(t')]\bar{\dot{N}}(t') >_{t'} . \tag{42}$$

In the homogeneous linear case with the particle number density N as the only variable one has the following simplified form with partial derivatives instead of the functional ones as in the inhomogeneous case of eq.(19) [8]

$$P\mathcal{B} = \frac{\partial\sigma}{\partial N}Tr(N\mathcal{B}\sigma). \tag{43}$$

Thus, the projection operator $P(t)$ (eq.(19)) becomes time independent and using eq.(41) one gets for any operator $\mathcal{C}$

$$Tr[\mathcal{C}P(t')L(t')\sigma(t')] = Tr[\mathcal{C}\frac{\partial\sigma}{\partial N}]Tr[NL(t')\sigma(t')] = 0. \tag{44}$$

Thus, $P(t')$ can be dropped in the expectation value of eq.(42) the rate equation has the form

$$K(t,t') = <\dot{N}(t)T(t,t')\bar{\dot{N}}(t')>_{t'} \tag{45}$$

with the linearized form of

$$T(t,t') = exp\{-i(t-t')(1-P)L\} \tag{46}$$

one finally has

$$\frac{\partial<N>_t}{\partial t} = \int dt' <\dot{N}(t)exp\{-i(t-t')(1-P)L\}\bar{\dot{N}}(t')>_{t'} \mathcal{A}'(t'). \tag{47}$$

This equation is similar to a result of Yamamoto [11], but gives the correct time behaviour. The result of Yamamoto can be derived by dropping P.

3.3 Comparison of reactions in diluted and dense systems

Selecting the observables in the following manner

$$\{A_n\} = f'\{i,k\} = \{a_i^+ a_k, \ldots, H\} \quad \{A_n\} = \{N_A, N_B, N_C, N_D, H\} \tag{48}$$

with the creation (annihilation) operators $a_i^+(a_k)$ of the species $i(k)$ one gets a kinetic equation of the Boltzmann type for the single particle density and in the next approximation - taking into account hard sphere diameters for the reacting particles and the increase of the collision frequency measured by the radial pair distribution $g(r)$ - a corresponding kinetic equation with a corrected collision term $J_E(ff)$ of the Enskog type ([8], [15], [16])

$$\frac{\partial f}{\partial t} + v\frac{\partial f}{\partial r} = J_E(ff) = g(r)J_B(ff) \tag{49}$$

for the single particle density $f(i,k,t) = Tr[f'\{i,k\}\sigma(t)]$.
Without going into the details here only the comparison for the rate coefficients in the diluted

$$k_f \sim \int da\,db\,dc\,dd| < Aa, Bb|T|Cc, Dd > |^2$$

$$* \exp-\beta\left(\frac{a^2}{2m_A} + \frac{b^2}{2m_B}\right)$$

$$*\delta\left(a^2 + b^2 - c^2 - d^2\right) \tag{50}$$

and more dense case

$$k_f \sim Im \int dadbdcddda'db'$$
$$* < Aa, Bb|T|Cc, Dd >< Cc, Dd|T^+|Aa', Bb' >$$
$$* \exp -\beta \left(\frac{a^2}{2m_A} + \frac{b^2}{2m_B} \right) g_{AB}(a, b, a', b', t)$$
$$* \left(\frac{a^2}{2m_A} + \frac{b^2}{2m_B} - \frac{c^2}{2m_C} - \frac{d^2}{2m_D} + i\epsilon \right)^{-1} \tag{51}$$

is given indicating that in the second case the rate constant is not any more a constant but due to the presence of $g(r, t)$ depends on concentration and time.

References

[1] D.Napier, B.D.Shizgal this volume

[2] J.Popielawski in Teoria Kinetyki Chemicznej, Edt. J.Popielawski, Warszawa 1990, p. 25

[3] A.S.Cukrowski, S.Fritzsche, J.Popielawski, in Far From Equilibrium Dynamics of Chemical Systems, Edts. J.Popielawski, J.Gorecki, World Scientific, Singapore, New Jersey, London, Hong Kong 1991, p. 91

[4] J.Gorecki, ibid., p. 133

[5] A.S.Cukrowski, J.Gorecki, J.Popielawski, S.Fritzsche, ibid., p. 390

[6] E.T.Jaynes, Statistical Physics (Brandeis Lectures 3,181), Benjamin, New York Amsterdam 1963

[7] B.Robertson, Phys.Rev. 144 (1966) 151

[8] R Haberlandt, Nova Acta Leopoldina NF 49 (1979) Nr. 233, 27

[9] R.Kubo, J.Phys.Soc.Japan 12 (1957) 570

[10] R. Haberlandt, Ann.Physik 43 (1986) 213

[11] T Yamamoto, Prog.Theor.Phys.(Kyoto) 10 (1953) 11

[12] R Der, S.Fritzsche Chem.Phys.Lett. 121 (1985) 177

[13] R.Haberlandt, W.Stiller, Ber.Bunsenges. 94 (1990) 1331

[14] S.Fritzsche, R.Haberlandt, J.Kärger, H.Pfeifer, K.Heinzinger, Chem.Phys. 174 (1993)

[15] R.Der, R.Haberlandt, A.Merkel, Chem.Phys. **53** (1980) 437

[16] R.Der, R.Haberlandt, H.J.Czerwon, Mol.Phys. **35** (1978) 1609, 229

KINETICS OF RAREFIED GASES FAR FROM EQUILIBRIUM

Valeri Rudyak

Novosibirsk State Academy of Civil Engineering, 630008, Novosibirsk, Russia

ABSTRACT

The influence of nonlocality effects on the kinetics of the rarefied gases is studied. Nonlocal solution of the Boltzmann equation and generalized hydrodynamic equations are discussed. It is shown that the Chapman-Enskog description of the hydrodynamic processes is not complete. Finally, the new equation of the Brownian particle motion is obtained. It is assumed that the carrier gas is in the local-equilibrium state.

1. INTRODUCTION

Kinetics of the rarefied gases is described by the Boltzmann equation for the distribution function f

$$df/dt = J(ff)$$

$$d/dt = \partial/\partial t + \mathbf{v} \cdot \nabla, \qquad J(ff) = \int d\mathbf{v}_1 \, d\sigma \, g(f'f_1' - ff_1) \ . \qquad (1.1)$$

If the Knudsen number K is small then eq. (1.1) has a small parameter. Its left hand part is of the order of K. For such a case we can seek a solution of eq. (1.1) in the form of a series. Such a solution obtained by Chapman and Enskog is named normal. The time-space dependence of this solution is determined by the evolution of the hydrodynamic variables

$$f(\mathbf{v}, \mathbf{r}, t) = f(\mathbf{v}, n(\mathbf{r}, t), p(\mathbf{r}, t), \mathbf{u}(\mathbf{r}, t)) \ , \qquad (1.2)$$

where n, $\mathbf{u}$, p are the density, velocity and pressure of the gas, respectively.

Thus, if initial conditions for the hydrodynamic variables are given we can determine the distribution function in any approximation of the Knudsen number. The variables n, $\mathbf{u}$, p satisfy the hydrodynamic equations. These are the Euler, Navier-Stokes, Burnett and etc.

284

equations.

The described solution is derived if $\partial f / \partial t \sim \mathbf{v} \cdot \nabla \sim K \ll 1$. Therefore, it is not applicable for the description of initial or boundary layers, which sizes are of the order of the free path time τ_λ or free path λ, respectively. To solve the initial or boundary problem by using the normal solution one needs to give the initial data of the hydrodynamic variables outside the initial or boundary layers. In general case these initial and boundary conditions are different from true conditions at the value of the order of K.

The normal methods of the Boltzmann equation solution are applied if the Knudsen number is small. What must we do if the Knudsen number is not very small? We have the flows far from equilibrium. In this case the transport phenomena become nonlocal. What is the significance of the nonlocality effects? Finally, how then nonlocality effects can be taken into account in the rarefied gases? These questions are investigated in the present paper.

2. NONLOCAL SOLUTION OF THE BOLTZMANN EQUATION

Usually when the hydrodynamic level of description is used it is assumed that the change of the gradients of hydrodynamic variables affects instantly the corresponding fluxes. We have the local and Markovian constitutive relations

$$\mathbb{P}(\mathbf{r}, t) = - 2\mu \overset{o}{\nabla} \mathbf{u}(\mathbf{r}, t) , \qquad \mathbf{q}(\mathbf{r}, t) = - \lambda \nabla T(\mathbf{r}, t) , \qquad (2,1)$$

where $\overset{o}{AB}$ is the traceless symmetric tensor, $\mathbb{P}$ is the stress tensor and $\mathbf{q}$ is the heat flux, μ and λ are the viscosity and heat conduction coefficients.

The Markovian character of the constitutive relations (2.1) is the consequence of the assumption of the infinite velocity of disturbance propagation in a medium. In reality this velocity is finite and the change of the hydrodynamic variables never affects the corresponding currents instantly. There is a lag in response.

The existence of the fluxes is caused by the interaction of physically small volumes of the fluid. Therefore, the flux at the point $\mathbf{r}$ is in general determined by some space region of the fluid that surrounds $\mathbf{r}$, including $\mathbf{r}$ itself. There is the nonlocal dependence of the gradients and fluxes. Thus, in general case nonlocal and delaying constitutive relations

$$\mathbb{P}(\mathbf{r},t) = -2 \iint d\mathbf{r}' \, dt_1 \, M(\mathbf{r}',\mathbf{r},t,t_1) \, \nabla^0 \mathbf{u}(\mathbf{r},t_1) \, ,$$

$$\mathbf{q}(\mathbf{r},t) = -\iint d\mathbf{r}' \, dt_1 \, \Lambda(\mathbf{r}',\mathbf{r},t,t_1) \, \nabla T(\mathbf{r},t_1) \tag{2,2}$$

need to be used. Here M and Λ are the relaxation transport kernels. We have written the linear constitutive relations but they can be easily generalized for nonlinear systems too.

The nonlocal relations of the type of eqs. (2.2) are well known and derived by the methods of nonequilibrium statistical mechanics (see for example [1,2]) In paper [3] the similar relations were obtained from the solution of the linearized Boltzmann equation.

Let us seek the solution of eq. (1.1) in the form of nonlocal functional of the local-Maxwellian function f_0

$$f(\mathbf{r},t) = \frac{1}{V'} \int d\mathbf{r}' \, f_0(\mathbf{r}+\mathbf{r}',t) \, \omega(\mathbf{r}',t) \, ,$$

$$f_0(\mathbf{v},\mathbf{r},t) = n(\mathbf{r},t) \left(\frac{m}{2\pi kT(\mathbf{r}.t)}\right)^{3/2} \exp\left(-\frac{mc^2(\mathbf{r},t)}{kT(\mathbf{r},t)}\right) \, , \tag{2.3}$$

where V' is a physically small volume and $\mathbf{c} = \mathbf{v} - \mathbf{u}$.

By substituting this expression into eq. (1.1) we get the equation

$$\frac{\partial\omega(\mathbf{r}')}{\partial t} + \frac{d\ln f_0(\mathbf{r}+\mathbf{r}')}{dt} \, \omega(\mathbf{r}') =$$

$$= \int d\mathbf{r}'' \, J\left[f_0(\mathbf{r}+\mathbf{r}') \, \omega(\mathbf{r}') \, f_{01}(\mathbf{r}+\mathbf{r}'') \, \omega(\mathbf{r}'')\right] f_0^{-1}(\mathbf{r}+\mathbf{r}') \, . \tag{2.4}$$

Let us further introduce $\omega = 1 + \Omega$. Then, the function Ω satisfies

$$\frac{\partial\Omega}{\partial t} - \frac{d\ln f_0}{dt} - \frac{d\ln f_0}{dt} \Omega + L\,\Omega + N(\Omega\,\Omega) \, , \tag{2.5}$$

where the linearized L and nonlinear N operators are easily determined. Integrating eq. (2.5) we find

$$\Omega(\mathbf{r}',t) = S(t,t_0)\,\Omega(\mathbf{r}',t_0) -$$

$$- \int_{t_0}^{t} dt_1 S(t,t_1)\left(-\frac{d\ln f_0}{dt_1} - \frac{d\ln f_0}{dt_1}\Omega - N(\Omega\,\Omega) \right)_{\mathbf{r}'}, \qquad (2.5a)$$

where the evolution operator satisfies

$$\partial S(t,t_1)/\partial t = L\, S(t,t_1)\,, \qquad\qquad S(t,t_1) = 1$$

We are solving eq. (2.5a) by iterations

$$\Omega^{(0)} = 0\,,$$

$$\Omega^{(n+1)}(\mathbf{r}',t) = S(t,t_0)\,\Omega(\mathbf{r}',t_0) -$$

$$- \int_{t_0}^{t} dt_1 S(t,t_1)\left(-\frac{d\ln f_0}{dt_1} - \frac{d\ln f_0}{dt_1}\Omega^{(n)} + \sum_{i=1}^{n} N(\Omega\,\Omega) \right)_{\mathbf{r}'},$$

In particular we can show that the first iteration gives the following function f

$$f(\mathbf{r},t) = \frac{1}{V'} \int d\mathbf{r}'\, f_0(\mathbf{r}+\mathbf{r}',t)\left\{ 1 + S(t,t_0)\,\Omega(\mathbf{r},t_0) - \right.$$

$$\left. - \int_{t_0}^{t} dt_1 S(t,t_1)\left[\mathbf{c}(C^2 - \tfrac{5}{2})\nabla\ln T + 2\,(\mathbf{CC} - \tfrac{1}{3}\,C^2\mathbb{U}):\nabla\mathbf{u} \right] \right\}_{t_1,\mathbf{r}+\mathbf{r}'}$$

$$\mathbf{C} = \mathbf{c}\,(m/2kT)^{1/2} \qquad\qquad (2.6)$$

Calculating the pressure tensor and the heat flux by means of this function yields the nonlocal constitutive relations (3.2) (of course, if we ignore the initial data).

3. GENERALIZED HYDRODYNAMIC EQUATIONS

The generalized constitutive relations (2.2) are nonlocal and delaying. One should distinguish the two types of nonlocality. The first is the nonlocality connected with the correlation of the dissipative fluxes and gradients and the second is the nonlocality resulting from the nonlocality of the relaxation transport kernels (RTK) . The latter is caused by the nonlocality of the evolution operator S which is a nonlocal functional of the hydrodynamic variables. As we see the hydrodynamic equations of the rarefied gases are integro-differential. These equations are reduced to the Navier-Stokes equations in the continuum limit when RTK is the delta function.

The generalized hydrodynamic equations can be reduced to the differential ones by localizing the constitutive relations (2.2). In order to achieve this goal we expand the gradients in the Taylor series near the point (r,t). Thus, obtained differential hydrodynamic equations contain the derivatives with respect to space and time of the order of $n = 1,2,\dots$. For example, the momentum equation takes the form

$$\rho \frac{Du}{Dt} + \rho u \nabla u + \nabla p = \nabla \sum_{k=0}^{\infty} \left[m_k \frac{\partial^k}{\partial t^k} n_k \nabla^k \right] (\nabla^o u) , \qquad (3.1)$$

where m_k and n_k are the local transport coefficients. The first terms in the right hand side of eq. (3.1), (k=0), determine the ordinary Navier -Stokes fluxes. The next terms give the corrections resulting from the nonlocality of the transport laws. The terms proportional to $\partial/\partial t$ or ∇ (k=1) determine the corrections of the order of K^2. Thus, these corrections are of the order of the Burnett terms. It is not easy to see that $n_1=0$ [2]. The space nonlocality does not change the transport laws in this approximation. The space nonlocality effects give additional terms only in the hydrodynamic equations of the third and higher orders. In particular, the third order stress tensor has the term

$$\frac{1}{2} \left[\int_{t_0}^{t} dt_1 \int dr' M(t,t_1,r,r')(r'-r)(r'-r) \right] : \nabla \nabla \nabla u \qquad (3.2)$$

On the other hand, on the Burnett level of description the hydrodynamic equations contain additional terms

$$2 \mu \, \tau_\lambda \gamma_1 \frac{\partial}{\partial t} (\nabla^\circ u) \quad \text{and} \quad \lambda \tau_\lambda \gamma_2 \frac{\partial}{\partial t} (\nabla T) \tag{3.3}$$

where γ_1 and γ_2 are some constants, for the Maxwellian molecules $\gamma_1 = \gamma_2 = 1$.

Thus, the space-time nonlocality determines the contributions into the hydrodynamic equations which cannot principally be obtained by the Chapman-Enskog method. In this sense the Chapman-Enskog method does not give the full description of a moderately rarefied gas.

Certainly, the differential approximation of type (3.1) is generally not equivalent to the initial integro-differential equations. The transition from the integro-differential equations to the differential ones is possible if the fields of the hydrodynamic variables are smooth enough. At the same time for some models of RTK the generalized hydrodynamic equations are equivalently reduced to the differential equations. This question has been studied in detail in our monograph [2]. We need to have in mind that the used exponential model of the rarefied gas RTK is applicable only for description of the weakly inhomogeneous gas. In general case, RTK are the nonlinear functionals of hydrodynamic variables and their gradients. There are not the universal RTK for the highly inhomogeneous flows. For example, for the quasihomogeneous flows ($\nabla q = 0$, $\nabla P = 0$, $\nabla M_n = 0$, M_n are the highest moments of the distribution function, $M_2 = P$) the functional M has the form

$$M(t, t_1) = \frac{\mu}{\tau_\lambda} \exp \left[- \frac{t - t_1}{\tau_\lambda} \right] \exp \left[- \int_{t_1}^{t} dt_2 \, \frac{7}{3} \frac{\partial u(t_2)}{\partial x} \right] \tag{3.4}$$

Therefore, if we describe the transport processes by means of the local transport coefficients they depend in general on gradients of hydrodynamic values.

4. ABOUT CORRELATION OF KINETIC AND HYDRODYNAMIC PROCESSES

When the Boltzmann equation is derived it is supposed that the gas relaxation proceeds in three stages. At the first stage there is a rapid relaxation process of the change of s-particle distribution function F_s ($s \geq 2$) with the characteristic time scale of the order of the collision duration τ_0. At the second stage the gas evolution is defined by the change in time of a single distribution function F_1 (this function differs from f_1 by a normalizing factor only). The time scale of this stage is τ_λ. Finally, at the third stage the gas evolution is determined by the change in time of the hydrodynamic variables. The characteristic time scale of this stage is τ_2 and usually $\tau_2 \gg \tau_\lambda$. However, for the highly nonequilibrium processes $\tau_2 \geq \tau_\lambda$. Therefore, for such processes the function F_s at the times $t \gg \tau_0$ would be a functional not only of F_1 but also of the hydrodynamic variables ϕ_α

$$F_s (x_1, \ldots, x_\sigma, t) = F_s(x_1, \ldots, x_s | \Gamma_1, \psi_\alpha) , \qquad s \geq 2 \qquad (4.1)$$

It is very important for a dense gas because the energy of such a gas is the moment of binary distribution function. In this sense the energy can be the "fast" variable itself. Thus, at least for a dense gas far from equilibrium it is impossible to distinguish between the kinetic and hydrodynamic stages of the system evolution.

Here we shall derive the kinetic equation of a moderately dense gas (see also [4-6]). In this case the first two equations of the BBGKY hierarchy have the form

$$\frac{dF_1}{dt} = \frac{N-1}{V} \iint dv_2 dr_2 \theta_{12} F_2 , \qquad (4.2)$$

$$\frac{dF_2}{dt} - \theta_{12} F_2 = \frac{N-2}{V} \iint dv_2 dr_2 (\theta_{12} + \theta_{23}) S_{-\tau}^{(3)} F_1 F_1 F_1 , \qquad (4.3)$$

$$\theta_{ij} = \frac{\partial \phi_{ij}}{\partial r} \left(\frac{\partial}{\partial p_i} - \frac{\partial}{\partial p_j} \right) ,$$

where Φ_{ij} is the interaction potential of molecules and $S_{-\tau}^{(s)}$ is the s-particles streaming operator. To take into account the existence of the "hydrodynamic mode" of the function F_2 we shall seek it in the form of a sum

$$F_2 = F_{20} + F_{21} \qquad (4.4)$$

of the quasi-equilibrium function

$$F_{20} = V^2 \int dr_3 dv_3 \cdots dr_N dp_N F_{N0} \qquad (4.5)$$

$$F_{NO} = \theta_N^{-1} \exp \{ - \, d\mathbf{r'}\beta(\mathbf{r'})[\hat{E}(\mathbf{r'}) - \mu(\mathbf{r'})\hat{n}]\}$$

and some function F_{21} . Here $\beta = 1/kT$, $\hat{n}$, $\hat{E}$ are the densities of the number of particles and energy of the system, respectively; μ is the chemical potential. Further the solution of eq. (4.3) is commonly built in the form

$$F_{21}(t) = S^{(2)}_{-(t-t_0)} [\, F_1(t_0) \, F_1(t_0) - F_{20}(t_0) \,]$$

$$- \int_{t_0}^{t} dt_1 S^{(2)}_{-(t-t_1)} [\, TF_1F_1 + L_2F_{20} - I(F_1F_1F_1)] \tag{4.6}$$

where I is the three-particle collision operator, T is some nonlocal projection operator and L_2 is the two particle Liouville operator. Substituting function (4.6) into eq. (4.1) we obtain the kinetic equation of the moderately dense gas which has an additional collision integral as compared with the usual equation. This collision integral is a function of the hydrodynamic variables gradients.

5. NONLOCALITY EFFECTS IN BROWNIAN MOTION

The Brownian motion gives another interesting example where taking into account of the nonlocal effects leads to the qualitatively new results. It is well known that the evolution of the Brownian particle can be described by the Fokker-Planck equation for the distribution function f_P

$$\frac{df_P}{dt} = \frac{\partial}{\partial \mathbf{V}} \left[\gamma(\mathbf{V} - \mathbf{u}) + kT\gamma \frac{\partial}{\partial \mathbf{V}} \right] f_P \quad , \tag{5.1}$$

where $\mathbf{R}, \mathbf{V}$ are the radius-vector and velocity of the particles and γ is the friction constant. Montgomery [7] ascertained that eq. (5.1) can be obtained from the Boltzmann equation if the collision integral is expanded in powers of the ratio m/M , where M is the mass of the Brownian particle. He has assumed that the distribution function f of a carrier gas is Maxwellian. Subsequently it was shown [8] that using the Navier-Stokes distribution function of the gas one obtains an additional term in the kinetic equation. This term is proportional to the local temperature gradient. However, the conclusions of these papers [7,8] are unsatisfactory at least in two points. Deriving the Boltzmann equation we assume that (i) the dimensions of the colliding molecules are negligibly small and (ii) the colliding molecule and particle are statistically independent. It may be shown that these conditions are not fulfilled even for enough small particles [9,10]. Therefore, the Boltzmann equation can describe the Brownian particle dynamics only if the carrier gas is very rarefied.

On the other hand, the Brownian particle is sufficiently large and its characteristic radius R_0 can be of the order of λ or larger. Then the interactions of molecules with the particle are not local. Therefore, the dynamics of interaction of molecules with the particle would be described by the kinetic equation with the nonlocal collision integral. If the molecules interact with the particle as hard spheres then the collision integral of such an equation is of the form

$$T = \int dv d\sigma \; [f_p(\mathbf{V}',\mathbf{r})f(\mathbf{v}',\mathbf{r}+R_0\mathbf{k}) - f_p(\mathbf{V},\mathbf{r})f(\mathbf{v},\mathbf{r}-R_0\mathbf{k})] \quad , \qquad (5.2)$$

where $\mathbf{k}$ is the unit vector in the direction from the center of the particle to the molecule at the moment of collision.

Expanding this integral in the series of gradients we find

$$T = \int dv d\sigma \; [f' f_p' - f f_p + R_0 k_0 (f_p' \, \nabla f' + f_p' \nabla f)] \quad . \qquad (5.3)$$

Further reduction can be achieved by using the known transformations [7]. Then on the right hand side of eq. (5.3) the following terms appear

$$\frac{\partial}{\partial \mathbf{V}} \left[\gamma(\mathbf{V} - \mathbf{u}) + kT\gamma \frac{\partial}{\partial \mathbf{V}} \right] f_p + \left[\alpha_1 \nabla n + \nabla T \right] \frac{\partial f_p}{\partial \mathbf{V}} \qquad (5.4)$$

Thus, the existence of the hydrodynamic variables gradients of the carrier medium has an influence on the kinetics of the Brownian particles even if the distribution function of the carrier gas is locally Maxwellian. Furthermore, in contrary to [8] we see that not only a thermal-diffusion term but also the diffusion one appears in the Fokker-Planck equation.

6. CONCLUSION

This paper is a short review of the subject of nonlocality in hydrodynamics. The main point is that the hydrodynamics of the gas far from equilibrium is nonlocal. The local equations correspond to small deviations from equilibrium. In general case we know very little about the relaxation transport kernels and need to use some models. However, it is impossible to obtain good results using the very simple models. The simplicity of the nature is not trivial. In conclusion I can say that the nonlocality is a very difficult problem in the modern hydrodynamics. We cannot like it but it exists.

ACKNOVLEDGEMENTS

I would like to thank the Organizing Committee of the Third International Symposium on "Far-from-Equilibrium Dynamics of Chemical Systems" and especially Dr. A. Cukrowski who suggested to present this work.

REFERENCES

1. D. Zubarev, Noneqilibrium Statistical Thermodynamics, Consultants Bureau, New York (1974).

2. V. Ya. Rudyak, Statistical Theory of Dissipative Processes in Gases and Liquids, (Science, Novosibirsk 1987), (in Russian).

3. M. Bixon, J.R. Dorfman, K.C. Mo, Phys. Fluids, 14. 1049 (1971).

4. V. Ya. Rudyak, Zhurnal Techn. Phys. 51, 2236 (1981), (in Russian).

5. V. Ya. Rudyak, N.N. Yanenko., Dokl. Akad. Nauk SSSR, 264, 1336 (1982) (in Russian).

6. V. Ya. Rudyak, N.N. Yanenko, in Rarefied Gas dynamics, Eds.: O.M. Belotserkovskii, M.N. Kogan, S.S. Kutateladze, A.K. Rebrov, (Plenum Press, New York, 1985), V.1, p.107.

7. D. Montgomery, Phys. Fluids, 14, 2088 (1971).

8. J. Fernandez de la Mora, J.M. Mercer, Phys. Rev., A26, 2178 (1982).

9. V. Ya. Rudyak, Pisma v Zhurnal Techn. Phys., 18 No 20, 77 (1982), (in Russian).

10. V. Ya Rudyak, M. Yu. Gladkov, Kinetic Equations of Molecular Gas Suspension and Fine dusty Gas, Preprint 2-93 NISI, Novosibirsk (1993).

THE KINETIC THEORY OF THE EFFECT OF CHEMICAL REACTION ON DIFFUSION

Bogdan Nowakowski and Jan Popielawski[†]

Institute of Physical Chemistry, Polish Academy of Sciences,

01-224 Warsaw, Kasprzaka 44/52, Poland

† deceased

Abstract

Simultaneous diffusion and reaction of a trace gas in a carrier gas is studied by means of the Boltzmann-Lorentz equation (BLE). It is assumed that the reaction term is relatively small and can be treated as a perturbation. The kinetic equation for the spatial Fourier transform of the distribution function is solved in the hydrodynamic regime by the Resibois method. The diffusive flux of molecules is calculated from the solution of BLE involving up to the second order perturbative terms. The diffusion coefficient contains a correction to the corresponding classical Chapman-Enskog result, accounting for the deformation of the distribution function by chemical reaction. The effect of chemical reaction on diffusion is calculated for the model of hard spheres reacting chemically with a finite activation energy. This influence can be appreciable if the diffusing gas features much lower molecular mass than the carrier gas.

I. Introduction

The nonequilibrium effects of chemical reactions in gaseous phase have attracted a research interest for over four decades. Prigogine and Xhrouet[1] initiated these investigations by proposing the Boltzmann equation complimented by a term accounting for a chemical reaction. If the chemical process is relatively slow and the reaction can be treated as a perturbation, the well-known Chapman-Enskog method[2] can provide the normal solution of this Boltzmann equation. A number of authors[3-10] used the normal solution to calculate the nonequilibrium contributions to the rate constant of chemical reaction in a spatially homogeneous gaseous system.

Additional effects of interaction of chemical reaction and

transfer processes are expected in inhomogeneous systems. The effect of chemical processes on the transport coefficients is of primary significance for the theory of dissipative structures, the formation of which involves the interplay of both above kind of processes. The Curie principle excludes the possibility of coupling of chemical reaction and transport processes within the scope of the linear nonequilibrium thermodynamics[11]. The corresponding conclusion can be deduced from the first approximation of the normal solution of the Boltzmann equation[8a]. These results are not valid beyond the linear approximation. In fact, a system of dilute gases presents one of few cases for which calculations of such cross effects can be effectively performed basing on the well-grounded nonequilibrium statistical theory.

Interaction of chemical reaction and transfer processes has been studied within the framework of the kinetic theory of gases by applying two methods of solution of the Boltzmann equation. Xystris and Dahler[12] and Eu and Li[13] calculated the corrections to the linear transport coefficients induced by the chemical reaction using the method of moments[14]. On the other hand, the Chapman-Enskog method was applied in the second (Burnett) approximation to calculate the viscosity coefficient in a system with chemical reaction[15, 16a]. Calculations of the numerical values of the corrections to the viscosity coefficient have shown[16b] that the two methods yield rather close but not completely coinciding results.

We present the study of diffusion and concomitant slow chemical reaction of a trace gas in a thermalizing carrier gas. Such a kinetic process is governed by the single linear Boltzmann-Lorentz equation for the distribution function of the inhomogeneous, trace component. We intend to calculate the correction to the diffusion coefficient induced by the chemical reaction. For linear(ized) kinetic equations Resibois[17] developed the appealing method of derivation of the linear transport coefficients, which adopts the perturbation technique of quantum theory. It provides results equivalent to the Chapman-Enskog method, but is more convenient and transparent in calculations.

In the following we define the physical system and formulate the corresponding kinetic model for which the perturbation (Resibois) method is applied to provide the solution in the hydrodynamic regime. The general expression for the diffusion coefficient can be extracted from the usual definition of a flux which involves subsequent corrections to the distribution function, obtained by the perturbation technique. The first order perturbation contribution yields the familiar formula obtained by the Chapman-Enskog method, whereas the next order term presents the correction to this classical result induced by the chemical process. As a numerical example, the effect of chemical reaction on the diffusion coefficient is calculated for the Present model[3a] of reactive hard spheres. The results obtained are discussed and compared with the corresponding results provided by the moment method[12a,13].

II. The kinetic model

We consider a dilute binary gaseous mixture consisting of the trace species A and the carrier gas C, i.e. their number densities satisfy the inequality

$$n_A \ll n_C \qquad (1)$$

Both the species are involved in a bimolecular chemical reaction A + C which is assumed to be relatively slow to be treated as a perturbation of the corresponding nonreative system. Under such conditions, Eq.(1) ensures that for the carrier gas the effect of chemical reaction is negligible, so that the component C maintains its equilibrium state and acts as a thermal bath. In effect, we can consider only one nonequilibrium component of the system – the inhomogeneously distributed trace (Lorentzian) gas A which diffuses through the medium. In addition, the feedback from the reverse reaction is neglected – this condition is achieved, for example, in early stages of reactions[18-20].

In the system considered, the distribution function $f(\underline{r}, \underline{v}, t)$ of position $\underline{r}$ and velocity $\underline{v}$ of molecules A at time t satisfies the suitably modified Boltzmann-Lorentz equation[1]

$$\frac{\partial f}{\partial t} + \underline{v}\,\frac{\partial f}{\partial \underline{r}} = J(f) + R(f) , \qquad (2)$$

where $J(f)$ is the Boltzmann-Lorentz term for the collisions with the total (elastic and reactive) cross section $d\sigma$

$$J(f) = \int \left(f' f_C'^{(0)} - f f_C^{(0)} \right) |\underline{v} - \underline{v}_C| d\sigma d\underline{v}_C \quad , \tag{3}$$

and $R(f)$ represents the correction due to the reactive collisions with the cross section $d\sigma^*$

$$R(f) = -\int f' f_C'^{(0)} |\underline{v} - \underline{v}_C| d\sigma^* d\underline{v}_C \quad . \tag{4}$$

In Eqs. (3,4) the primed distribution functions are calculated for postcollisional velocities, and $f_C^{(0)}$ denotes the equilibrium distribution of C, that is the Maxwellian velocity distribution at temperature T, multiplied by the uniform density n_C

$$f_C^{(0)}(v_C) = n_C \left(\frac{m_C}{2\Pi kT} \right)^{3/2} exp\left(- \frac{m_C v_C^2}{2kT} \right) \quad , \tag{5}$$

Above m_C is the mass of a molecule C and k is the Boltzmann constant. (We try to simplify notation by omitting the subscript "A" for symbols referring to the species A.) Because of condition (1), the elastic collisions A-A have been neglected in Eq. (2). We consider the inhomogeneous distribution function that depends only on one spatial variable, $f(\underline{r}, \underline{v}, t=0) = f(x, \underline{v}, t=0)$. The collision and reaction operators, (3) and (4) respectively, are isotropic and retain this symmetry all the time. The important feature is that the Boltzmann-Lorentz Eq. (2) is the linear integro-differential equation for the distribution function $f(\underline{r}, \underline{v}, t)$. In this case, the Fourier transform can be applied to Eq. (2), yielding the following equation

$$\frac{\partial \phi}{\partial t} + iqv_x \phi = J(\phi) + R(\phi) \quad , \tag{6}$$

where ϕ is the spatial Fourier transform of the distribution function which with the above assumptions has the specific form

$$\phi(q, \underline{v}, t) = \int e^{-iqx} f(x, \underline{v}, t) dx \quad . \tag{7}$$

The macroscopic reaction-diffusion equation is recovered from Eq. (2) in the hydrodynamic regime, in which the scale of the spatial inhomogeneities is large (relative to the mean free path of the gas molecules) and accordingly the wave vector q in Eq. (6) is small. Under these conditions, equation (6) can be treated as composed of the principal kinetic equation for the homogeneous evolution of the velocity distribution

$$\frac{\partial \phi}{\partial t} = J(\phi) \ , \tag{8}$$

and the perturbation Q formed by the reactive and convective terms of Eq. (6)

$$Q(\phi) = R(\phi) - iqv_x\phi \ . \tag{9}$$

The collision operator (3) is self-adjoint for the suitably defined scalar product of functions

$$\langle \chi | \phi \rangle = \int [\psi_0(v)]^{-1} \chi^*(\underline{v})\phi(\underline{v})d\underline{v} \ . \tag{10}$$

Above, ψ_0 is the equilibrium Maxwellian velocity distribution of A

$$\psi_0(v) = \left(\frac{m}{2\pi kT}\right)^{3/2} exp\left(-\frac{mv^2}{2kT}\right) \ , \tag{11}$$

which is the only stationary solution of Eq. (8). The perturbation Q is not exactly the self-adjoint operator because the convective term is antihermitian. However, the simple form of this non-hermicity allows to apply the formalism for the hermitian operators if the right and left eigenfunctions are used, respectively.

In the hydrodynamic regime the long-time behaviour of the solution of Eq. (6) is studied. It is well-known that except $\lambda_0 = 0$ all the eigenvalues of the collision operator (3) are negative, and moreover for the intermolecular potential $V(r) \sim r^{-k}$, $k>2$, λ_0 is the isolated eigenvalue[21]. (The eigenfunctions are generally not known explicitly, with the prominent exception of the Maxwellian molecules, for which $k=5$.) The small perturbation (9) shifts slightly the eigenvalues (and introduces the dependence on q), but their ordering is retained also for the perturbed equation (6). The basic perturbed eigenvalue λ_0' is a small, but finite negative quantity, while the relative magnitudes of the other eigenvalues are much greater, $\lambda_i'/\lambda_0' \gg$ 1 for $i \geq 1$. Consequently, the prevailing contribution to the solution of the perturbed equation (6) in the hydrodynamic regime is given by the perturbed eigenfunction ψ_0', associated with the perturbed eigenvalue λ_0'

$$\phi(q,\underline{v},t) \simeq N_q\psi_0'(q,\underline{v}) \ exp(\lambda_0't) \ , \quad t \geq |\lambda_0'|^{-1} \ . \tag{12}$$

298

The coefficient N_q is related to the Fourier transform of the density of particles, n_q

$$n_q(t) = \int \phi d\underline{v} = \langle \psi_0 | \phi \rangle = N_q exp(\lambda_0' t) . \tag{13}$$

The other important hydrodynamic variable, the flux j_q , is calculated from the equation

$$j_q(t) = \int v_x \phi d\underline{v} = \langle v_x \psi_0 | \phi \rangle . \tag{14}$$

The components y and z of the flux vanish because of the assumed symmetry.

Contrary to ψ_0, the perturbed eigenfunction ψ_0' depends on the variable q of the Fourier transform, and so provides the distribution function f nonuniform in space.

III. The perturbation solution

The equation for the basic eigenfunction

$$(J+Q)\psi_0' = \lambda_0' \psi_0' \tag{15}$$

can be solved by the perturbation method. The eigenfunction and eigenvalue are expanded in the series

$$\psi_0' = \psi_0 + \psi_0'^{(1)} + \psi_0'^{(2)} + \ldots \tag{16}$$

$$\lambda_0' = \lambda_0 + \lambda_0'^{(1)} + \lambda_0'^{(2)} + \ldots , \tag{17}$$

where $\psi_0'^{(j)}$ and $\lambda_0'^{(j)}$ are the j-th order perturbation contribution to ψ_0' and λ_0' , respectively. In the standard manner, the equations for these corrections are obtained from Eq. (15) using expansions (16, 17), treating the perturbation Q as the first order term, and grouping the terms of the same order. The well-known solutions for the first two orders can be presented in the form

$$\lambda_0'^{(1)} = \langle \psi_0 | Q(\psi_0) \rangle \tag{18}$$

$$\psi_0'^{(1)} = -J^{-1} \widetilde{\widetilde{Q(\widetilde{\widetilde{\psi_0}})}} \tag{19}$$

$$\psi_0^{\prime\,(2)} = J^{-1}Q\widetilde{J^{-1}Q(\psi_0)} - \lambda_0^{\prime\,(1)}J^{-2}\widetilde{Q(\widetilde{\psi_0})} \tag{20}$$

In the above equations tilde denotes the functions which do not contain the components of the kernel of J, i.e. are orthogonal to ψ_0

$$\tilde{\chi} = \chi - \langle\psi_0|\chi\rangle\psi_0 \tag{21}$$

For functions of form (21) the inverse operator J^{-1} is well defined. The higher order contributions involve increasing number of the operators $J^{-1}Q$ (precisely $j-1$ of them in $\lambda_0^{(j)}$ and $\psi_0^{(j)}$), the norm of which is the scale of the relative perturbation. For this reason, the magnitude of the perturbative terms is rapidly diminishing function of the order j.

Using functions (19), (20) in the expansion (16), one can calculate from Eq. (14) the flux including up to the second order terms

$$j_q = \langle v_x\psi_0 \mid n_q(\psi_0 + \psi_0^{\prime\,(1)} + \psi_0^{\prime\,(2)})\rangle$$

$$= \Big(-\langle v_x\psi_0 |J^{-1}\widetilde{Q(\widetilde{\psi_0})}\rangle + \langle v_x\psi_0 |J^{-1}Q\widetilde{J^{-1}Q(\psi_0)}\rangle$$

$$-\langle\psi_0|Q(\psi_0)\rangle\langle v_x\psi_0 |J^{-2}\widetilde{Q(\widetilde{\psi_0})}\rangle \Big)\ n_q \tag{22}$$

The flux can be expressed explicitly in terms of the reactive and convective components of the perturbation operator Q of Eq. (9). This transformation shows that some contribution to j_q vanish for symmetry reason. The function $\langle v_x\psi_0|$ is antisymmetric with respect to the inversion of v_x and can produce finite terms only when combined in the scalar product with the function of the same symmetry. The collision J and reaction R operators retain the symmetry of the transformed function in the velocity space; that means that their action on the isotropic function ψ_0 yields an isotropic function of $\underline{v}$. Consequently, the convective component iqv_x of the perturbation Q is the only factor of $\psi_0^{\prime\,(j)}$ antisymmetric with respect to v_x. Evidently, only the terms which involve odd number of the convective operators provide nonvanishing contributions to the flux j_q. One obtains from Eq. (22)

$$j_q = \Big(\ \langle v_x \psi_0 | J^{-1} (v_x \psi_0) \rangle$$

$$- \langle v_x \psi_0 | J^{-1} v_x J^{-1} \widetilde{R(\widetilde{\psi_0})} \rangle - \langle v_x \psi_0 | J^{-1} R J^{-1} (v_x \psi_0) \rangle$$

$$+ \langle \psi_0 | R(\psi_0) \rangle \langle \psi_0 | v_x J^{-2} (v_x \psi_0) \rangle \Big)\ i q n_q\ . \tag{23}$$

IV. Diffusion coefficient - general formulae

The inverse Fourier transform of Eq. (23) yields the familiar equation

$$j_q = - (\ D' + D'')\ \frac{\partial n}{\partial x}\ , \tag{24}$$

where $D'+D''$ stands for the diffusion coefficient consisting of the parts obtained from the first and second order terms of the perturbation solution, respectively

$$D' = - \langle v_x \psi_0 | J^{-1} (v_x \psi_0) \rangle \tag{25}$$

$$D'' = \langle v_x \psi_0 | J^{-1} v_x J^{-1} \widetilde{R(\widetilde{\psi_0})} \rangle + \langle v_x \psi_0 | J^{-1} R J^{-1} (v_x \psi_0) \rangle$$

$$- \langle \psi_0 | R(\psi_0) \rangle \langle \psi_0 | v_x J^{-2} (v_x \psi_0) \rangle\ . \tag{26}$$

The term D' involves only the convective operator and does not account for the chemical reaction - it is exactly the classical diffusion coefficient D_{CE} calculated by the Chapman-Enskog method for a nonreactive trace gas[2]. The first order of the perturbation solution corresponds to the first approximation of the normal solution of the Boltzmann equation; at this level of approximation the interaction of chemical reaction and diffusion is not predicted.

The effect of chemical reaction on D is obtained only from the next order contribution, D'', which is related to the second (Burnett) approximation of the Chapman-Enskog solution. Although Eq. (26) involves the formal superpositions of integral operators, like e.g. $J^{-1} R J^{-1}$, which seem to be untreatable in practical calculations, it can be easily transformed to the much more convenient form using the hermicity of the operators J, R, and v_x. The first component of D'' is recasted as

$$\langle v_x \psi_0 | J^{-1} v_x J^{-1} \widetilde{R(\widetilde{\psi}_0)} \rangle = \langle v_x J^{-1}(v_x \psi_0) | J^{-1} \widetilde{R(\widetilde{\psi}_0)} \rangle \ , \tag{27}$$

The second term of D'' we present in the more symmetric form

$$\langle v_x \psi_0 | J^{-1} R J^{-1}(v_x \psi_0) \rangle = \langle J^{-1}(v_x \psi_0) | R J^{-1}(v_x \psi_0) \rangle \ , \tag{28}$$

and respectively the fourth term

$$\langle v_x \psi_0 | J^{-2}(v_x \psi_0) \rangle = \langle J^{-1}(v_x \psi_0) | J^{-1}(v_x \psi_0) \rangle \ . \tag{29}$$

It is useful to introduce the following functions

$$\omega(v) = -J^{-1} \widetilde{R(\widetilde{\psi}_0(v))} \tag{30}$$

$$v_x \chi(v) = J^{-1}(v_x \psi_0(v)) \ . \tag{31}$$

ω and χ are isotropic functions of v, because the integral operators J and R are isotropic. Using the above functions in Eqs. (26–29), we can present the correction to the diffusion coefficient induced by the chemical reaction, Eq. (26), in the form

$$D'' = -\langle \omega | v_x^2 \chi \rangle + \langle v_x \chi | R(v_x \chi) \rangle - \langle \psi_0 | R(\psi_0) \rangle \langle \chi | v_x^2 \chi \rangle \ . \tag{32}$$

The compact form of Eq. (32) is easily used for calculations once the functions ω and χ are found for a specific molecular model. The main advantage of the performed transformations is the lowering of the order of the terms involved in the equations for D''. While this correction is obtained originally only from the second order perturbation contribution, its transformed form of Eq. (32) is expressed in terms of ω and χ, which are the solutions of the kinetic equations for the first order corrections in the distribution function $\psi_0'(v, t)$, Eq. (19). The applied symbolic formalism can be appreciated when compared with the complexity of the corresponding transformation performed in the traditional manner, presented in Chap. XV of Ref. 2.

V. The effect of chemical reaction on diffusion for the model of reactive hard spheres

To calculate D'' for a specific molecular interaction we have to solve the equations for the functions ω and χ

$$J(\omega(v)) = -\tilde{\tilde{R}}(\tilde{\tilde{\psi}}_0(\tilde{\tilde{v}})) \tag{33}$$

$$J(v_x\chi(v)) = v_x\psi_0(v) \quad . \tag{34}$$

We use the operators J, Eq.(3), and R, Eq.(4), for the model of hard spheres[2] of diameters d_A and d_C for the components A and C, respectively. For the chemical reaction A+C we adopt the line-of-centers model[3a], in which the colliding hard spheres are reacting with the probability s_f (steric factor) if their relative velocity

$$g = \underline{v} - \underline{v}_B \tag{35}$$

satisfies the condition

$$\underline{e} \cdot g \geq g^* \ , \tag{36}$$

where $\underline{e}$ is the unit vector along the line connecting the centers of the molecules at the instant of impact. The threshold relative velocity g^* determines the activation energy of the reaction which we define in the dimensionless form

$$\varepsilon^* = \frac{\mu g^{*2}}{2kT} \ , \tag{37}$$

where μ denotes the reduced mass

$$\mu = \frac{m \cdot m_C}{m + m_C} \ , \tag{38}$$

We apply for ω and χ the standard expansions

$$\omega(v) = \psi_0 \left(a_1 S_{1/2}^{(1)}(c^2) + a_2 S_{1/2}^{(2)}(c^2) + \ldots \right) \tag{39}$$

$$\chi(v) = \psi_0 \left(b_0 S_{3/2}^{(0)}(c^2) + b_1 S_{3/2}^{(1)}(c^2) + \ldots \right) \ , \tag{40}$$

where $S_i^{(p)}$ are the Sonine polynomials of the dimensionless velocity

$$c^2 = \frac{mv^2}{2kT} \ . \tag{41}$$

The equations for the coefficients a_p and b_p are readily obtained from Eqs. (33) and (34), respectively, noting that the scalar products involving the operator J are calculated as the brackets in Ref.2

$$\langle \psi_0 \underline{c}^{\delta} S_i^{(p)}(c^2) | J(\psi_0 \underline{c}^{\delta} S_i^{(q)}(c^2)) \rangle = -\left[\underline{c}^{\delta} S_i^{(p)}(c^2), \underline{c}^{\delta} S_i^{(q)}(c^2) \right]_{AC} \tag{42}$$

where $\delta = 0, 1$ is an exponent. Useful formulae for calculation of the products involving the operator R are given in Ref. 8a .

It is convenient to express the solutions of Eqs. (33, 34) in terms of the characteristic time scales related to the frequency of elastic collisions

$$(\tau_J)^{-1} = n_C d_{AC}^2 \left(\frac{8\Pi kT}{\mu} \right)^{1/2} , \tag{43}$$

and reactive collisions

$$(\tau_R)^{-1} = n_C d_{AC}^2 \left(\frac{8\Pi kT}{\mu} \right)^{1/2} s_f \exp(-\varepsilon^*) , \tag{44}$$

where the collisional diameter is $d_{AC} = (d_A + d_C)/2$. From Eq. (1) follows the important relation $\tau_J \ll \tau_R$, ensuring that in the Boltzmann-Lorentz equation (2) the reaction term R is much smaller than the collision term J. The function ω, in the form of expansion (39) consisting of the first two terms, is given by

$$a_1 = \frac{\tau_J}{4\tau_R} \frac{m+m_C}{m} \left[G(M_C)\left(\frac{1}{2}+\varepsilon^*\right) + \frac{1}{2}M_C^2\left(\frac{3}{4}+2\varepsilon^*-\varepsilon^{*2}\right) \right] / G(M_C) \tag{45}$$

$$a_2 = \frac{\tau_J}{4\tau_R} \frac{m+m_C}{m} M_C\left[\frac{3}{4}+2\varepsilon^*-\varepsilon^{*2}\right] / G(M_C) , \tag{46}$$

with

$$M_C = \frac{m_C}{m + m_C} , \tag{47}$$

$$G(M_C) = 5-6M_C+7M_C^2 . \tag{48}$$

Similarly, the coefficients of expansion (40) of the function χ are[2]

$$b_0 = -\frac{3}{4}\tau_J \left[M_C[1-M_C H(M_C)] \right]^{-1} \tag{49}$$

$$b_1 = -\frac{3}{2}\tau_J \left[\frac{H(M_C)}{M_C[1-M_C H(M_C)]} \right] , \tag{50}$$

304

where

$$H(M_C) = \frac{M_C}{30-44M_C+27M_C^2} \ .$$ (51)

Using the functions ω and χ of Eqs. (39, 40), we can express the effect of the chemical reaction on the diffusion coefficient in the following form

$$D^{(3)} = -D_{CE} \left[a_1(1-5H) + 5a_2 H \right.$$

$$- \frac{\tau_J}{\tau_R[1-M_C H]} \left(H^2 M_C^2 \left(\frac{225}{4}+\frac{105}{2}\varepsilon^*+7\varepsilon^{*2}-2\varepsilon^{*3}\right) \right.$$

$$+ HM_C \left(1+11H+\frac{11}{2}H^2+(1+10H+11H^2)\varepsilon^*-(1+7H-11H^2)\varepsilon^{*2}\right)$$

$$\left. \left. + \frac{1}{8}-\frac{3}{2}H-\frac{11}{2}H^2+(\frac{1}{4}+H-65H^2)\varepsilon^*-2H\varepsilon^{*2} \right) \right] \ .$$ (52)

The Chapman-Enskog diffusion coefficient D_{CE} can be calculated from Eq. (25) with the use of the function χ

$$D_{CE} = -\frac{1}{3}\langle\psi_0|v^2\chi\rangle = \frac{3kT}{4\mu}\tau_J[1-M_C H(M_C)] \ .$$ (53)

The result for D calculated in Refs. 12a, 13 by means of the moment method, if adopted to the system considered, has the following form in the notation of this paper (cf. Eqs. (69, 70) of Ref. 13)

$$D = \frac{kT}{2m} \left(\frac{\langle \underline{c}S_{3/2}^{(1)}(c^2)\psi_0|(J+R)(\underline{c}\psi_0)\rangle^2}{\langle \underline{c}S_{3/2}^{(1)}(c^2)\psi_0|(J+R)(\underline{c}S_{3/2}^{(1)}(c^2)\psi_0)\rangle} - \langle \underline{c}\psi_0|(J+R)(\underline{c}\psi_0)\rangle \right)^{-1} .$$ (54)

It has been verified[12a, 13] that for a nonreactive system, $R=0$, Eq. (54) provides the proper result (53).

VI. Discussion

The Boltzmann-Lorentz equation for the trace gas has been solved by the Resibois method in order to derive the expression for the correction to the diffusion coefficient induced by the chemical reaction, Eq. (32). This result is obtained from the terms of the perturbative series one order higher than those providing the usual formula for the diffusion coefficient. While the adopted method of solution features some conveniences in calculations, it is limited to linear equations - this is the obstacle in applying this approach to the mixture with more balanced concentrations, which involves nonlinear terms.

In general terms, the influence of chemical process on transfer phenomena is a result of the deformation of the molecular velocity distribution by the proceeding reaction. We have calculated the relative correction to the classical diffusion coefficient

$$\gamma = \frac{D - D_{CE}}{D_{CE}} = D^{(3)}/D_{CE} \tag{55}$$

using Eqs. (52) and (54) with the steric factor $s_f = 0.02$. This value assigned, γ is only a function of the relative molecular mass M_C, and the activation energy ε^*. The latter parameters expose the relevant dependence of γ on the rate of the process of relaxation and the intensity of chemical reaction.

Figure 1 presents the dependence of γ on the mass M_C, obtained for $\varepsilon^* = 1$. The relative correction γ calculated with Eq. (52) is generally small, but increases considerably (as the absolute value) for $M_C \to 1$,

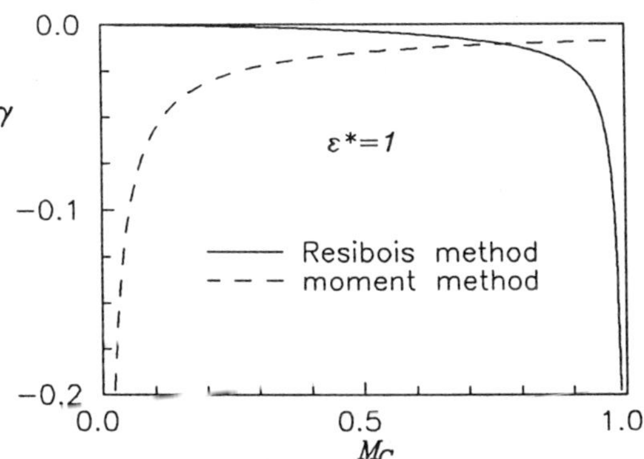

Figure 1. The relative correction γ to the diffusion coefficient as a function of the relative molecular mass M_C.

that is when the molecular mass of the diffusing species is much lower than of the carrier gas. Such slope can be explained in terms of the competition of the relaxation process and chemical reaction. If the masses of molecules A and C differ substantially, the exchange of

306

energy in elastic A-C collisions is not effective. The mechanism of elastic scattering is then not enough efficient to restore the equilibrium velocity distribution perturbed by the chemical reaction. However, this perturbation itself decays when molecules of species A become much heavier than those of the carrier gas, i.e. in the range $M_C \cong 0$. In this limit the condition for a reactive collision, Eq. (36), practically does not depend on the velocity $\underline{v}$ of the heavier molecule A. In such case the molecules of the reactant A are depleted uniformly by the proceeding reaction, but the shape of their velocity distribution remains nearly unperturbed. Consequently, the correction γ is decaying as $M_C \to 0$. The result of the moment method[12a, 13], Eq. (54), but does not predict the explained above correct dependence for γ in the extreme ranges of M_C.

In order to get an idea what magnitude of the effect can be expected in the most favorable conditions $M_C \approx 1$, in Figure 2 we present γ calculated for $M_C = 0.95$ by means of Eq. (52) as a function of the

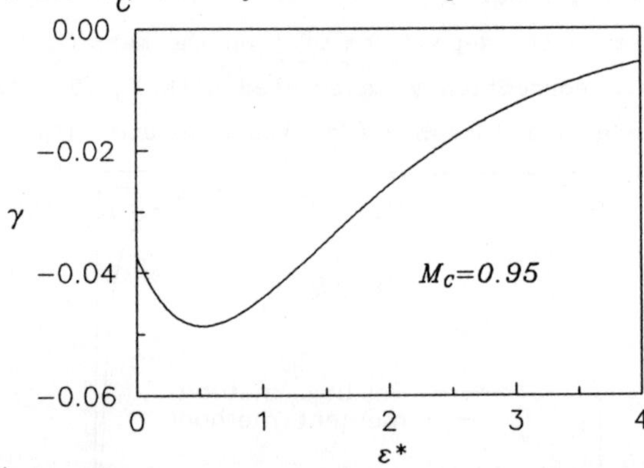

Figure 2. The relative correction γ to the diffusion coefficient, calculated by means of Eq. (52), as a function of the dimensionless activation energy ε^*.

activation energy ε^*. The relative correction reaches the extreme around $\varepsilon^*=0.52..$. For higher ε^*, the correction tapers off, because the reactive collisions become less frequent. On the other hand, for small values of ε^* the molecules are consumed relatively more uniformly from the whole range of velocities and the deformation of the molecular energy distribution is weaker.

For the model of reactive hard spheres the correction to the diffusion coefficient is always negative, because in this case the reaction removes preferably the molecules with a higher energy. Diffusion is slowed down, since the mean velocity of molecules of the reactant is lower than the mean velocity calculated for the temperature of the mixture. Shizgal and Karplus[8b,c] pointed out the importance of the depression of the intrinsic reactant temperature and indicated that it correlates with the nonequilibrium effects in chemical reaction. In the simplest manner, the modified diffusivity can be calculated by introducing into the Chapman-Enskog formula (53) the first order approximation for the corrected temperature T'[8b,c]

$$T' = T(1-a_1) \tag{56}$$

However, expanding the obtained result for small a_1 one cannot recover correction (52), nor even the numerical factor for a_1 in that equation.

As follows from Eq.(52), the nonequilibrium effects of chemical reaction are proportional to the ratio τ_J/τ_R, which approximately determines the relative magnitude of the reaction term against the collision term. The form of the expressions obtained suggests quite pronounced effects for the more intensive chemical reaction. However, in extending of the obtained results to this range one encounters limits of validity of the perturbation solution. On the other hand it has been shown previously[9,10], that at least for the studies of nonequilibrium effects of chemical reaction, the perturbative result provides the reliable basis for quantitative conclusions even outside its conventional range of applicability

We are not aware of any experimental measurements of the effect of chemical reaction on diffusion. While experimental data are lacking, the results obtained in this paper could be verified by appropriate numerical simulations. The available methods include the direct simulation of solution of the Boltzmann equation[22,23], which has been adopted for reactive systems[24], as well as molecular dynamics technique. In the specific but important range $M_C \approx 1$, the numerical solution of the kinetic equation specialized for the Lorentz gas[10] can be applied.

References

1. I. Prigogine and E. Xhrouet, Physica **15**, 913 (1949).

2. S. Chapman and T.G. Cowling, *The Mathematical Theory of Nonuniform Gases* (Cambridge, London, 1953).

3. R.D. Present, (a) J. Chem. Phys. **31**, 747 (1959); (b) *ibid.* **48**, 4875 (1968).

4. R.D. Present and M. Morris, J. Chem. Phys. **50**, 151 (1969).

5. J. Ross and P. Mazur, J. Chem. Phys. **35**, 19 (1961).

6. C.W. Pyun and J. Ross, J. Chem. Phys. **40**, 2572 (1964).

7. C.W. Pyun, J. Chem. Phys. **48**, 1306 (1968).

8. B. Shizgal and M. Karplus, (a) J. Chem. Phys. **52**, 4262 (1970); (b) *ibid.* **54**, 4345 (1971); (c) *ibid.* **54**, 4357 (1971).

9. A.S. Cukrowski, S. Fritzsche, and J. Popielawski, Acta Phys. Pol. **A84**, 369 (1993).

10. A.S. Cukrowski, J. Popielawski, W. Stiller, and R. Schmidt, J. Chem. Phys. **89**, 197 (1988).

11. S.R. De Groot and P. Mazur, *Nonequilibrium Thermodynamics* (North-Holland, Amsterdam 1962).

12. N. Xystris and J.S. Dahler, (a) J. Chem. Phys. **68**, 354 (1978); (b) *ibid.* **68**, 374 (1978); (c) *ibid.* **68**, 387 (1978).

13. B.C. Eu and K.-W. Li, Physica **88A**, 135 (1977).

14. H. Grad, Commun. Pure Appl. Math. **2**, 231 (1949).

15. J. Popielawski, J. Chem. Phys. **83**, 790 (1985).

16. A.S. Cukrowski and J. Popielawski, (a) Acta Phys. Pol. **A70**, 321 (1986); (b) *ibid.* **A71**, 853 (1987).

17. P. Resibois, J. Stat. Phys. **2**, 21 (1970); Bull. Cl. Sci. Acad. Roy. Belg. **56**, 160 (1970).

18. B. Shizgal, J. Chem. Phys. **55**, 76 (1971).

19. A.S. Cukrowski, J. Popielawski, L. Qin, and J.S. Dahler, J. Chem. Phys. **97**, 9086 (1992).

20. J. Górecki and B.C. Eu, J. Chem. Phys. **97**, 6695 (1992).

21. Y. Pao, Commun. Pure Appl. Math. **27**, 407 (1974).

22. G.A. Bird, *Molecular Gas Dynamics* (Clarendon Press, Oxford 1976).

23. K. Nanbu, J. Phys. Soc. Japan **49**, 2042 (1980); *ibid.* **49**, 2050 (1980).

24. F. Baras, M. Malek-Mansour, Physica **188A**, 253 (1992).

NONEQUILIBRIUM CONTRIBUTIONS TO THE RATE OF CHEMICAL REACTION IN BIMOLECULAR GAS PHASE REACTIONS

A.S. Cukrowski

Institute of Physical Chemistry, Polish Academy of Sciences, Pl-01224 Warsaw, Poland

ABSTRACT

Thermally activated bimolecular reactions in gas phase are analyzed. Results for the rate constant of chemical reaction, obtained from the perturbation method of solution of the Boltzmann equation, are compared with corresponding results obtained from the other methods . For one reaction as well as for two competitive reactions in the Lorentz gas, results obtained from numerial solution of the Fokker-Planck equation are used for such comparisons. For a dilute gas numerical results obtained from Monte Carlo simulations are used for this purpose. Additional comparisons with the results, obtained from simplified Boltzmann equations in which velocity distribution functions of components are Maxwellian but nonequilibrium temperatures of components are time dependent, are also presented. It is shown that the analytical results for the nonequilibrium corrections to the rate constant obtained from the perturbation method, in which the products and reverse reaction are neglected, are accurate for very slow reactions only. For early stages of relatively faster reactions the results obtained from perturbation method show a qualitative agreement with the above mentioned results used for comparisons. For further stages of reactions the role of reverse reaction may be very important. The effects of diminishing of the rate constant of chemical reaction can be very large in the Lorentz gas and in systems in which competitive reactions proceed.

1. INTRODUCTION

This lecture is a survey of the last results which have been obtained due to a joint international scientific research initiated by Professor Jan Popielawski in the problems connected with nonequilibrium effects in gas phase reactions.

Prigogine et al [1,2] were the first to analyze a deformation of the Maxwellian velocity distribution function due to proceeding of a bimolecular chemical reaction in a dilute gas. As a result of such a deformation the nonequilibrium rate of chemical reaction is smaller than that calculated with the use of the Maxwellian distribution function. This effect has been analyzed in many papers [3-24] in which appropriate Boltzmann equation has been solved by approximate methods such as the perturbation method and the moments method. Popielawski has taken into consideration that in contrary to the results obtained by these methods for transport processes [25,26] the results for rate of chemical reaction obtained from the perturbation method could not have been verified by experimental results. Therefore, Popielawski suggested that for a system in which a chemical reaction proceeds the analytical results obtained by the perturbation method of solution of the Boltzmann equation could be verified by comparison with numerical computer results. For this purpose we used three types of results for comparisons:

(a). For the Lorentz gas we obtained accurate results from the appropriate numerical solution of the Fokker-Planck equation.

(b). For a dilute gas we performed two types of simulations, i.e. first Monte Carlo simulations and then for a relatively more dense gas the molecular dynamics simulations.

(c). For a dilute gas we performed also calculations in which we assumed that in the Boltzmann equation the velocity distribution function were Maxwellian and all the nonequilibrium effects were associated with a change of temperature in time only. In this case a set of obtained differential equations could be solved exactly.

This lecture is organized as follows. In section 2 we characterize shortly the Boltzmann equation for a chemically reacting gas and after description of the model of reactive cross section we discuss such quantities like nonequilibrium correction to the rate of chemical reaction in quasistationary state and corresponding Shizgal-Karplus temperature of reagent in this state. In sections 3, 4 and 5 we discuss the results for comparisons obtained as mentioned in points (a), (b) and (c), respectively. In section 6 we additionally discuss the case of competitive reactions. In section 7 we present the final conclusions.

2. THE BOLTZMANN EQUATION AND DEFINITIONS OF THE QUANTITIES USED FOR COMPARISONS OF RESULTS

We consider the following bimolecular reaction in a dilute gas:

$$A + B \; \xrightleftharpoons \; C + D \tag{1}$$

The components A, B, C, and D are the reagents and in some cases an additional component G playing a role of a carrier gas can be introduced. We assume that the molecular masses of reagents do not change when the reaction proceeds, i.e.

$$m_A = m_C \qquad m_D = m_D \tag{2}$$

We neglect all heat effects of this reaction, i.e. we assume that this reaction is neither exothermal nor endothermal one. However, we will take into consideration that, even in the simplest models of reactive collisions, the particles of reagents need not have the same average kinetic energies.

We write down the Boltzmann equation [12] in the following form

$$\frac{\partial f_A}{\partial t} = - I_{el} - I_{re} \tag{3}$$

where

$$
\begin{aligned}
I_{el} = - &\iint (f_A' f_{A_1}' - f_A f_{A_1}) \; \sigma_{AA_1} \; g_{AA_1} \; d\Omega_{AA_1} \; dc_{A_1} \\
- &\iint (f_A' f_B' - f_A f_B) \; \sigma_{AB} \; g_{AB} \; d\Omega_{AB} \; dc_B \\
- &\iint (f_A' f_C' - f_A f_C) \; \sigma_{AC} \; g_{AC} \; d\Omega_{AC} \; dc_C \\
- &\iint (f_A' f_D' - f_A f_D) \; \sigma_{AD} \; g_{AD} \; d\Omega_{AD} \; dc_D \\
- &\iint (f_A' f_G' - f_A f_G) \; \sigma_{AG} \; g_{AG} \; d\Omega_{AG} \; dc_G \tag{4}
\end{aligned}
$$

$$I_{re} = -\iint (f_C\, f_D - f_A\, f_B) \, \sigma^*_{AB} \, g_{AB} \, d\Omega_{AB} \, dc_B \qquad (5)$$

where f_i and f_i' are the velocity distribution functions for i-th component before and after collisions, respectively, t is time, σ_{ij} and σ^*_{ij} are the differential elastic and reactive cross sections for colliding molecules i and j , c_i and c_j are their velocities, $g_{ij}= c_i - c_j$, Ω_{ij} denotes the solid angles, the index A_1 is introduced to distinguish the second colliding particle of A. In the same way after changing the appropriate indices only additional three Boltzmann equations for change of f_i (i = B,C,D) in time may be written.

For special simplified cases described in sections 3 and 4 using the perturbation method of solution we have solved the Boltzmann equation for the component A. Neglecting products and the reverse reaction we have introduced the nonequilibrium velocity distribution function as

$$f_A = f_A^{(0)}(T)\, (1 + \psi_A) = f_A^{(0)} + f_A^{(1)} \qquad (6)$$

where $f_A^{(0)}(T)$ is the Maxwellian velocity distribution function at temperature T

$$f_A^{(0)}(T) = n_A \left(\frac{m_A}{2\pi k_B T} \right)^{3/2} \exp \left(- \frac{m_A\, c_A^{\,2}}{2 k_B T} \right) \qquad (7)$$

and ψ_A is expanded in the Sonine polynomials

$$\psi_A = \sum_i a_A^{(i)}\, S_{1/2}^{(i)}(\mathscr{C}_A^2) \qquad (8)$$

where

$$\mathscr{C}_A^2 = \frac{m_A\, c_A^{\,2}}{2\, k_B T} \qquad (9)$$

We have introduced the nonequilibrium value of the rate of chemical reaction in an usual way

$$v_A = -\frac{dn_A}{dt} = \iint f_A \, f_B \, \sigma^*_{AB} \, g_{AB} \, d\Omega_{AB} \, dc_B \qquad (10)$$

and compared v_A with its equilibrium value

$$v_A^{(0)} = \iint f_A^{(0)} \, f_B^{(0)} \, \sigma^*_{AB} \, g_{AB} \, d\Omega_{AB} \, dc_B \qquad (11)$$

The quantity η defined as

$$\eta = 1 - v_A / v_A^{(0)} \qquad (12)$$

is very convenient for a description of nonequilibrium corrections.

We introduce as Shizgal and Karplus [16] the temperatures of the components

$$T_i = \frac{2}{3n_i k_B} \int f_i \, \frac{1}{2} \, m_i c_i^2 \, d\,\sigma_i \qquad (i = A, D, C, D, G) \qquad (13)$$

and the temperature of the system by

$$T = \sum_i \frac{n_i}{n} \, T_i \qquad (14)$$

As shown by Shizgal and Karplus [16] for the solution of the Boltzmann equation the temperature of the reagent A can be calculated as

$$T_A^{SK} = T \, (1 - a_A^{(1)}) \qquad (15)$$

Therefore, the change of temperature T_A due to nonequilibrium effects can be calculated as

$$\Delta T = T - T_A^{SK} \qquad (16)$$

The quantities η (see eq. (12)) and ΔT have been used for comparisons of the results obtained from the perturbation theory with the numerical results. We have calculated the nonequilibrium rate of chemical reaction v_A introducing nonequilibrium value of f_A (see eqs. (6) - (8)) into eq. (10).

In all the comparisons we have used the line-of-centers model [3] for the differential reactive cross section

314

$$\sigma^*_{AB} = \begin{cases} 0 & E < E^* \\[2ex] \frac{1}{4} s_F d^2_{AB} (1 - E^*/E) & E > E^* \end{cases} \qquad (17)$$

where s_F is the steric factor, d_{AB} is the average diameter of the colliding spheres, E is the relative kinetic energy of the spheres and E^* denotes the threshold energy. In the Monte Carlo simulations as well as for a molecular dynamics simulations, in which we analyzed reactive collisions between A and A_1 only, equivalent definition of this model has been used

$$\sigma^*_{AA_1} = \begin{cases} 0 & k \cdot g_{AA_1} \leq g^* \\[2ex] \frac{1}{4} s_F d^2 & k \cdot g_{AA_1} > g^* \end{cases} \qquad (18)$$

where k is the unit vector in the line of centers of colliding molecules A modelled as hard spheres with the diameter d , g^* is threshold relative velocity connected with the threshold energy of the reaction

$$E^* = mg^{*2}/ 4 \qquad (19)$$

We have introduced the dimensionless reduced threshold energy ε^* as

$$\varepsilon^* = E^* / k_B T \qquad (20)$$

The equilibrium rate of chemical reaction for the line-of-centers model is

$$v^{(0)}_A = 2 s_F n_A n_B d^2_{AB} \left(2\pi k_B T/m_{AB}\right)^{1/2} \exp\left(-\varepsilon^*\right) \qquad (21)$$

where m_{AB} is the reduced mass.

In some cases we have additionally introduced the quantity η calculated directly from the Shizgal-Karplus temperature T_{SK}. Then we calculated

$$\eta(T^{SK}_A) = 1 - v(T^{SK}_A) / v^{(0)} \qquad (22)$$

where $v(T^{SK}_A)$ is calculated as in eq. (21) but with T replaced by T^{SK}_A.

3. ANALYSIS OF THE PERTURBATION METHOD FOR THE LORENTZ GAS MODEL

In this case (see Refs. [27-29]) we limited to analysis of a simplified form of chemical reaction (1) discussed for early stages of the process when the concentration of products is small enough to be neglected, i.e.

$$A + B \longrightarrow products \tag{23}$$

We additionally assumed that

$$n_A \ll n_B \ll n_G \tag{24}$$

$$m_A \ll m_B$$

$$m_A \ll m_G \tag{25}$$

Because of the conditions (24) and (25) the integral for elastic collisions in eq. (4) simplifies to

$$I_{el} = - \iint \left(f_A' f_G^{(0)'} - f_A f_G^{(0)} \right) \sigma_{AG} \, g_{AG} \, d\Omega_{AG} \, d\mathbf{c}_G \tag{26}$$

We obtained the following results from the perturbation method of solution of the Boltzmann equation using the first Sonine polynomial in eq. (8)

$$\eta(1) = \frac{1}{4} s_F \frac{n_B}{n_G} \frac{m_G}{m_A} \left(\frac{d_{AB}}{d_{AG}} \right)^2 \left(\varepsilon^* + \frac{1}{2} \right)^2 \exp\left(-\varepsilon^*\right) \tag{27}$$

and

$$\Delta T = \frac{1}{4} s_F \frac{n_B}{n_G} \frac{m_G}{m_A} \left(\frac{d_{AB}}{d_{AG}} \right)^2 \left(\varepsilon^* + \frac{1}{2} \right) \exp\left(-\varepsilon^*\right) T \tag{28}$$

We have compared the results from eqs. (27) and (28) with results obtained from a numerical solution of the following differential equation derived by Desloge and Mathysse [30]

$$\frac{\partial f_A}{\partial t} = \frac{m_A n_G}{m_G c_A^2} \frac{\partial}{\partial c_A} \left[c_A^3 \, \pi d_{AG}^2 \left(\frac{k_B T}{m_A} \frac{\partial f_A}{\partial c_A} + c_A f_A \right) \right] - 4\pi n_B c_A \sigma_{AB}^* f_A \tag{29}$$

Eq. (29) is the corresponding kinetic equation for the Lorentz gas for which $m_A \ll m_G$. Also equations corresponding to eqs. (27) and (28) (in which two Sonine polynomials were introduced) were used for such comparisons. From eqs. (27) and (28) it can be seen that for very large values of $n_B m_G / n_G m_A$ the quantity η would be larger than 1 and ΔT would be larger than T. This would correspond to an unphysical situation, namely to negative rate of chemical reaction and to negative temperature of A. This shows that the perturbation solution of the Boltzmann equation can not be used if the chemical reaction is too fast. From accurate numerical solutions of the Lorentz-Fokker-Planck equation (29) it follows that: (a) for $\eta > 0.2$ the results obtained from the perturbation method can be treated as a rough approximation only, (b) for $0.03 < \eta < 0.2$ a qualitative agreement with the results obtained from eqs. (27) and (28) can be observed, (c) for $\eta < 0.03$ the results obtained from the perturbation method can be treated as accurate.

Discussing the Lorentz gas I would like to mention about another case of chemical reaction in dilute gas. Shizgal and Fitzpatrick [19] analyzed the reaction

$$A + B \longrightarrow \text{products} \tag{30}$$

and additionally assumed that

$$n_A \ll n_B \tag{31}$$

$$m_A = m_B \tag{32}$$

$$d_A = d_B \tag{33}$$

We can see that in this case the elastic term in eq. (4) simplifies to

$$I_{el} = - \iint \left(f_A' \, f_B^{(0)'} - f_A \, f_B^{(0)} \right) \, \sigma_{AB} \, g_{AB} \, d\Omega_{AB} \, dc_B \tag{34}$$

We can derive in the same way as for the Lorentz gas in Ref. [27] the following formula for η

$$\eta(1) = \frac{1}{4} s_F \left(\varepsilon^* + \frac{1}{2} \right)^2 \exp\left(-\varepsilon^*\right) \tag{35}$$

and for ΔT connected with the Shizgal-Karplus temperature (see eqs. (15) and (16))

$$\Delta T = \frac{1}{4} s_F \left(\varepsilon^* + \frac{1}{2} \right) \exp\left(-\varepsilon^*\right) T \tag{36}$$

It is interesting that the results obtained from simple analytical eq. (35) are nearly the same as the numerical results obtained by Shizgal and Fitzpatrick (see Fig. 7 in Ref. [19]) with the use of perturbation solution of the Boltzmann equation.

4. ANALYSIS OF THE PERTURBATION METHOD FOR A DILUTE GAS

We consider (see Refs. [31-33]) the simplest bimolecular chemical reaction in a dilute gas :

$$A \;+\; A \;\longrightarrow\; B \;+\; B \tag{37}$$

in order use the perturbation solution of the Boltzmann equation we analyze the reaction in a simplified form

$$A \;+\; A \;\longrightarrow\; \text{"negligible" products} \tag{38}$$

In this case the elastic integral (see eq. (4)) simplifies to

$$I_{el} = - \iint \left(f'_A \, f'_{A_1} - f_A \, f_{A_1} \right) \sigma_{AA_1} \, g_{AA_1} \, d\Omega_{AA_1} \, dc_{A_1} \tag{39}$$

and the reactive integral (see eq. (5)) has a form

$$I_{re} = \iint f_A \, f_{A_1} \, \sigma^*_{AA} \, g_{AA} \, d\Omega_{AA} \, dc_{A_1} \tag{40}$$

The Boltzmann equation (3) with the collision integrals in the form of eqs. (39) and (40) have been already solved by many authors. The perturbation method of solution have been used by many authors without taking into consideration that the temperature of reagents can be different from the temperature of the system [3,9,15,34]. We decided to solve this equation for an isolated system and to take into consideration that the temperature of reagents can be different from the temperature of the system. We compared the analytical results obtained with the results obtained from the Nanbu Babovsky Monte Carlo computer simulations (see Refs. [35-38]) performed for a system in which the chemical reaction (37) proceeds. In order to make such comparisons we assumed that the reaction (37) is in its early stage in which the concentration of products may be neglected. We took into consideration that the most energetic molecules A are consumed and although the number density of products may be neglected the energy of component A transferred to component B had to be taken into account. We disregarded the density of products but not the energy consumed by them. Usually in an analysis of reaction (38) it is assumed that the energy of A is conserved. The assumption of energy transfer to product

B permitted to calculate $a_A^{(1)}$ (see eq. (8)) which in the case of energy conservation of component A equals zero. We performed calculations in a similar way as for the Lorentz gas. We obtained an expression for η an also calculated the Shizgal-Karplus temperature (see eq. (15)) and ΔT (see eq. (16)). In the first approximation, i.e. using only the first Sonine polynomial we obtained:

$$\eta(1) = \frac{1}{2} s_F \left(\varepsilon^* + \frac{1}{2}\right)^2 \exp(-\varepsilon^*) \tag{41}$$

$$\Delta T = \frac{1}{2} s_F \left(\varepsilon^* + \frac{1}{2}\right) \exp(-\varepsilon^*) T \tag{42}$$

It is interesting to observe that the effects described by eqs. (41) and (42) are two times larger than those in eqs. (35) and (36). The results obtained from eq. (41) are different from those obtained from the Present's result [3] in which products have been totally neglected

$$\eta_P(2) = \frac{1}{32} s_F \left(\varepsilon^{*4} - 2\varepsilon^{*3} + \frac{1}{2}\varepsilon^{*2} + \frac{1}{2}\varepsilon^* + \frac{1}{16}\right) \exp(-\varepsilon^*) \tag{43}$$

To verify eqs. (41) and (42) obtained for the reaction (38) we performed the Nanbu-Babovsky Monte Carlo simulations (see Refs. 35–38) for the reaction (37). We performed these simulations for a dilute gas represented by N = 500 spheres only. In order to have better accuracy in each simulation we performed many runs (more than 1000) and for each time step Δt we obtained average values of numbers of spheres A and B (N_A and $N_B = 500 - N_A$) and their temperatures (T_A and T_B). With the increase of time N_A decreased. The temperature of T_A which in beginning was $T_A = T = 300K$ also decreased but after reaching minimum it increased. For slow reactions this temperature minimum was represented as a plateau. We treated the time region corresponding to this temperature plateau as a quasistationary state and calculated the nonequilibrium rate v_A of chemical reaction in this region. We treated this temperature minimum as the Shizgal-Karplus temperature. We calculated the quantities η and ΔT (see eqs. (12) and (16)) and compared their values with those calculated from eqs. (41) and (42).

As the reaction (38) may be treated as an approximate simplification of the reaction (37) only if the reaction (37) is very slow we tried to introduce small steric factors. However, we could not obtain sufficient accuracy in the simulations if s_F was smaller than 0.1 (for calculations of η) and smaller than 0.05 (for calculations of ΔT). It is interesting that even for relatively fast reactions the shapes of curves representing η as a function of ε^* as well as ΔT as a function of ε^* for both the types of results (obtained from eq. (38) with perturbation method and from eq. (37) with the simulations) were the same. The smaller was s_F the smaller were the differences between the types of results mentioned. As η and ΔT (see eqs. (40) and (41)) are linear functions of s_F we analyzed in the comparisons η/s_F and $\Delta T/s_F$. For relatively slower reactions characterized by $3 < \varepsilon^* < 5$ and $s_F = 1$ or by $\varepsilon^* < 3$ and $s_F = 0.1$ we obtained better agreement between the results obtained from eqs. (41) and (42) and those from the Monte Carlo simulations. It was observed that if s_F became smaller (e.g. $s_F = 1$, $1/2$, $1/3$, $1/5$, $1/10$, $1/20$ for ΔT results) the results for η/s_F and $\Delta T/s_F$ obtained from the Monte Carlo simulations were closer to the analytical results calculated from eqs. (41) and (42). The same tendency could be observed from the molecular dynamics simulations performed by Gorecki [39] (for $s_F = 1$, 0.5 and 0.2). The results for η obtained from analytical expression derived with use of two Sonine polynomials (for $0 < \varepsilon^* < 5$) only slightly differed from results obtained from eq. (41). The results for $\eta(T_A^{SK})$ obtained from the Shizgal-Karplus temperature (see eqs. (22), (16) and (42)) were also similar to those discussed above. Additional comparisons have been made with the molecular dynamics simulations for relatively more dense gas. The molecular dynamics results and the Monte Carlo results obtained by Gorecki (see Refs. 39-41) were nearly the same (with the exception of a very fast reaction characterized by $\varepsilon^* = 0$).

5. SIMPLIFIED TIME DEPENDENT SOLUTION OF THE SET OF BOLTZMANN EQUATIONS

In an analysis of the energy relaxation in binary nonreacting gas we have already obtained an analytical expression for a relaxation time and verified this result by Monte Carlo simulations [42]. We had similar experience with analysis of relaxation translational energy in perpendicular directions in one component dilute gas [43]. It is interesting that the mentioned result for binary gas could be also derived from the Sather–Dahler theory of relaxation [44] and from the Shizgal theory [45] after simplifying assumption that the velocity distribution functions of components are Maxwellian. Analyzing nonequilibrium effects associated with chemical reaction in the Lorentz gas we also observed that the shape of a nonequilibrium velocity distribution function is very similar to that of Maxwellian one (see Fig. 1 in Ref. 28). Also the way of calculation of $\eta(T_A^{SK})$ from $v(T_A^{SK})$ (see eqs. (21) and (22)) convinced us that introduction of nonequilibrium temperature to an expression for rate of chemical reaction (derived with the use of Maxwellian velocity distribution function) can give reasonable results. That is why we decided in Ref. [46] to solve the Boltzmann equations for reagents assuming that the velocity distribution functions are Maxwellian and the nonequilibrium effects are associated with time dependence of temperatures of the reagents.

After introduction of such an assumption the collision integrals I_{el} and I_{re} can be integrated and instead of the set of the Boltzmann equations we obtained a set of partial differential equations. We decided additionally to analyze the effect of the neglecting of the reverse reaction. Therefore, we analyzed the following chemical reaction

$$ A \; + \; A \; \rightleftharpoons \; B \; + \; B \tag{44} $$

We obtained the following set of differential equations

$$ \frac{d\,n_A}{d\,t} = -\,4\,s_F\,n_A^2\,d^2\left(\pi k_B T_A / m_A\right)^{1/2}\exp\left(-E^*/k_B T\right)\left(1 - F_{nBB}\right) \tag{45} $$

$$ \frac{d\,T_A}{d\,t} = -\,\frac{4}{3}\,d^2 n_B\left(\frac{\pi k_B(T_A + T_B)}{m}\right)^{1/2}(T_A - T_B)(1 - F_{rel}) $$

$$ -\,\frac{4}{3}\,s_F d^2\,n_A T_A\left(\frac{\pi k_B T_A}{m}\right)^{1/2}\left[\frac{E^*}{k_B T_A} + \frac{1}{2}\right]\exp\left(-\frac{E^*}{k_B T_A}\right)(1 - F_{TBB}) \tag{46} $$

where

$$F_{nBB} = \left(\frac{n_B}{n_A}\right)^2 \left(\frac{T_B}{T_A}\right)^{1/2} \exp\left[-\left(\frac{E^*}{k_B T_B} - \frac{E^*}{k_B T_A}\right)\right] \tag{47}$$

$$F_{nBB} = \left(\frac{n_B}{n_A}\right)^2 \left(\frac{T_B}{T_A}\right)^{3/2} \frac{E^*/k_B T_B + 1/2}{E^*/k_B T_A + 1/2} \exp\left[-\left(\frac{E^*}{k_B T_B} - \frac{E^*}{k_B T_A}\right)\right] \tag{48}$$

$$F_{rel} = \frac{3}{2}\frac{n_B}{n_A} s_F\left(-\frac{2T_B}{T_A + T_B}\right)^{1/2}\left(-\frac{E^*}{k_B T_B}\right) \tag{49}$$

we did not introduced analogous equations for the component B because
of the conditions

$$n_A + n_B = n = const \tag{50}$$

$$n_A T_A + n_B T_B = nT = const \tag{51}$$

After solution of eqs. (45) and (46) we could obtain numerical results
for η/s_F and $\Delta T/s_F$ (see eqs. (12) and (16)) also for very small steric
factors (even for $s_F = 0.001$). For such small steric factors we
obtained results which confirmed the analytical results from eqs. (41)
and (42) for the dimensionless threshold energy $0 < \varepsilon^* < 5$. We have
additionally analyzed the role of the reverse reaction. For very slow
reactions this role can be practically neglected. However, for fast
reactions this role is very important. The values of η/s_F and ΔT are
smaller than those from the corresponding results obtained with
neglecting of the reverse reaction. The time necessary to obtain a
quasistationary state and the Shizgal-Karplus temperature is also
smaller. The fundamental differences can be observed if the reaction
approaches to equilibrium because with taking into account the
reverse reaction one obtains $n_A \longrightarrow n/2$ and $n_B \longrightarrow n/2$, whereas after
neglecting of the reverse reaction one obtains $n_A \longrightarrow 0$ and $n_B \longrightarrow n$.
Naturally the last great differences could have been expected because
the neglecting of products and of the reverse reaction is usually
introduced for the early stages of chemical reaction only.

6. NONEQUILIBRIUM EFFECTS IN SYSTEMS IN WHICH COMPETITIVE REACTIONS CAN PROCEED

Systems in which competitive reactions proceed have been already

322

analyzed by Shizgal and Fitzpatrick [19] as well as by Gorensek and Kostin [47,48]. In Ref. 49 we examined the Lorentz gas with the following two reactions

$$A + B \longrightarrow products \qquad (52)$$

$$A + C \longrightarrow products \qquad (53)$$

for which the reactive cross sections were described by the line-of-centers model. We assumed for molecular masses and densities

$$m_A \ll m_B \qquad m_A \ll m_C \qquad (54)$$

$$n_A \ll n_B \qquad n_C \ll n_B \qquad (55)$$

For the threshold energies we assumed

$$E^*_{AB} \ll E^*_{AC} \qquad (56)$$

We performed analysis in a similar way as in Ref. 27. We used perturbation solution of the Boltzmann equation for the fastest reaction (52). We calculated the Shizgal-Karplus temperature T_A^{SK} which was smaller than the equilibrium temperature T. We showed that the decrease of temperature of common reactant A can have a large effect on the rate of the slower reaction (53) if the threshold energy E^*_{AC} is large. We obtained this large effect for η (the relative decrease of the rate of chemical reaction) after replacing T by T_A^{SK} in a typical expression for the rate of chemical reaction (see eqs. (20) - (22)). As reaction (52) is much faster than reaction (53) the Shizgal-Karplus temperature may be expressed as

$$T_A^{SK} = \left[1 - \frac{1}{4} s_F \frac{m_B}{m_A} \left(\frac{d_{AB}}{d_{AE}} \right)^2 \left(\frac{E^*_{AB}}{k_B T_A} + \frac{1}{2} \right) \exp \left(- \frac{E^*_{AB}}{k_B T_A} \right) \right] T \qquad (57)$$

and the rate of the slower reaction (2) as

$$v_{AC} = 2 s_F n_A n_C d_{AC}^2 \left(2\pi k_B T_A^{SK} / m_{AB} \right)^{1/2} \exp \left(- \frac{E^*_{AC}}{k_B T_A^{SK}} \right) \qquad (58)$$

As in Ref. 27 we compared such results with numerical results obtained from the numerical solution of appropriate Fokker-Planck equation for reaction (52). The slower reaction can be even nearly suppressed by the faster one.

7. FINAL CONCLUSIONS

We have used the results obtained in our previous papers [27-29,

31-33, 40, 42, 43, 46, 49] in order to show what nonequilibrium effects associated with bimolecular chemical reaction in gas can appear and how they can be compared with the results obtained within perturbation solution of the Boltzmann equation.

As shown in section 3 in the case of the Lorentz gas the rate of chemical reaction can be diminished very much (see eqs .(27), (12) and (21)). It can be diminished even nearly to 100% if the masses of components differ very much because the relaxation of energy of components is very slow in this case. The temperature of reacting component A may fall down nearly to zero for hard spheres (in which quantum effects are not taken into account). Perturbation method of solution of the Boltzmann equation can overestimate such effects for fast reactions. Unphysical negative values of the rate of chemical reaction and of temperature of the component A can be obtained. (see eqs. (27) and (28)). The results obtained from the perturbation method are accurate only if the relative correction η to the rate of chemical reaction does not exceed 3%. For 3 $<\eta$ < 20% such results are only in qualitative agreement with the exact results obtained from the Lorentz-Fokker-Planck equation and with further increase of η the perturbation results become inaccurate and may be even unphysical if η > 1.

Additionally, new simple results has been presented for the case in which the masses and molecular diameters of the reagents are the same and a small amount of component A reacts with component B treated as a heat bath (see eqs. (35) and (36)). The results obtained from the simple analytical eq. (35) are nearly the same as numerical results obtained by Shizgal and Fitzpatrick [19].

In section 4 discussing the simplest bimolecular reaction in a dilute gas (see eq. (38)) we have shown very simple analytical results for η and ΔT (see eqs. (41) and (42)). These results were obtained from the perturbation method of solution of the Boltzmann equation after assumption that in early stages of chemical reaction even if the number density of products can be neglected the energy consumed by them should be taken into account. These results have been verified by comparisons with the Nanbu-Babovsky Monte Carlo simulations (additionally confirmed by molecular dynamics simulations) performed

324

for reaction (37). Similarly as in the case of the Lorentz gas, the differences between the analytical results (eqs. (41) and (42)) describing reaction (38) and those obtained from the simulations describing reaction (37) became smaller and smaller if the rate of chemical equation decreased. We obtained the smaller chemical reaction rate by introduction of small steric factor. Because of an accuracy of simulations we could not introduce the steric factors smaller than 0.05.

We have overcome this difficulty by solution of the set of differential equations in section 5. We obtained these equations after assumption that the velocity distribution functions of the reagents are Maxwellian and the nonequilibrium effects are connected with time dependence of number densities and temperatures of the reacting components (see eqs. (45) and (46)). We have additionally observed the effects connected with the reverse reaction as we performed this analysis for reaction (44). A solution of eqs. (45) and (46) permitted to get results even for steric factor equal to 0.001 and to get a verification of eqs. (41) and (42).

It is important that in the case of reaction described by eqs. (37) and (38) the perturbation analytical results have been verified by Nanbu-Babovsky Monte Carlo computer simulations and additionally by molecular dynamics. These results have been additionally verified by a solution of the set of differential equations obtained from the Boltzmann equations (after introduction of the Maxwellian velocity distribution functions with nonequilibrium time dependent temperatures and number densities). It is interesting that in a similar way (but without analysis of the reverse reaction) Górecki [50] in the limiting case of $s_F \longrightarrow 0$ has also confirmed the analytical form of eqs. (41) and (42).

It is very interesting that in the case of slow reactions for which the perturbation method should give accurate results the nonequilibrium corrections η to the rate constant of chemical reaction calculated from eq. (41) should be two times larger than η calculated from eq. (35). This can be explained as follows. In chemical reactions (30) and (38) the reactants A and B have the same masses and diameters. Therefore, we can simply discuss the problem of nonequilibrium effects as a competition of two processes: (I) the process of destroying of the Maxwell-Boltzmann distribution (connected with reactive collisions) (II) the process of maxwellization (connected with elastic collisions). If the reaction is very slow the majority of collisions of A with A in reaction (38) or of collisions A with B in reaction (30) is elastic. It is easy to compare nonequilibrium effects associated with reactions (38) and (30) in the case in which the total density of the systems is the same. It is important that in these reactions $d_A = d_B$ and $m_A = m_B$. If a small fraction of collisions, say δ percent, is reactive then in reaction (38) the process of demaxwellization of component A is two times more effective than in the case of reaction (30). It follows from the fact that in each reactive collision in reaction (38) two spheres A react and disturb the Maxwellian distribution of component A whereas in each collision in reaction (30) only one sphere A reacts. In both the cases $(100 - \delta)$ percent of collisions is elastic and there is no difference in efficiency of maxwellization of the component A because the efficiency of translational energy relaxation is the same. As the demaxwellization in reaction (38) is two times more effective than in reaction (30) and the process of maxwellization proceeds with the same efficiency for both the reactions the nonequilibrium effects associated with reaction (38) are two times larger than those associated with reaction (30). We can see this tendency if we compare eqs. (35) and (41) as well as eqs. (36) and (42).

According to Shizgal [51] for a rigorous derivation of such equations as eqs. (41) and (42) it is necessary to solve by the perturbation method a set of two Boltzmann equations for components A and B and analyze the limiting case in which $n_B \longrightarrow 0$ in stationary state. Dahler [52] is also interested in such an analysis. I hope that this will be possible in near future.

In section 6 we have shown that in the case of two competitive reactions proceeding in the Lorentz gas the decrease of the rate of the slower reaction can be very large. It is interesting that in this case the reason of this large effect is different than in the case of one reaction in the Lorentz gas. Namely, in the case of one reaction the process of maxwellization connected with energy relaxation can be very slow because of the great difference between the molecular masses. Therefore, the process of demaxwellization connected with proceeding of chemical reaction can be much faster than the first process. In the case of two competitive reactions the molecules possessing the most appropriate molecular velocities for proceeding the slower reaction can be consumed in the faster reaction.

The competition of two reactions and the effect of suppression of a slower reaction by a faster one have been recently analyzed and confirmed by molecular dynamics by Gorecki and Kawczynski [53] as well as by Gorecki and Hanazaki [54].

The problems discussed in this paper may be interesting because nonequilibrium effects in chemical reactions (see Refs. 55 and 56) are still being investigated.

Acknowledgements. This work was partially supported by KBN grant No: 2P 0830 91 01.

REFERENCES

[1] I. Prigogine, E. Xhrouet, Physica 15, 913 (1949).

[2] I. Prigogine, M. Mahieu, Physica 16, 51 (1950).

[3] R.D. Present, J. Chem. Phys. 31, 747 (1959).

[4] J. Ross, P. Mazur, J. Chem. Phys. 35, 19 (1964).

[5] C.W. Pyun, J. Ross, J. Chem. Phys. 40, 2572 (1964).

[6] M.D. Kostin, J. Chem. Phys. 43, 2679 (1965).

[7] M.D. Kostin, J. Chem. Phys. 46, 1316 (1967).

[8] C.W. Pyun, J. Chem. Phys. 48, 1306 (1968).

[9] R.D. Present, J. Chem. Phys. 48, 4875 (1968).

[10] R.D. Present, Morris M., J. Chem. Phys. 50, 151 (1969).

[11] D.M. Chapin, M.D. Kostin, J. Chem. Phys. 48, 3067 (1968)

[12] D.M. Chapin, M.D. Kostin, J. Chem. Phys. 52, 5317 (1970)

[13] L. Monchick, J. Chem. Phys. 53, 2091 (1970).

[14] L. Monchick, J. Chem. Phys. 53, 4307 (1970).

[15] B. Shizgal, J. Karplus, J. Chem. Phys. 52, 4262 (1970).

[16] B. Shizgal, J. Karplus, J. Chem. Phys. 54, 4345 (1971).

[17] B. Shizgal, J. Karplus, J. Chem. Phys. 54, 4357 (1971).

[18] B. Shizgal, J. Chem. Phys. 55, 76 (1971).

[19] B. Shizgal, J. M. Fitzpatrick, Phys. Rev. 18, 267 (1978).

[20] B. Ch. Eu, K.W. Li, Physica A 88, 135 (1977).

[21] N. Xystris., J.S. Dahler, J. Chem. Phys. 68, 354 (1978).

[22] N. Xystris., J.S. Dahler, J. Chem. Phys. 68, 374 (1978).

[23] N. Xystris., J.S. Dahler, J. Chem. Phys. 68, 387 (1978).

[24] J. Popielawski, in Dynamics of Systems with Chemical Reactions, Ed. J. Popielawski, Proc. Internat. Symp. Swidno (Poland), June 1988, World Sci. Publ., Singapore, New Jersey 1989, p. 95.

[25] S. Chapman, T.G. Cowling, The Mathematical Theory of Nonuniform Gases, Cambridge University, Cambridge 1960.

[26] J.H. Ferziger, H.G. Kaper, Mathematical Theory of Transport Phenomena in Gases, North Holland, Amsterdam 1972.

[27] A.S. Cukrowski, J. Popielawski, R. Schmidt, W. Stiller, J. Chem. Phys. 89, 197 (1988).

[28] W. Stiller, J. Schmidt, J. Popielawski, A.S. Cukrowski, J. Chem. Phys. 93, 2425 (1990).

[29] A.S. Cukrowski, J. Popielawski, W. Stiller, R. Schmidt, J. Chem. Phys. 95, 6192 (1991).

[30] E.A. Desloge, W. Mathysse, Am. J. Phys. 28, 1 (1960).

[31] A.S. Cukrowski, S. Fritzsche, J. Popielawski, in Far-from-Equilibrium Dynamics of Chemical Systems, Eds. J. Popielawski and J. Gorecki, Proc. Second Internat. Symp. Swidno (Poland), September 1990, World Sci. Publ., Singapore, New Jersey 1991, p. 91.

[32] J. Popielawski, A.S. Cukrowski, S. Fritzsche, Physica A 188, 344 (1992).

[33] A.S. Cukrowski S. Fritzsche, J. Popielawski, Acta Phys. Pol. 82, 369 (1993).

[34] J.M. Fitzpatrick, E.A. Desloge, J. Chem. Phys. 59, 5527 (1979).

[35] K. Nambu, J. Phys. Soc. Japan 49, 2042 (1980).

[36] K. Nambu, J. Phys. Soc. Japan 49, 2050 (1980).

[37] K. Nambu, J. Phys. Soc. Japan 52, 3382 (1980).

[38] H. Babovsky, Math. Methods Appl. Sci. 8, 229 (1986).

328

[39] J. Gorecki,   in Far-from-Equilibrium Dynamics of Chemical
      Systems, Eds. J. Popielawski and J. Gorecki, Proc. Second
      Internat Symp. Swidno (Poland), September 1990, World Sci. Publ.,
      Singapore, New Jersey 1991, p. 133.

[40] J. Gorecki, J. Popielawski, A.S. Cukrowski, Phys. Rev. A 44,
      3791 (1991).

[41] J. Gorecki, J. Chem. Phys. 95, 2041 (1991).

[42] A.S. Cukrowski, S. Fritzsche, Ann. Phys. (Leipzig) 48, 377
      (1991).

[43] S. Fritzsche, A.S. Cukrowski, Acta Phys. Polon. A 74, 811
      (1988).

[44] N.F. Sather, K.F. Dahler, J. Chem. Phys. 35, 2029 (1961).

[45] B. Shizgal, J. Chem. Phys. 72, 3156 (1980).

[46] A.S. Cukrowski, J. Popielawski, Lihong Qin, J.S. Dahler, J.
      Chem. Phys. 97, 9086 (1992).

[47] M.B. Gorensek, M.D. Kostin, J. Chem. Phys. 81, 1277 (1984).

[48] M.B. Gorensek, M.D. Kostin, J. Chem. Phys. 83, 2280 (1985).

[49] A.S. Cukrowski, A.L. Kawczyński, J. Popielawski, W. Stiller,
      R. Schmidt, Chem. Phys. 159, 39 (1992).

[50] J. Gorecki, Polish J. Chem. 66, 1183 (1992).

[51] B. Shizgal, private communication.

[52] J.S. Dahler, private communication.

[53] J. Gorecki, A.L. Kawczyński, J. Chem. Phys. 96, 4647 (1992).

[54] J. Gorecki, I. Hanazaki, Chem. Phys. 180 (1994) in press.

[55] F. Baras, M. Malek Mansour, Phys. Rev. Lett. A 63, 2429 (1991).

[56] M. Malek Mansour, F. Baras, Physica A 188, 253 (1992).

ON NONEQUILIBRIUM CORRECTIONS TO THE RATE CONSTANT OF A THERMALLY ACTIVATED BIMOLECULAR CHEMICAL REACTION IN A GAS PHASE

A.S. Cukrowski* , J. Gorecki* and S. Fritzsche**

* Institute of Physical Chemistry, Polish Academy of Sciences,
Pl-01224 Warsaw, Poland
** Research Group Statistical Theory of Non-Equilibrium
Processes, WIP, KAI e.V., D-04303 Leipzig, Germany

ABSTRACT

Nonequilibrium corrections to the rate constant of a chemical reaction, proceeding according to the line-of-centers model, are analyzed for threshold energies in the range from $4.5\ k_B T$ to $8\ k_B T$ (k_B - Boltzmann constant, T - temperature of the system). Results obtained from molecular dynamics and from Nanbu-Babovsky Monte Carlo simulations are compared with a perturbation solution of the Boltzmann equation. The comparison indicates that a good approximation expression for the rate constant can be obtained considering first order terms in the expansion of velocity distribution function in Sonine polynomials. It is also shown that the concept of nonequilibrium temperature (Shizgal-Karplus temperature) of reactant leads to an accurate formula for the nonequilibrium rate constant.

I. INTRODUCTION

In this paper we compare different analytical formulae for the nonequilibrium rate constant for a model thermally activated reaction with the results of microscopic simulations using both Monte-Carlo and molecular dynamics techniques. We restrict our attention to reactions characterized by high activation energies, because for such reactions a stationary nonequilibrium state of reactant is achieved at low concentrations of product. Therefore, the assumptions of perturbation approach within which the analytical expression for rate constant is derived, are more justified, than in the case of low activation energies.

Prigogine and his coworkers Xhrouet and Mahieu [1,2] were the first who analyzed a nonequilibrium velocity distribution function generated by a thermally activated bimolecular chemical reaction. These authors have shown that if a chemical reaction is slow enough then the velocity distribution function and associated with it change in the rate constant of a chemical process can be evaluated from perturbation solution of the Boltzmann equation. The nonequilibrium distribution function has an important influence on chemical kinetics and transport processes (see, e.g. survey elaborated by Popielawski [3] and references therein). Most of the papers are concerned with the line-of-centers model for reactive collisions introduced by Present [4]. Within this model molecules of reactant are represented by spheres. The differential cross section for reactive collisions in a bimolecular process

$$A + A \longrightarrow B + B \tag{1}$$

is defined as

$$\sigma^*_{AA_1} = \begin{cases} 0 & k \cdot g_{AA_1} \leq g^* \\ \\ \dfrac{1}{4} s_F d^2 & k \cdot g_{AA_1} > g^* \end{cases} \tag{2}$$

where k is the unit vector in the line of centers of colliding molecules A (modeled by hard spheres with the diameter d), s_F is the steric factor and g^* is threshold relative speed connected with the threshold energy E^* of the reaction (1) in the following way:

$$E^* = mg^{*2}/ 4 \tag{3}$$

The rate v_A of chemical reaction (1) can be calculated as:

$$v_A = - \frac{dn_A}{dt} = k_A n_A^2 = \iint f_A f_{A_1} \sigma^*_{AA_1} g_{AA_1} d\Omega_{AA_1} dc_{A_1} \tag{4}$$

where: n_A and t are the number density and time, k_A denotes the rate constant of reaction (1), g_{AA_1} is the relative speed of colliding molecules A and A_1, f denotes the velocity distribution functions, σ^* stands for the reactive cross section, Ω is the solid angle, c the velocity of one sphere and the indices A and A_1 are introduced to distinguish each of two spheres A involved in (1).

If the velocity distribution function of the reactant is Maxwellian

$$f_A^{(0)}(T) = n_A \left(\frac{m_A}{2\pi k_B T}\right)^{3/2} \exp\left(-\frac{m_A \, c_A^2}{2k_B T}\right) \tag{5}$$

where k_B and T are the Boltzmann constant and temperature respectively, then for the line-of-centers model the equilibrium value of rate constant reads:

$$k_A^{(0)} = 4 \, s_F \, n_A^2 \, d_{AB}^2 \, \left(\pi k_B T/m\right)^{1/2} \exp\left(-\varepsilon^*\right) \tag{6}$$

where the dimensionless reduced threshold energy ε^* is

$$\varepsilon^* = E^* / k_B T \tag{7}$$

Using the perturbation solution of the Boltzmann equation [4] one obtains an expression for a nonequilibrium form of f_A. Next, introducing f_A to eq. (4) one gets a corresponding formula for the nonequilibrium value of rate constant k. It is convenient to define the quantity η describing a relative change in the rate constant due to the nonequilibrium effects

$$\eta = 1 - k_A / k_A^{(0)} \tag{8}$$

In this work we compare various analytical formulae for η obtained from the perturbation solution of the Boltzmann equation (section II) with the results obtained from computer simulations (section III). Discussion of the results is presented in section IV.

II. THE ANALYTICAL FORMULAE FOR THE NONEQUILIBRIUM CORRECTIONS TO THE RATE CONSTANT OF A CHEMICAL REACTION

In order to derive a simple analytical expression for η it is convenient to analyze reaction (1) in a state characterized by a negligible small concentration of product B [1,3]. In this case reaction (1) may be written in a simplified form

$$A + A \longrightarrow \text{"negligible" products} \tag{9}$$

Such simplified description is justified only if the reaction (1) is slow. For the line-of-centers model it means that either ε^* is large or s_F is small (see eqs. (4) and (6)). In the case of large ε^* analytical expression for η have been derived by Present [4,5] and later analyzed and reconfirmed by Shizgal and Karplus [6]. These authors have shown that:

$$\eta_P(2) = \frac{1}{32} s_F \left(\varepsilon^{*4} - 2\,\varepsilon^{*3} + \frac{1}{2}\,\varepsilon^{*2} + \frac{1}{2}\,\varepsilon^* + \frac{1}{16} \right) \exp\left(-\varepsilon^*\right) \tag{10}$$

$$\eta_P(2,3) = \frac{1}{7680} s_F \left(\frac{32}{9}\,\varepsilon^{*6} - \frac{128}{3}\,\varepsilon^{*5} + \frac{1184}{3}\,\varepsilon^{*4} - \frac{1904}{3}\,\varepsilon^{*3} \right.$$

$$\left. + 138\,\varepsilon^{*2} + 140\,\varepsilon^* + 17 \right) \exp\left(-\varepsilon^*\right) \tag{11}$$

where the numbers (2) and (2,3) are introduced in order to indicate the Sonine polynomials which are used to approximate f_A in the perturbation solution of the Boltzmann equation. It has been shown recently that for an adiabatic system characterized by a small ε^* the energy transfer from the reagent A to product B plays an important role even if the concentration of product may be neglected. Adopting the method of Shizgal and Karplus [7] for the case of the adiabatic system we obtained new formulae for η [8]:

$$\eta(1) = \frac{1}{2} s_F\,\varepsilon_1^{\,2} \exp\left(-\varepsilon^*\right) \tag{12}$$

$$\eta(1,2) = \left[\frac{1}{2} s_F \, \varepsilon_1 [\varepsilon_1 + \frac{1}{30}(\varepsilon_1 - \varepsilon_2)] - \frac{1}{60} s_F \varepsilon_2 (\varepsilon_1 - \varepsilon_2) \right] \exp(-\varepsilon^*)$$

$$- \frac{1}{16} s_F^2 \left[\varepsilon_1 + \frac{1}{30}(\varepsilon_1 - \varepsilon_2) \right]^2 \varepsilon_2 \left[\exp(-\varepsilon^*) \right]^2 \qquad (13)$$

where

$$\varepsilon_1 = \varepsilon^* + \frac{1}{2} \qquad (14)$$

$$\varepsilon_2 = \varepsilon^{*\,2} - \varepsilon^* - \frac{1}{4} \qquad (15)$$

We have derived the following formula for the nonequilibrium temperature of A (so called Shizgal-Karplus temperature T_A^{SK}) for reaction (1) (see ref. [8]):

$$T_A^{SK} = \left[1 - \frac{1}{2} s_F \, \varepsilon_1 \exp(-\varepsilon^*) \right] T \qquad (16)$$

Substituting T_A^{SK} instead of T in eq. (8) one gets:

$$k\left(T_A^{SK}\right) = 4 \, s_F d_A^2 \left(\pi k_B T_A^{SK}/m_A \right)^{1/2} \exp\left(-\varepsilon^* T/ T_A^{SK} \right) \qquad (17)$$

$$\eta(T_A^{SK}) = 1 - k(T_A^{SK})/k^{(0)} \qquad (18)$$

It worthwhile to notice that the results obtained from eqs. (12) and (18) are nearly the same.

In the case of small ε^* the analytical formulae for η were compared with computer simulations based on the Monte Carlo Nanbu-Babovsky algorithm [8] (the method is described in refs. [9-12]) and on molecular dynamics technique [13]. For $\varepsilon^* < 4$ the results obtained from eqs. (10) and (11) are significantly different from those coming from eqs. (12) and (13) (see Refs. 8 and 13). This result can be easily understood because these reactions are too fast to be considered as a small perturbation and the decrase in the reagent's energy is significant. But for $s_F = 1$ and $\varepsilon^* > 4$ the differences between various formulae become smaller. It encouraged us to perform more accurate simulations in order to find

334

which approximation is the best for the range of activation energies where the assumptions of perturbation method seem to be satisfied.

III. COMPARISON OF SIMULATIONS AND ANALYTICAL FORMULAE

In this section we compare the formulae (10), (11), (12) and (13) with computer simulations which use the Monte Carlo Nanbu–Babovsky method [8-12] and molecular dynamics technique for reactive hard spheres [13-15].

As both the methods have been extensively described in the literature [8,13-17], here we outline only the systems analyzed and the results obtained.

In our Monte Carlo simulations the system consisted of $N_A = 500$ spheres representing A, which were placed in a volume $V = 1666.7$ nm^3 and characterized by temperature 300 K. The spheres A which took part in reaction (1), proceeding with the reactive cross section described by eq. (2), were changed to spheres B. The masses and diameters of spheres were $m_A = m_B = 16$ g/mole and $d_A = d_B = 0.35$ nm which corresponded to the packing fraction 0.0067. The changes in N_A and N_B ($N_B = 500 - N_A$) and the temperatures T_A and T_B calculated from the averaged kinetic energies of A and B as well as the numbers of reactive and elastic collisions were calculated in time increments characterized by time steps $\Delta t = 0.6941\ 10^{-13}$ s. In order to improve accuracy 3000 runs were performed for every reactive cross section and the average values of the quantities mentioned above were calculated for each time step. After a few hundred time steps the temperature T_A stabilized at relatively long lasting stationary value (see Refs. 16 and 17) which was regarded by us as a quasistationary nonequilibrium one. Using this temperature we calculated the nonequilibrium rate constant of chemical reaction from eq. (17) and obtained η from eq. (18).

Molecular dynamics simulations for a thermally activated binary reaction in a system of reactive hard spheres are described in Refs. 13 - 15. The results presented below were obtained for systems in which both A and B are represented by spheres with mass 32 a.u.

and diameter 0.5 nm. The periodic boundary conditions were used in all simulations. Few systems characterized by different packing density and number of spheres were considered (1000 spheres in the case of packing fractions 0.14, 0.19, 0.27 and 500 spheres in the case of packing fraction 0.08).

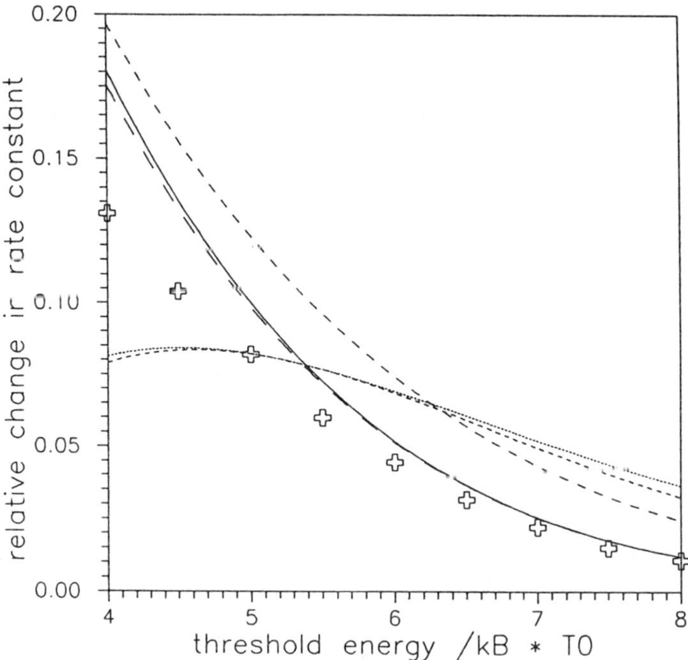

Fig. 1. Nonequilibrium correction η (see eq. (8)) for $s_F = 1$ as a function of dimensionless threshold energy ε^* (eq. (7)) calculated with the use of eqs. (12), (18), (13), (10) and (11) and represented by solid, long dashed, dashed, short dashed and dotted lines, respectively. Crosses represent the results obtained from the Monte Carlo simulations.

In Fig. 1. the results of the Monte Carlo simulations (denoted as crosses) are compared with those obtained from eqs. (10)-(13) and (18). The molecular dynamics data are shown in Figs. 2 and 3. The ratio $k_A / k_A^{(0)}$ is plotted as a function of concentration of B (in molar fractions) for two different values of ε^*. The points denote values obtained from the average time between consecutive

336

reactive collisions, as it is described in [13]. The solid line shows the result calculated as in Eq. (18), in which the temperature of A corresponds to the average kinetic energy of spheres representing A at a given concentration of product measured in simulations.

The accuracy of molecular dynamics calculations decreases with increasing ε^* because less different reaction paths may be created using the same equilibrium trajectory. Let us notice that the nonequilibrium rate constant, after a rapid decrease immediately after reaction starts, remains nearly constant. If one assumes that the rate constant does not depend on time for $t \in [t_o, t_o + \Delta t]$, than the rate equation (eq. (4)) for reaction (1) may be transformed to the form:

$$k\, \Delta t = (\, n_A(t_o + \Delta t))^{-1} - (\, n_A(t_o))^{-1} \tag{19}$$

Eq. (19) may be applied to calculate the average rate constant in the time interval $[t_o, t_o + \Delta t]$ and the statistical errors should be reduced if compared with the method based on average time between consecutive reactive collisions. The rate constants calculated using the average transition time between $n_A(t_o) = 0.95$ and $n_A(t_o + \Delta t) = 0.9$ are shown in Fig. 4. The calculations performed for other intervals of concentrations from the range $0.9 < n_A < .97$ lead to almost identical results. The rate constants are shown for three various densities of the system and we think that the difference between the results may be regarded as a measure of the statistical error. The results of Monte Carlo simulations (crosses) and those calculated from analytical formulae are also presented on this figure.

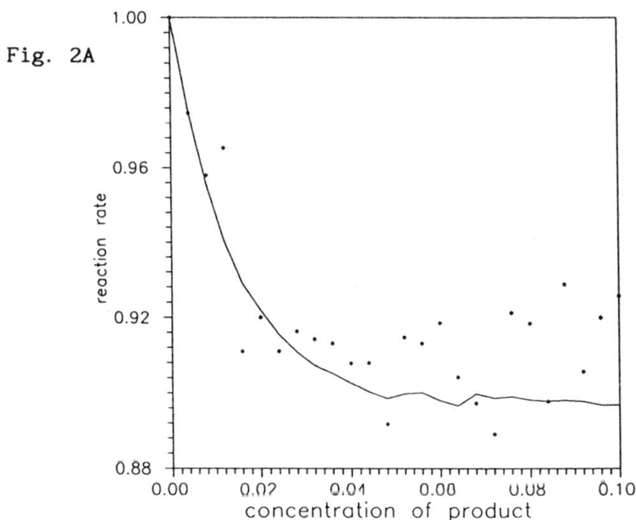

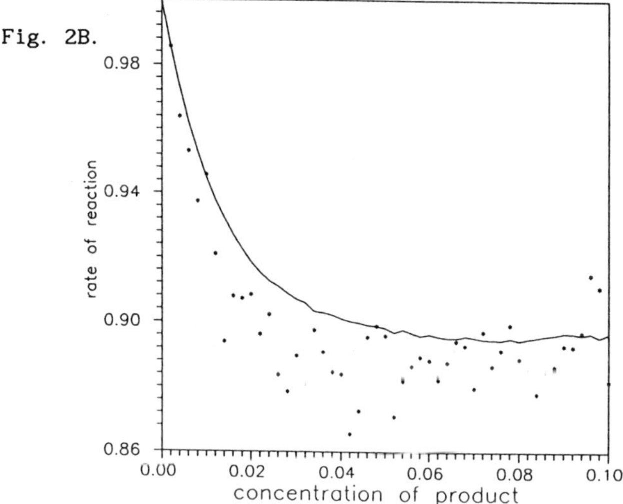

Fig. 2. The reaction rate constant k normalized to its value for a thermalized system $k^{(0)}$ as a function of concentration of products for $\varepsilon^* = 4.5$. Points show the values calculated directly from the average time between consecutive reactive collisions, the solid line uses eq. (17) in which the average kinetic energy of A observed in simulations is substituted as T_{SK}. The packing fractions are: 0.083 (Fig 2A), 0.27 (Fig 2B).

338

Fig. 3A.

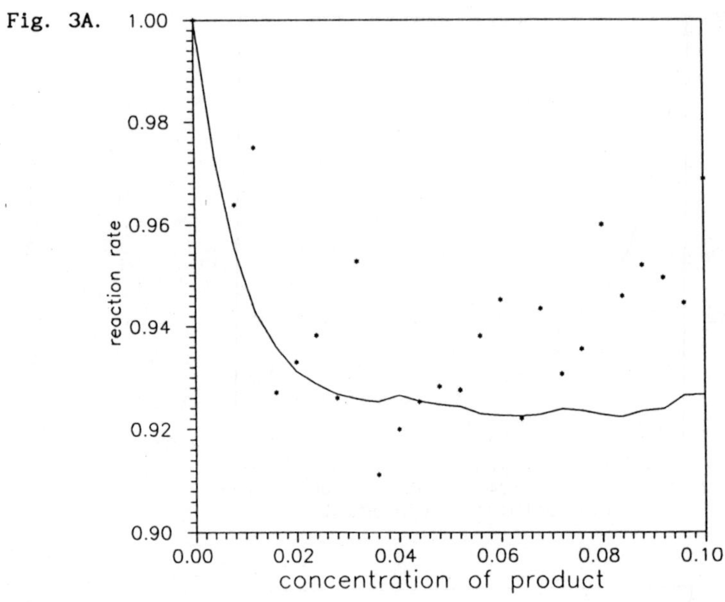

Fig. 3B.

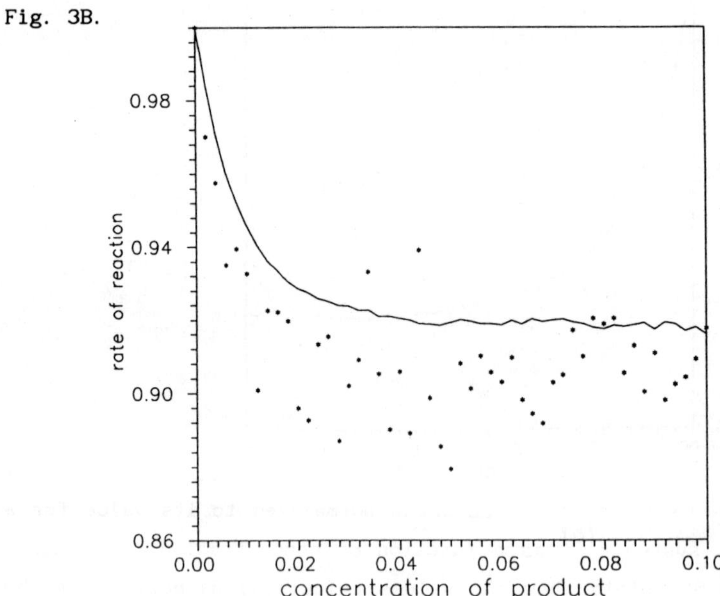

Fig. 3. The reaction rate constant k normalized to its value for a thermalized system $k^{(0)}$ as a function of concentration of products for $\varepsilon^* = 5.0$. Notation as in Fig. 2.

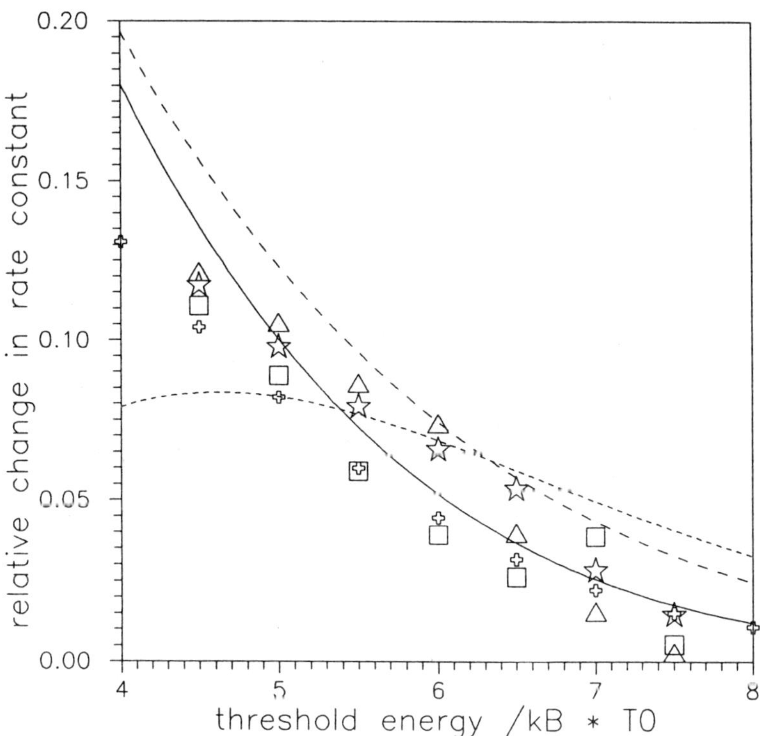

Fig. 4. The average quasistationary values of η obtained in molecular dynamics simulations; the packing fractions are: squares - 0.14, stars - 0.19, triangles - 0.27. For comparison the results of analytical formulae ((12) - solid, (13) - dashed, (10) - short dashed lines) and of Monte Carlo simulations (crosses) are shown.

IV. DISCUSSION

The results presented in section III seem to be accurate enough to show that in the examined range of ε^* the best description of nonequilibrium rate constant in an adiabatic system is obtained from formula (12). Also the method based on the nonequilibrium temperature (eq. (17)) gives a fairly good answer. It should be emphasized that η obtained with the use of the Shizgal-Karplus temperature may be also derived from the solution of a set of phenomenological kinetic equations for concentration and the energy density of A (see [18,19]). On the other hand it is surprising that an expansion based on two Sonine polynomials (eq.(13)), which is expected to be more accurate than that based on the first Sonine polynomial, happens to be worse. It may suggest that the convergence of the Sonine series is slow and more terms need to be considered. We plan to study this problem in future. As it may be expected, the formulae, in which the contribution from the first Sonine polynomial was neglected (eqs.(10) and (11)), cannot be literally applied in the case of an adiabatic system.

ACKNOWLEDGMENT

This work was supported by KBN grant No: 2P 0830 91 01.

REFERENCES

[1] I. Prigogine, E. Xhrouet, Physica **15**, 913 (1949).

[2] I. Prigogine, M. Mahieu, Physica **16**, 51 (1950).

[3] J. Popielawski, in Dynamics of Systems with Chemical Reactions, Ed. J. Popielawski, Proc.Internat. Symp. Swidno (Poland), 1988, World Scientific, Singapore, 1989; see also A.S. Cukrowski in this volume.

[4] R.D. Present, J. Chem. Phys. **31**, 747 (1959).

[5] R.D. Present, J. Chem. Phys. **48**, 4875 (1968).

[6] B. Shizgal, J. Karplus, J. Chem. Phys. **52**, 4262 (1970).

[7] B. Shizgal, J. Karplus, J. Chem. Phys. **54**, 4345 (1971).

[8] A.S. Cukrowski, S. Fritzsche, J. Popielawski, in Far-from-Equilibrium Dynamics of Chemical Systems, Eds. J. Popielawski and J. Gorecki, Proc. Second Internat. Symp. Swidno (Poland), 1990, World Scientific, Singapore, 1991.

[9] K. Nanbu, J. Phys. Soc. Japan **49**, 2042 (1980).

[10] K. Nanbu, J. Phys. Soc. Japan **49**, 2050 (1980).

[11] K. Nanbu, J. Phys. Soc. Japan **52**, 3382 (1980).

[12] H. Babovsky, Math. Methods Appl. Sci. **8**, 229 (1986).

[13] J. Gorecki, J. Popielawski, A.S. Cukrowski, Phys. Rev. **A 44**, 3791 (1991).

[14] J. Gorecki, in Far-from-Equilibrium Dynamics of Chemical Systems, Eds. J. Popielawski and J. Gorecki, Proc. Second Internat Symp. Swidno (Poland), 1990, World Scientific, Singapore, 1991.

[15] J. Gorecki, J. Chem. Phys. **95**,, 2041 (1991).

[16] J. Popielawski, A.S. Cukrowski, S. Fritzsche, Physica **A 188**, 344 (1992).

[17] A.S. Cukrowski S. Fritzsche, J. Popielawski, Acta Phys. Pol. **82**, 369 (1993).

[18] A.S. Cukrowski, J. Popielawski, Lihong Qin, J.S. Dahler, J. Chem. Phys. **97**, 9086 (1992).

[19] J. Gorecki, Polish J. Chem. **66**, 1183 (1992).

SPHERULITIC GROWTH OF POLYMER SYSTEMS
IN CONVECTION FIELD UNDER NON-EQUILIBRIUM CONDITIONS

Adam Gadomski[*] and Jerzy Łuczka[†]

[*] Department of Polymer Physics, Silesian University,
Śnieżna St. 2, PL 41-200 Sosnowiec, Poland

[†] Department of Theoretical Physics, Silesian University,
Bankowa St. 14, PL 40-007 Katowice, Poland

ABSTRACT. *The growth process of spherulites in polymer systems is studied. An external convection field is considered. Non-equilibrium regime near interface is postulated. The influence of a kinetic coefficient and a saturation parameter on the growing process is demonstrated.*

1. INTRODUCTORY REMARKS

Many phenomena in nature such as erosion, turbulence, fracture, pattern formation, and growth are typical problems of far-from-equilibrium processes. The last example, growth processes are intensively studied from both theoretical as well as experimental points of view (cf. Ref. [1]). Among many growth processes like aggregation (especially diffusion-limited aggregation), solidification, precipitation, etc., the spherulitic growth is also the subject of studies [2]. There exist many experimental reports devoted to some most important aspects of that growth process (e.g., kinetics, morphology, structure peculiarities, thermodynamics, etc.) [3], but a theoretical description at least of the spherulitic kinetics is so far unsatisfactory [4].

Before embarking on this not completely understood field of research, a few basic informations concerning the spherulitic growth are needed. Firstly, spherulites are sphere-like patterns consisting of a dendrite-like skeleton (whose arms are branched at noncrystallographic angles) and amorphous phase (mostly consisting of macromolecules and/or small crystallites) dispersed in between. Secondly, they emerge frequently in many polymeric systems, being basically grown from either supersaturated solution [2,4,5] or undercooled melt [2,6,7,8]. A few examples can be listed immediately, e.g., poly-(phenylene sulfide) [2,4] or pol-(oxyethylene) [5] spherulites grown from solution, and poly-(propylene), poly-(ethylene) [2] or nylon 6 spherulites [6] grown from undercooled melt. Thirdly, there exists a theoretical (analytical as well as simulational) description of the spherulitic growth which is based on methods used in the description of a diffusion controlled pattern formation process [1] in which the internal boundary condition with the so-called Gibbs-Thomson term has been completed by a non-equilibrium term, first proposed in [4].

In this paper we consider a model of the spherulitic growth that is based on a mass convection field instead of a diffusion field. We would like to state clearly that we consider the growth of spherulites from solution and in our simple approach spherulites are represented by spherical objects. Note that a spherulite is 3d-system (there also exist some 2d-objects commonly known as cylindrulites) [2]. The non-equilibrium character of the process can quantitatively be manifested by at least

(i) external field feeding the growing object,

(ii) internal boundary condition prescribed at the

interface: spherulite-surroundings.

One should also realize that spherulites can be understood as some competition systems.

It is worth to recall the following experimental observation: the growth velocity $v \equiv \frac{dR}{dt}$ (R is the sphere radius; $R \equiv R(t)$, where t is time) is only a parametric function of temperature T (but the process in question is isothermal!) and slightly depends upon a particular system under study [2]. Then, in this case we may solely expect that asymptotically

$$R \propto t \qquad (1)$$

Note that it substantially differs from the well-known relationship $R \propto t^{1/2}$, characteristic for purely diffusive processes. Finally, one may notice that spherulites are obviously nonfractals as a whole, but the skeleton ("dendrite") may have a fractal character (cf. Ref.[1] for going into details).

344

2. DESCRIPTION OF THE MODEL

An evolution equation for spherulites can be formulated in the following form [10]:

(i) the mass conservation law as a fundamental evolution equation for growing spherulite, with an initial condition which is an initial shape (a surface) of the growing spherulite,

(ii) specification of the concentration field of medium and of the spherulite at the interface; it in general should be associated with non-equilibrium boundary conditions,

(iii) specification of fluxes through the interface and its connection with the concentration field of surroundings; in general it allows to introduce not exclusively diffusive but also others fluxes of particles or objects.

It has been shown that an evolution equation for growing objects with an ideal or perturbed spherical symmetry has the form [10-12]

$$\left[C - c(\tilde{r}, \vartheta, \phi)\right]\frac{d\tilde{r}}{dt} = -\vec{j}[c(\tilde{r}, \vartheta, \phi)] \cdot \vec{n}_o \tag{2}$$

where $\tilde{r} \equiv \tilde{r}(\vartheta, \phi; t)$ is an equation of a surface of the growing object in the spherical coordinate system (r, ϑ, ϕ), C is its density which may depend on space variables and can generally be of stochastic nature [2, 11]. A quantity $c(r, \vartheta, \phi)$ stands for concentration of external particles at a point (r, ϑ, ϕ), $\vec{j}[c(r, \vartheta, \phi)]$ is the flux of particles outside of the object which depends functionally on concentration at a point (r, ϑ, ϕ) and $\vec{n}_o$ is the outer normal to the surface of the object [10].

As regarding point (ii), the concentration of the particles at the surface of the growing object is determined by thermodynamic conditions and geometry of the surface. Under assumption of local thermodynamical equilibrium near the interface, it has the form of the Gibbs-Thompson relation [3]. In a more realistic model, the surface is away from local equilibrium and deviation from this state is proportional to the growth velocity of the interface [4, 12],

$$c(\tilde{r}, \vartheta, \phi) = c_o\left[1 + \Gamma K(\tilde{r}, \vartheta, \phi) - \beta\left(\frac{d\tilde{r}}{dt}\right)^{\nu}\right] \tag{3}$$

where c_o is the concentration field at a flat interface, Γ is the capillary constant, β is a positive kinetic coefficient, ν is constant and may manifest a possible non-Newtonian character of external medium[13], $K(\tilde{r}, \vartheta, \phi)$ is twice the mean curvature of the object surface and the last term describes a deviation from the thermodynamic equilibrium [4, 12, 13]. When $\beta=0$, one gets the well known Gibbs-Thomson condition [3].

Let us consider point (iii). If the feed of the growing object is purely convective then [10]

$$\vec{j}[c(r, \vartheta, \phi)] = c(r, \vartheta, \phi)\, \vec{v}(r, \vartheta, \phi)\ , \tag{4}$$

where $\vec{v}(r, \vartheta, \phi)$ is a convection velocity.

A set of Eqs. (2)-(4) will be applied for describing the evolution of spherulites. Here, we will study a simplified model of an ideal sphere of radius $R \equiv R(t)$ and of constant density C immersed in a radial constant convective field with attraction to a central point and a linear case of dependence of the concentration c in (3) upon interface velocity, $\nu = 1$. It means that (2)-(4) take the form

$$\left[C - c(R)\right]\frac{dR}{dt} = -\vec{j}[c(R)] \cdot \hat{e}_r \tag{5}$$

($\hat{e}_r$ is a unit radial vector),

$$c(R) = c_o\left[1 + \frac{2\Gamma}{R} - \beta\frac{dR}{dt}\right] \tag{6}$$

$$\vec{j}[c(R)] = c(R)\,\vec{v}(R, t), \quad \vec{v}(R, t) = -v_o\,\hat{e}_r \tag{7}$$

where v_o is a positive constant (mean velocity of particles).

Eqs. (5)-(7) lead to the basic evolution equation of the form

$$\left[C - c_o\left(1 + \frac{2\Gamma}{R} - \beta\frac{dR}{dt}\right)\right]\frac{dR}{dt} = c_o v_o\left[1 + \frac{2\Gamma}{R} - \beta\frac{dR}{dt}\right] \tag{8}$$

with an initial condition $R(t=t_o) = R_o > 0$.

3. ANALYSIS OF EVOLUTION EQUATION

The growth process described by Eq. (8) is influenced by five parameters, C, c_o, Γ, β and v_o. But in fact only two parameters are physically important,

$$\Delta = (C - c_o)/c_o \ , \qquad \beta_o = \beta v_o \qquad (9)$$

The quantity Δ is a measure of the saturation in the system [3] and β_o is a rescaled kinetic coefficient. Indeed, rescaling the variables R and t to dimensionless quantities $r = r(\tau)$ and τ via the relations

$$r = (1/2\Gamma)R \qquad\qquad \tau = (v_o/2\Gamma)t \qquad (10)$$

One obtains from Eq. (8) the following nonlinear ordinary differential equation

$$\beta_o\left(\frac{dr}{d\tau}\right)^2 + (\Delta + \beta_o - \frac{1}{r})\ \frac{dr}{d\tau} - \frac{1}{r} - 1 = 0 \qquad (11)$$

This equation is an algebraic quadratic equation with respect to $dr/d\tau$. One of its root has to be ruled out. It is determined by the limiting case $\beta_o \rightarrow 0$ in (11). Thus, from (11) one gets

$$\frac{dr}{d\tau} = \frac{d(r) + 1}{2\beta_o r} - \frac{\Delta + \beta_o}{2\beta_o} \qquad (12)$$

where

$$d(r) = \left([\ (\Delta + \beta_o)r - 1]^2 + 4\beta_o r(r + 1) \right)^{1/2} \qquad (13)$$

A solution of Eq. (12) cannot be expressed by known functions. Numerical results and properties of solutions of Eq. (12) are presented in Figs. (1)-(2). The dependence of $r(\tau)$ on the rescaled kinetic coefficient β_o is sketched in Figs. 1 . Our results show that when β_o increases then growth becomes slower. At early stage of evolution, the influence of β_o on $r(\tau)$ is strongly nonlinear. For long times, the growth is linear in time according to (1) and β_o influences only on the slope of the curve $r(\tau)$. Fig. 2 shows the influence of the saturation parameter Δ on the growing process.

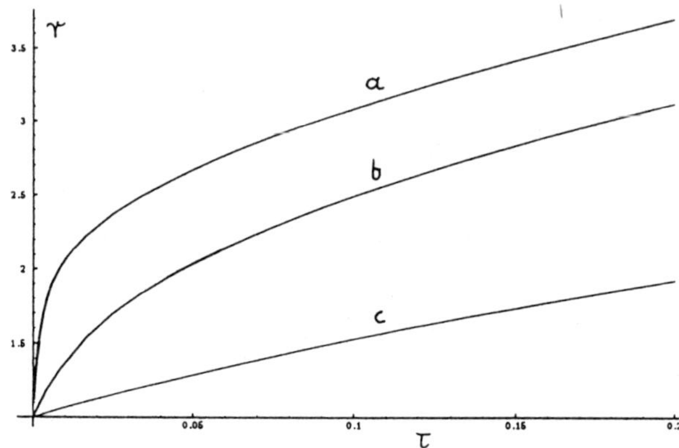

Fig.1. The time dependence of the rescaled spherulite radius $r(\tau)$ for the saturation parameter $\Delta=0.5$ and a few representative values of a kinetic coefficient: (a) $\beta_o = 0.001$; (b) $\beta_o = 0.01$; (c) $\beta_o = 0.1$.

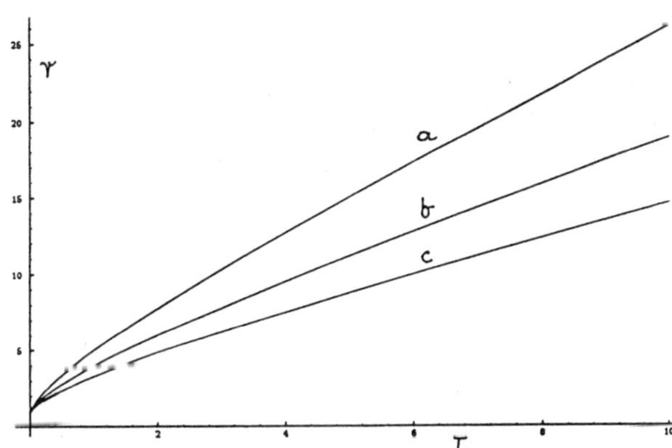

Fig.2. The time-dependence $r(\tau)$ for $\beta_o = 0.1$ and for : (a) $\Delta = 0.2$; (b) $\Delta = 0.5$; (c) $\Delta = 0.8$.

4. CONCLUSIONS

In this work we tried to show that spherulites, being some semicrystalline patterns, manifest not only some unusual structural but also theoretically unexpected kinetic behavior which is, in fact, represented by the asymptotic formula given by Eq.(1). We proposed some mechanism which is in agreement with experimental evidences (cf., [2], [4], [6]) and can reproduce the linear R(t)-dependence reported. In our opinion the mechanism proposed in [4] is not the only one possible in many experimental systems. To enlarge the theoretical set-up we proposed the model which takes into account some new possibilities. It is worth to be done because the interest concerning spherulites has some practical aspects e.g., piezoelectric features of polymeric materials or mechanical durability of plastics (cf., [14] and references therein).

ACKNOWLEDGEMENT

We are very indebted to the Organizers of the Symposium "Far-from-Equilibrium Dynamics of Chemical Systems", September 6-10, Borki near Tomaszów Mazowiecki (Poland), for giving us the possibility to present our ideas concerning the spherulitic growth. We thank Katarzyna Jarząbek for help and criticism during preparation of this work. One of us (A.G.) is thankful to Ewa Mielcarzewicz for useful discussions.

The work supported in part by KBN (Poland).

REFERENCES

1. T.Vicsek, *Fractal Growth Phenomena*, World Scientific, Singapore, 1991.

2. H. D. Keith and F. J. Padden, Jr., J.Appl.Phys. **34** 2409 (1963).

3. A.A.Chernov, *Modern Crystallography III. Crystal Growth*, Springer-Verlag, Berlin, 1984.

4. N.Goldenfeld, J.Crystal Growth **84** 601 (1987);
    F.Liu and N.Goldenfeld, Phys.Rev. **A42** 895 (1990).

5. D.K.Carpenter, G.Santiago and A.H.Hunt, J.Polymer Sci.: Polymer Symposium **44** 75 (1974).

6. A. Gałęski, A. S. Argon and R. E. Cohen, Macromol. Chem. **188** 1195 (1987).

7. A. Atanasov, Polymer Bull. (Berlin) **17(5)** 445 (1987).

8. A. Sanches, C. Marco J. G. Fatou and A. Bello, Macromol. Chem. **188** 1205 (1987).

9. I. Prigogine and I. Stengers, *Order out of Chaos. Man's New Dialogue with Nature*, Flamingo (An Imprint of Haroper Collins Publishers), London, 1985, chap. 4-6.

10. J. Łuczka, A. Gadomski and Z. J. Grzywna, *Growth Driven by Diffusion*, IMA Preprint Series # 869 (1991), University of Minnesota;

    J. Łuczka, A. Gadomski and Z. J. Grzywna, Czech. J. Phys. **42** 577 (1992);

    A. Gadomski, Z. J. Grzywna and J. Łuczka, Chem. Engng. Sci. **48** 3713 (1993).

11. A. Gadomski and J. Łuczka, Acta Phys. Pol. **B24** 725 (1993).

12. A. Gadomski and J. Łuczka, Inter. J. Quantum Chem. **49** (1994).

13. E. Ben-Jacob and P. Garik, Physica **D38** 16 (1989).

14. D. W. van Krevelen, *Properties of Polymers. Correlations with Chemical Structure*, Elsevier Publishing Company, Amsterdam , 1972.

ON INVESTIGATION OF RELAXATION PHENOMENA IN THE POLY(4-METHYL-1-PENTENE) MEMBRANE MATERIAL WITH PARTICULAR ATTENTION TO THE SOLVENT CONTENT

Adam Danch, Katarzyna Jarząbek

Silesian University, Institute for Technological Problems, Department of Polymer Physics, 41-200 Sosnowiec, Śnieżna 2, POLAND.

ABSTRACT

The dielectric relaxation properties of opaque and transparent polymer membranes have been investigated. The measurements were carried out in the frequency range from 800Hz to 10MHz and at temperatures from $20^{\circ}C$ to $160^{\circ}C$ using a capacitance bridge and circuit magnification meter. Main object of this study are membranes obtained by casting on glass from solution of 4% poly(4-methyl-1-pentene) (PMP) in cyclohexane (C_6H_{12}). However, some data for other membranes of different PMP concentrations and carbon tetrachloride (CCL_4) as a solvent were obtained as well. A few relaxation processes have been found and the interpretation of their origin is given. An influence of the solvent content on relaxation effects has also been considered.

INTRODUCTION

For most applications of polymeric materials the dielectric constant (ε') and dielectric loss (ε'') are important practical parameters. Also studies of the dielectric properties provide a great deal of information for understanding the molecular structure [1,2] and the mechanism of electric polarization in dielectric or ferroelectric materials [3,4]. A relaxation process comprise the approach to equilibrium of a system, which is initially out of thermodynamic equilibrium. Application of irreversible thermodynamics for description of relaxation processes [5] helps one to explain the meaning of some of concepts of the relaxation, and shows (in particular) that the Debye equations give a good description of a relaxation process in the first approximation.

More than two independent variables usually exist in a relaxing system. Usually, such a system contains a source of energy which is different when compared to that connected with elastic properties. This source of energy may be contributed by polarization in electric or magnetic fields.

In this study of poly(4-methyl-1-pentene) (PMP) membranes we show that the information on the influence of forming conditions, solvent content and heat treatment on relaxation effects is useful in understanding of polymer superstructure. PMP belongs to the class of polyolefines and its thermal, transport and mechanical properties have been studied using different laboratory technics for many years [6-9]. Some other authors used this polymer as the permselective layer in composite membranes [10,11]. It is known that properties of polymeric materials do not exclusively depend upon chemical structure of polymer chain, but also the role of crystalline and amorphous phases (superstructure) is pronounced. Due to a structural rearrangement mostly connected to micro-(random behaviour of the mass center of the chain in its close vicinity) and macro-(a rotational and/or a translational diffusion) Brownian motion of the macromolecules caused by heat treatment, it is possible to obtain some crystalline modifications in polymer samples [12-16]. Of course, properties of a polymeric material depend on the amount of crystallinity and the distribution of crystallite sizes in the matrix. It is also worth to notice that existence of some more complicated polymer semicrystalline patterns can be related to differences in stereochemical structure of polymer chains. A various tacticity, in turn, may manifest not only in a crystalline form, but also in an amorphous state [17,6].

The purpose of this work is to determine the dielectric and thermodynamic parameters of relaxation processes in order to obtain certain useful information about dynamics of the phase transition (glass transition and/or other relaxation phenomena). A second aim of this study is to propose some possible explanation for differences found in opaque and transparent PMP membranes.

EXPERIMENTAL

Materials

Poly(4-methyl-1-pentene), trade name TPX, Mitsui & Co, LTD, used without purification, glass temperature (T_g) 30°C, melting point (T_m) 240°C, density 0.83g/cm^3.

Cyclohexane (C$_6$H$_{12}$), POCH, boiling temperature (T_b) 81.4°C, T_m 6.8°C, purity 99.7%, used as a solvent with no further purification.

Carbon tetrachloride (CCl$_4$), POCH, T_b 76.8°C, T_m -22.8°C, purity 99.7%, used as a solvent with no further purification.

Sample Characterization

Sheet membranes were obtained by casting on glass plate from solutions of 4% PMP in cyclohexane or carbon tetrachloride, dried at temperature range from -10°C to 20°C for two weeks. This way opaque and transparent polymer membranes of different thickness were obtained. It was established by Differential Thermal Analysis (DTA) that both types of membranes were semicrystalline and two glass transition temperatures (38°C and 120°C) were observed in each of them [17]. Thermogravimetric analysis (TG) showed about 1.5% weight content of solvent for transparent membranes after two weeks of drying (after 6 weeks no content of solvent was detected). Contrary to the transparent membranes the opaque ones did not contain solvent. For opaque membranes obtained from more diluted solutions 24 hours period of drying was long enough to get rid of solvent. It is worth to notice that the PMP opaque membranes have been successfully achieved only from cyclohexane solutions (a more detailed description of this process is presented, e.g. [17,18]).

Dielectric Measurements

The samples before dielectric measurements were silver-plated in argon atmosphere, the silver layer serving as electrodes. The capacitance bridge and the circuit magnification meter, covering the frequency 800Hz and the frequency range from 10^4Hz to 10^7Hz, were used for dielectric measurements. The sample was placed inside a chamber in which the temperature was controlled up to ±0.1°C.

RESULTS AND DISCUSSION

The temperature and frequency dependencies of dielectric loss tangent (tan6δ), and the dielectric constant ε' for the samples are shown in Figs 1-5.

Over the range from 20°C to 160°C two relaxation peaks at 800Hz can be observed. The literature data (elastic modulus and loss factor vs. temperature curve for unannealed PMP sample [6]) and earlier experience with PMP membranes [17] have shown that those changes in dielectric properties may be attributed to glass transition process. The glass transition temperatures (T_g) of transparent PMP membranes are of 62°C and 100°C, and opaque membranes of 59°C and 118°C, respectively. The differences between T_g values obtained by other authors depend on determination method used [6, 10,17].

The activation enthalpies were calculated from Arrhenius graph [19] using the least-squares method. It was found (see Fig.6) that four relaxation processes took place for opaque sample. One of these processes (α_i) had an activation enthalpy of 1.854 kJ/mol, the second (I) 1.111 kJ/mol, the third (α_a) 1.674 kJ/mol, and 0.412 kJ/mol for β relaxation process. A simple explanation for those processes ($\alpha_i, \alpha_a, \beta$) is given on the basis of the structure of a polymer. The origin of relaxation I is unknown and more detailed investigations than described in this paper are necessary to verify and elucidate all aspects of this process. However, it could be supposed that the relaxation I is due to a movement of space charges, which perform a local compensation of the polarization. The parameters of this relaxation process are difficult to be extracted from the measured spectra (Figs 1-3). It was possible to obtain more reliable values of the parameters using a fitting technique, if the existence of three peaks in the dielectric loss spectrum (α_i, I, α_a, in Fig.1) was assumed. Similar values for process I were obtained (like in opaque samples) for transparent membranes from frequency dependencies of dielectric loss factor after using of fitting technique. In relation to other dielectric relaxation processes (α_i, α_a) some differences were observed. The parameters for transparent membranes (enthalpy, height, width, and temperature of maximum value of peak) were different.

Unfortunately, the enthalpy values were obtained from four points for opaque and two points for transparent sample, which caused a big error of estimation. However, in our opinion, those measurements might serve as a confirmation of differences existing in superstructure between opaque and transparent membranes.

Both kinds of samples were annealed at $160^{\circ}C$ for 30 minutes. Measured values of tan6δ for those samples showed opposite trends. Transparent membranes after being annealed behaved as those polymers which are mostly known from literature [20,1,3], i.e., the glass transition peak in dielectric loss spectrum diminished and the glass temperature slightly shifted from $62^{\circ}C$ to $58^{\circ}C$. It is contrary to the behavior for opaque membranes, where the glass temperature has the same value $(T_g, 59^{\circ}C)$ but the height of tan6δ increased (see Fig.4). The reported X-ray photographs of the five PMP crystalline modifications showed that the cyclohexane- and carbon tetrachloride-cast membranes were mostly close to modification denoted as IV (hexagonal, lattice constants a=b=22.17tÅ and c=6.69tÅ [11]). It was noticed, that with increase of a and b, and decrease of c parameters, the permeability of PMP membrane increased. Mohr et al. [11] have shown that "...the crystallites are preferentially oriented with the chain axes perpendicular to the film surface..." but the distribution of those crystalline regions is mainly random and depends on forming conditions of membranes. A similar effect was observed on IR-spectra (polarized light), which could lead to the conclusion of increased orientation of crystallites in opaque sample [17], assumed. It is difficult to depict the degree of crystallinity, the size and crystal form of transparent and opaque sample basing on the dielectric measurement, but the investigations show that the ordering of opaque membranes in the amorphous region is most probably greater than in the case of transparent ones. It could be supposed that this state is due to interaction between the solvent molecules $(T_m$ is about $6.8^{\circ}C)$ and the polymer macromolecules. The opaque membranes were obtained at $-10^{\circ}C$ and it was significantly lower with respect to melting point of cyclohexane. It was noticed, when the solution concentration decrease, the temperature of membrane formation process could be higher. For example, at $0^{\circ}C$ the 0.5% weight opaque membranes (useless due to their

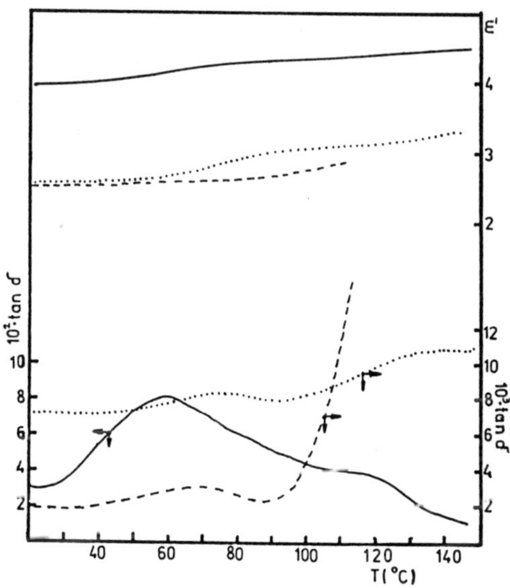

Fig. 1.   Dielectric   loss   tangent   and   dielectric   constant   vs. temperature at various frequencies for opaque membrane; (————) 800Hz, (.....) 50kHz, (— — —) 1MHz.

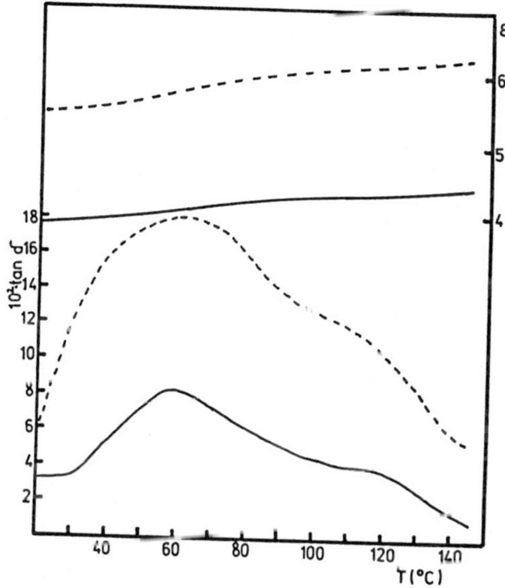

Fig.2. Dielectric loss tangent and dielectric constant vs. temperature at 800Hz for (————)opaque and (— — —)transparent membranes.

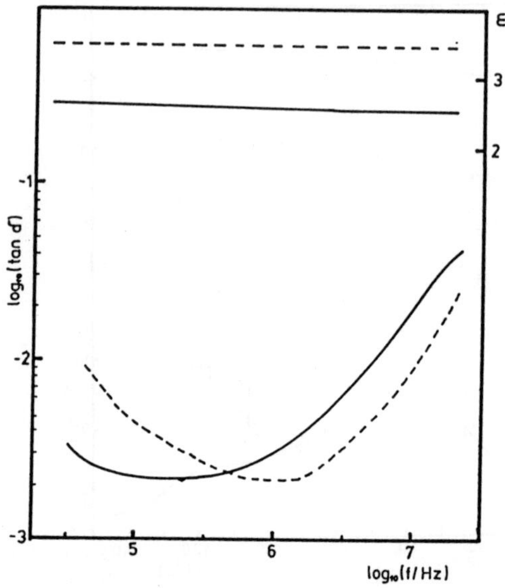

Fig. 3. Frequency dependence of tan δ and ε' for (————)opaque and
(– – –)transparent membranes at room temperature.

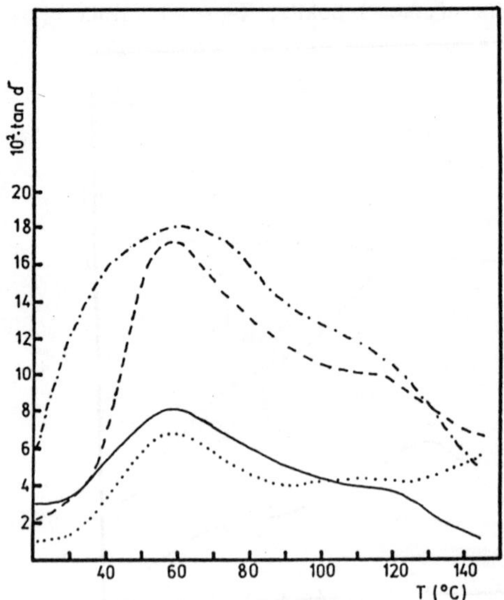

Fig. 4. Temperature dependence of tan δ, observed for annealed
(– – –)opaque, (.....)transparent and unannealed (————)opaque,
(. — . —-)transparent membranes at 800Hz.

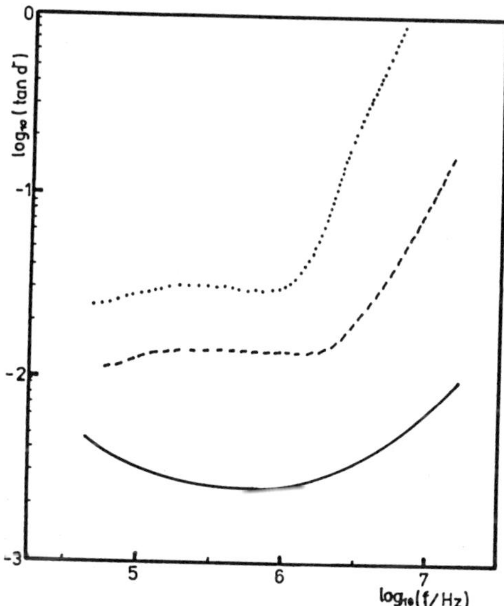

Fig.5. Frequency dependence of tan δ for transparent membranes obtained from carbon tetrachloride solution in different content of solvent at room temperature; (————) 1 5 weight% , (— — —) 6%, (. . . . . .) 9.5% .

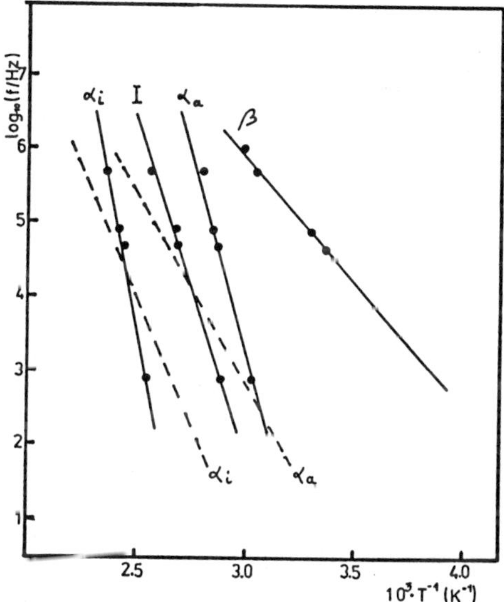

Fig.6. Log frequencies vs. reciprocal of temperature for (————)opaque and (— — —)transparent membranes.

mechanical properties) were obtained. Those observations need some kriometric measurements.

One should realize that the dielectric spectroscopy for annealed and unanneald sample gives evidence for some enforced and weak thermodynamic ordering in the structures of opaque membranes. A significantly lower value of $\tan 1\delta$ for opaque membranes, in contrary to those transparent ones, could ensure that the structure of opaque membrane is more stiff. It can be changed by a heat treatment in some adequately chosen temperature . The lower value of the activation enthalpy of glass transition in transparent membrane material is a consequence of the fact that changes among equilibrium states are facilitated.

It was noticed that some occluded amounts of solvent molecules act like a plasticizer in relaxation process. It facilitates the motion of polymer chains, which causes an increase of the loss factor. The kind of solvent (size and polarity of molecules) has some influence on this quantity . Although no more carefully done investigations have been carried out, it has been stated that a slight amount of solvent (3%) can cause the total overlapping of relaxation peaks (β and/or α). Some heat treatment above T_g-value, i.e., the evaporation of the slight amounts of solvent molecules occluded, reveals the peaks which have been overlapping before. Regarding to this fact $\tan 1\delta$ is significantly decreased, what is suitable for PMP membranes applications.

CONCLUSIONS

The dielectric relaxation measurements for PMP samples give us information about the phase transition at $58^{\circ}C$ (800Hz) and about other relaxation processes. The α_1- and the α_a- relaxation processes are due to the motion of the polymer chain and thus depend on the chain structure of the polymer (isotactic and atactic modifications of PMP). According to that fact two of the T_g-values for opaque as well as for transparent samples were observed.

Generally speaking, the membranes of both types showed different dielectric properties, which could be caused by different patterns of polymer superstructure. It was noticed, that some occluded amounts of solvent molecules (in transparent membranes) act like low-molecular-weight compounds on relaxation behaviour.

ACKNOWLEDGEMENTS

The authors would like to thank Dr. Adam Gadomski (Silesian University) for helpful discussions and useful comments. We are also grateful to Dr. Dionizy Czekaj (Silesian University) for the suggestions and helps in the experiments.

REFERENCES

1. N.G.McGrumm, B.E.Read and G. Williams, "Anelastic and Dielectric Effects in Polymer Solids.", Wiley, London 1967.
2. D.Sęk and A.Danch, J.Poly.Mater. **7**,313(1990).
3. J.Van Turnhout, "Thermal Stimulated Discharge of Polymer Electrets.", Elsevier, Amsterdam 1975
4. Y.Xu, "Ferroelectric Materials and Their Applications", North-Holland, Amsterdam 1991.
5. H.Fröhlich, "Theory of Dielectric.", Oxford Univ.Press, London 1958.
6. J.H.Griffith and B.G.Ranby, J.Polymer Sci. **44**,369(1960)
7. H.Ke, J.Polymer Sci. Part A, **1**,1453(1963).
8. G.Charlet and G.Delmas, Polymer **25**,1619(1984).
9. T.He and S.Porter, Polymer **28**,1321(1987).
10. A.C.Puleo, D.R.Paul and P.K.Wong, Polymer **30**,1357(1989)
11. J.M.Mohr and D.R.Paul, Polymer **32**,1236(1991).
12. M.Takayanagi and M.Kawasaki, J.Macromol.Sci.-Phys.(B) **1**,741(1967).
13. S.Aharoni, G.Charlet and C.Delmas, Macromolecules **14**,1390(1981).
14. G.Charlet and G.Delmas, Polym.Bull. **6**,367(1982).
15. G.Charlet, G.Delmas, J.Revol and R.Manley, Polymer **25**,1613(1984).
16. T.Tanigami, H.Suzuki, K.Yamaura and S.Matsuzawa, Macromolecules **18**,2595(1985).
17. A.Danch and A.Gadomski, sent to Editor.

18. A.Gadomski and A.Danch, in preparation.

19. J.M.Janik; ed.,"Fizyka Chemiczna.", PWN, Warszawa 1989.

20. P.Hedvig, "Dielectric Spectroscopy of Polymers.", Akademiai Kiado, Budapest 1977.